ADVANCED COMPUTATIONAL NANOMECHANICS

Microsystem and Nanotechnology Series

Series Editors – Ron Pethig and Horacio Dante Espinosa

Advanced Computational Nanomechanics

Silvestre, February 2016

Micro-Cutting: Fundamentals and Applications

Cheng, Huo, August 2013

Nanoimprint Technology: Nanotransfer for Thermoplastic and Photocurable Polymer

Taniguchi, Ito, Mizuno and Saito, August 2013

Nano and Cell Mechanics: Fundamentals and Frontiers

Espinosa and Bao, January 2013

Digital Holography for MEMS and Microsystem Metrology

Asundi, July 2011

Multiscale Analysis of Deformation and Failure of Materials

Fan, December 2010

Fluid Properties at Nano/Meso Scale

Dyson et al., September 2008

Introduction to Microsystem Technology

Gerlach, March 2008

AC Electrokinetics: Colloids and Nanoparticles

Morgan and Green, January 2003

Microfluidic Technology and Applications

Koch et al., November 2000

ADVANCED COMPUTATIONAL NANOMECHANICS

Edited by

Nuno Silvestre

University of Lisbon
Portugal

Library of Congress Cataloging-in-Publication Data

Advanced computational nanomechanics / edited by Nuno Silvestre.
     pages cm
  Includes bibliographical references and index.
  ISBN 978-1-119-06893-8 (cloth)
  1. Nanotechnology–Mathematics. 2. Nanoelectromechanical systems–Mathematical models.
3. Nanostructures–Mathematical models. 4. Micromechanics–Mathematics. I. Silvestre, Nuno, editor.
  T174.7.A385 2016
  620′.5–dc23

                                                                    2015035009

A catalogue record for this book is available from the British Library.

Typeset in 10/12pt TimesLTStd by SPi Global, Chennai, India

Printed and bound in Singapore by Markono Print Media Pte Ltd

1   2016

Contents

List of Contributors

Avinash Akepati, Graduate Research Assistant, Department of Aerospace Engineering and Mechanics, University of Alabama, Tuscaloosa, AL, USA

Ganesh Balasubramanian, Assistant Professor, Department of Mechanical Engineering, Iowa State University, Ames, IA, USA

Irene J. Beyerlein, Theoretical Division, Los Alamos National Laboratory, Los Alamos, NM, USA

Khoa Bui, School of Chemical, Biological and Materials Engineering, The University of Oklahoma, Norman, OK, USA

Steven W. Cranford, Assistant Professor, Department of Civil and Environmental Engineering, Northeastern University, Boston, MA, USA

Ghasem Ghadyani, Faculty of Mechanical Engineering, University Malaya, Kuala Lumpur, Malaysia

Ali Ghavamian, Research Assistant, School of Engineering, Griffith University, Gold Coast Campus, Southport, Queensland, Australia

Yuantong Gu, Professor, School of Chemistry, Physics and Mechanical Engineering, Queensland University of Technology, Brisbane, QLD, Australia

Mesut Kirca, Assistant Professor, Department of Mechanical Engineering, Istanbul Technical University, Istanbul, Turkey

Huong Nguyen, School of Chemical, Biological and Materials Engineering, The University of Oklahoma, Norman, OK, USA

Andreas Öchsner, Prof. Dr.-Ing., School of Engineering, Griffith University, Gold Coast Campus, Southport, Queensland, Australia

Dimitrios V. Papavassiliou, Professor, School of Chemical, Biological and Materials Engineering, The University of Oklahoma, Norman, OK, USA; Division for Chemical, Bioengineering and Environmental Transport Systems, National Science Foundation, Arlington, VA, USA

Nicola M. Pugno, Professor, Department of Civil, Environmental and Mechanical Engineering, University of Trento, Trento, Italy; Centre of Materials and Microsystems, Bruno Kessler Foundation, Trento, Italy; School of Engineering and Materials Science, Queen Mary University, London, UK

Moones Rahmandoust, Dr., Protein Research Center, Shahid Beheshti University, G.C., Velenjak, Tehran, Iran; School of Engineering, Griffith University, Gold Coast Campus, Southport, Queensland, Australia; Deputy Vice Chancellor Office of Research and Innovation, Universiti Teknologi Malaysia, Johor Bahru, Johor, Malaysia

Ruth E. Roman, Graduate Research Assistant, Department of Civil and Environmental Engineering, Northeastern University, Boston, MA, USA

Samit Roy, William D Jordan Professor, Department of Aerospace Engineering and Mechanics, University of Alabama, Tuscaloosa, AL, USA

Hiroyuki Shima, Associate Professor, Department of Environmental Sciences, University of Yamanashi, Kofu, Yamanashi, Japan

Yoshiyuki Suda, Associate Professor, Department of Electrical and Electronic Information Engineering, Toyohashi University of Technology, Toyohashi, Aichi, Japan

Keka Talukdar, HOD, Department of Physics, Nadiha High School, Durgapur, West Bengal, India

Albert C. To, Associate Professor, Department of Mechanical Engineering and Materials Science, University of Pittsburgh, Pittsburgh, PA, USA

Chien Ming Wang, Professor, Engineering Science Programme, Faculty of Engineering, National University of Singapore, Singapore, Singapore

Yu Wang, PhD Student, School of Computing, Engineering and Mathematics, University of Western Sydney, Sydney, NSW, Australia

Xi F. Xu, Professor, School of Civil Engineering, Beijing Jiaotong University, Beijing, China

Yingyan Zhang, Senior Lecturer, School of Computing, Engineering and Mathematics, University of Western Sydney, Sydney, NSW, Australia

Series Preface

Books in the Wiley's Microsystem and Nanotechnology Series are intended, through scholarly works of the highest quality, to serve researchers and scientists wishing to keep abreast of advances in this expanding and increasingly important field of technology. Each book in the series is also intended to be a rich interdisciplinary resource for not only researchers but also teachers and students of specialized undergraduate and postgraduate courses.

Past books in the series include the university textbook *Introduction to Microsystem Technology* by Gerlach and Dötzel, covering the design, production and application of miniaturized technical systems from the viewpoint that for engineers to be able to solve problems in this field, they need to have interdisciplinary knowledge over several areas as well as the capability of thinking at the system level. In their book *Fluid Properties at Nano/Meso Scale*, Dyson *et al.* take us step by step through the fluidic world bridging the nanoscale, where molecular physics is required as our guide, and the microscale where macro continuum laws operate. Jinghong Fan in *Multiscale Analysis of Deformation and Failure of Materials* provides a comprehensive coverage of a wide range of multiscale modelling methods and simulations of the solid state at the atomistic/nano/submicron scales and up through those covering the micro/meso/macroscopic scale. Most recently, *Nano and Cell Mechanics: Fundamentals and Frontiers*, edited by Espinosa and Bao, assembled through their own inputs and those of 47 other experts of their chosen fields of endeavour, 17 timely and exciting chapters that represent the most comprehensive coverage yet presented of all aspects of the mechanics of cells and biomolecules.

In this book edited by Professor Nuno Silvestre, who is well known for his own work in the modelling and simulation of carbon nanotube structures, he has assembled 11 chapters written by international experts that comprehensively cover the experimental, modelling and theoretical studies of the mechanical and thermal properties of carbon nanotubes, polymer nanocomposites and other nanostructures. The research literature on *Computational Nanomechanics* is spread among many journals specialising in mechanics, computer science, materials science and nanotechnology. The latest advances in this rapidly moving field of nanotechnologies have been collected together in a single book!

Although the chapters are primarily intended for established scientists, research engineers and PhD students who have knowledge in materials science and numerical simulation methods, the clarity of writing and pedagogical style of the various chapters make much of this book's content suitable for inclusion in undergraduate and postgraduate courses.

Ronald Pethig
School of Engineering
The University of Edinburgh

Preface

During the last decade, nanomechanics emerged on the crossroads of classical mechanics, solid-state physics, statistical mechanics, materials science, and quantum chemistry. As an area of nanoscience, nanomechanics provides a scientific foundation of nanotechnology, that is, an applied area with a focus on the mechanical properties of engineered nanostructures and nanosystems. Owing to smallness of the studied objects (atomic and molecular systems), nanomechanics also accounts for discreteness of the object, whose size is comparable with the interatomic distances, plurality, but finiteness, of degrees of freedom in the object, thermal fluctuations, entropic effects, and quantum effects. These quantum effects determine forces of interaction between individual atoms in physical objects, which are introduced in nanomechanics by means of some averaged mathematical models called interatomic potentials. Subsequent utilization of the interatomic potentials within the classical multibody dynamics provides deterministic mechanical models of nanostructures and systems at the atomic scale/resolution. This book focuses on a variety of numerical and computational methods to analyze the mechanical behavior of materials and devices, including molecular dynamics, molecular mechanics, and continuum approaches (finite element simulations). In resume, several computational methods exist to model and simulate the behavior of nanostructures. All of them present advantages and drawbacks. This book presents a survey on the computational modeling of the mechanical behavior of nanostructures, with particular emphasis on CNTs, graphene, and composites. It includes 11 chapters and each chapter is an independent contribution by scientists with worldwide expertise and international reputation in the technological area.

In Chapter 1, Y.Y. Zhang, Y. Wang, C.M. Wang, and Y.T. Gu present a state-of-the art review on the thermal conductivity of graphene and its polymer nanocomposites. It is known that some progress has been achieved in producing graphene–polymer nanocomposites with good thermal conductivity, but interfacial thermal resistance at the graphene/polymer interfaces still hinders this improvement. This chapter reports recent research studies, which have shown that covalent and noncovalent functionalization techniques are promising in reducing the interfacial thermal resistance. The authors argue that further in-depth research studies are needed to explore the mechanisms of thermal transport across the graphene/polymer interfaces and achieve graphene–polymer nanocomposites with superior thermal conductivity.

In Chapter 2, M. Kirca and A.C. To describe the mechanics of CNT network materials. Recently, the application of CNTs has been extended to CNT networks in which the CNTs are joined together in two- or three-dimensional space and CNT networks. In this chapter, a thorough literature review including the most recent theoretical and experimental studies are presented, with a focus on the mechanical characteristics of CNT network materials. As a supporting material, some recent studies of authors are also introduced to provide deeper understanding in mechanical behavior of CNT network materials.

In Chapter 3, H. Shima and Y. Suda present the "Helical carbon nanomaterial," which refers to exotic nanocarbons having a long, thin, and helical morphology. Their spiral shape can be exploited for the development of novel mechanical devices such as highly sensitive tactile nanosensors, nanomechanical resonators, and reinforced nanofibers in high-strain composites. Because the quantitative determination of their mechanical properties and performance in actual applications remains largely unexamined, advanced computational techniques will play an important role, especially for the nanomaterials that hold the promise for use in next-generation nanodevices that have been unfeasible by the current fabrication techniques. This chapter gives a bird's eye view on the mechanical properties of helical carbon nanomaterials, paying particular attention to the latest findings obtained by both theoretical and experimental efforts.

In Chapter 4, G. Ghadyani and M. Rahmandoust present a review of the fundamental concepts of the Newtonian mechanics including Lagrangian and Hamiltonian functions, and then the developed equations of motion of a system with interacting material points are introduced. After that, based on the physics of nanosystems, which can be applicable in any material phases, basic concepts of molecular dynamic simulations are introduced. The link between molecular dynamics and quantum mechanics is explained using a simple classical example of two interacting hydrogen atoms and the major limitations of the simulation method are discussed. Length and timescale limitation of molecular dynamic simulation techniques are the major reasons behind opting multiscale simulations rather than molecular dynamics, which are explained briefly at the final sections of this chapter.

In Chapter 5, a probabilistic strength theory is presented by X. Frank Xu to formulate and model probability distribution for strength of CNTs and CNT fibers. A generalized Weibull distribution is formulated to explain statistical features of CNT strength, and a multiscale method is described to show how to upscale strength distribution from nanoscale CNTs to microscale CNT fibers. The probabilistic theory is considered applicable to fracture strength of all brittle materials, and with certain modifications, to failure of non-brittle materials as well. The benchmark model on strength upscaling from CNTs to fibers indicates that the full potential of CNT fibers for exploitation is expected to be in the range between 10 and 20 GPa with respect to mean strength, due to universal thermodynamic effects and inherent geometric constraints.

In Chapter 6, A. Ghavamian, M. Rahmandoust, and A. Öchsner review and present the latest developments on nanomechanics of perfect and defective heterojunction CNTs. Homogeneous and heterojunction CNTs are of a great importance because of their exceptional properties. In this chapter, the studies on considerable number of different types of perfect and atomically defective heterojunction CNTs with all possible connection types as well as their constructive homogeneous CNTs of different chiralities and configurations are presented and their elastic, torsional, and buckling properties are numerically investigated based on the finite element method with the assumption of linear elastic behavior. They conclude that the atomic defects in the structure of heterojunction CNTs lead to an almost linear decrease in the mechanical stability and strength of heterojunction CNTs, which appeared to be considerably more in the models with carbon vacancy rather than Si-doped models.

In Chapter 7, S. Roy and A. Akepati present a methodology for the prediction of fracture properties in polymer nanocomposites. These authors propose a new methodology to compute J-integral using atomistic data obtained from LAMMPS (Large-Scale Atomic/Molecular Massively Parallel Simulator). As a case study, the feasibility of computing the dynamic atomistic J-integral over the MD domain is evaluated for a graphene nano-platelet with a central crack using OPLS (Optimized Potentials for Liquid Simulations) potential. For model verification, the values of atomistic J-integral are compared with results from linear elastic fracture mechanics (LEFM) for isothermal crack initiation at 300 K. Computational results related to the path-independence of the atomistic J-Integral are also presented. Further, a novel approach that circumvents the complexities of direct computation of entropic contributions in polymers is also discussed.

In Chapter 8, R.E. Roman, N.M. Pugno, and S.W. Cranford characterize the mechanical behavior of 2D nanomaterials and composites. They relate the recent isolation of graphene from graphite with the discovery and synthesis of a multitude of similar two-dimensional crystalline. These single-atom thick materials (2D materials) have shown great promise for emerging nanotechnology applications, and a full understanding of their mechanistic behavior, properties, and failure modes is essential for the successful implementation of 2D materials in design. R.E. Roman, N.M. Pugno, and S.W. Cranford focus on the fundamental nanomechanics of 2D materials, describing how to characterize basic properties such as strength, stiffness, bending rigidity, and adhesion using computational methods such as molecular dynamics (MD). Failure is discussed in terms of the local, atomistic interpretation of quantized fracture mechanics (QFM) and extended to bulk systems via Nanoscale Weibull Statistics (NWS).

In Chapter 9, K. Talukdar studied the effect of chirality on the mechanical properties of defective CNTs. In spite of many existing studies, the experimental values of the mechanical properties of the CNTs vary in magnitude up to one order as the actual internal structure of the CNTs is still not completely known and the properties vary largely with their chirality and size. For better understanding of the internal molecular details, computer modeling and simulation may serve as an important tool, and MD simulations provide the internal dynamical change that the system is running through in course of time under various external forces. In this chapter, a comprehensive investigation is presented to find out the influence of chirality on the defective CNTs.

In Chapter 10, G. Balasubramanian describes the mechanics of thermal transport in mass disordered nanostructures. Its scope is to provide a flavor of a set of computational approaches that are employed to investigate heat transfer mechanisms in mass disordered nanostructures, and in particular those containing isotope impurities. Carbon nanomaterials have been attractive case studies for their remarkable properties. Although the examples below demonstrate effects of isotopes mostly in carbon nanotubes and graphene, the approaches are generic and applicable to a wide array of nanomaterials. This chapter begins from the fundamentals of thermal transport in nanomaterials, moves to engineering material properties and their computation, and finally concludes with data-driven methods for designing defect-engineered nanostructures.

In Chapter 11, D.V. Papavassiliou, K. Bui, and H. Nguyen present a study on the thermal boundary resistance (TBR) effects in CNT composites. In this work, the authors review briefly predictive models for the thermal behavior of CNTs in composites, and simulation efforts to investigate improvements in Keff with an increase in CNT volume fraction. Emphasis is placed in the investigation of the TBR in CNTs coated with silica in order to reduce the overall TBR of the composites. The authors propose the use of multiwalled CNTs (instead of single-walled CNTs), because heat can be transferred mainly through the outer wall offering the advantage of a larger area of heat transfer. A better approach proposed by the authors might be to design composites where the CNT orientation, when it leads to anisotropic thermal properties, or thermal rectification could be explored.

Finally, the editor would like to thank all authors for having accepted the challenge of writing these chapters and also for they support to achieve this high-quality book. The editor also acknowledges Wiley professionals (Clive Lawson and Anne Hunt) and SPi Global (Durgadevi Shanmughasundaram) for their enthusiastic and professional support.

Nuno Silvestre
Lisbon, May 2015

1

Thermal Conductivity of Graphene and Its Polymer Nanocomposites: A Review

Yingyan Zhang[1], Yu Wang[1], Chien Ming Wang[2] and Yuantong Gu[3]

[1]*School of Computing, Engineering and Mathematics, University of Western Sydney, Sydney, NSW, Australia*
[2]*Engineering Science Programme, Faculty of Engineering, National University of Singapore, Singapore, Singapore*
[3]*School of Chemistry, Physics and Mechanical Engineering, Queensland University of Technology, Brisbane, QLD, Australia*

1.1 Introduction

In recent years, there have been many research studies conducted on the properties and applications of graphene. Experimental and theoretical studies have revealed that graphene is hitherto the best thermal conductor ever known in nature. This extraordinary thermal transport property of graphene makes it promising in various applications. One of the popularly pursued applications is its use as thermally conductive fillers in polymer nanocomposites. It is expected that the novel graphene–polymer nanocomposites possess superior thermal conductivity and thus hold an enormous potential in thermal management applications. In this chapter, we review the recent research progress of graphene, with an emphasis on its thermal conductivity and its polymer nanocomposites.

1.2 Graphene

1.2.1 Introduction of Graphene

Graphene is the latest discovered carbon allotrope in 2004, following the discoveries of fullerenes and carbon nanotubes in the past three decades. Before its discovery, graphene actually has, for many years, been deemed as an academic material that could not stand-alone in

Advanced Computational Nanomechanics, First Edition. Edited by Nuno Silvestre.
© 2016 John Wiley & Sons, Ltd. Published 2016 by John Wiley & Sons, Ltd.

the real world because of the thermal instability of two-dimensional structures at the nanoscale level [1, 2]. However, this perception was quashed in 2004 by the ground-breaking research of Geim and Novoselov [3–5]. For the first time, they succeeded in producing real samples of graphene. Owing to their pioneering work, they were awarded the 2010 Nobel Prize in Physics. Ever since the discovery of graphene, its extraordinary properties and huge potential applications have attracted immense attentions from the scientific world. This section gives a brief overview on graphene. In Sections 1.2.1.1 and 1.2.1.2, the structure and synthesis of graphene are explained. Section 1.2.2 gives an overview on its electronic, mechanical and optical properties. The emphasis is given on its thermal conductivity in Section 1.2.3.

1.2.1.1   Structure

Graphene is a single two-dimensional flat layer of carbon atoms, where the carbon atoms are arranged in a honeycomb-like hexagonal crystalline lattice [3–10]. Figure 1.1 shows an illustration of the lattice structure of graphene.

The properties of graphene are determined by its bonding mechanism. Carbon is the sixth element in the periodic table, and one carbon atom possesses six electrons. In the ground state, its configuration of electrons is $1s^2 2s^2 2p^2$, where s and p indicate different atomic orbitals, the numbers in front of the orbital represent different electron shells and the superscript numbers denote the number of electrons in the shells. This electronic configuration of carbon means that there are two electrons filling, respectively, in 1s, 2s and 2p orbitals. The two electrons in 1s orbital are close to the nucleus and are irrelevant for chemical reactions. The 2s orbital is approximately 4 eV lower than the 2p orbitals; thus, it is energetically favourable for a carbon atom to keep two electrons in the 2s orbital and the remaining two electrons in the 2p orbital [11]. The 2p orbital can be divided into $2p_x$, $2p_y$ and $2p_z$ orbitals for designation purposes. In the ground state, each of the two electrons in the 2p orbital occupies one single 2p orbital, that

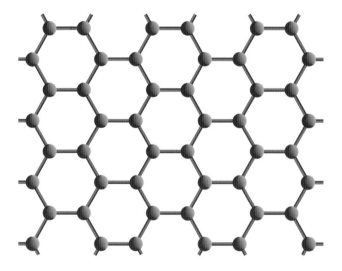

Figure 1.1   Lattice structure of graphene

is $2p_x$ and $2p_y$. While in a presence of other carbon atoms, it is favourable for a carbon atom to excite one electron from the 2s orbital to the third 2p orbital (i.e. $2p_z$ orbital) to form covalent bonds. It costs 4 eV for this excitation, but the energy gain from the covalent bond is actually higher than the cost. Therefore, in an excited state, a carbon atom has four equivalent orbitals 2s, $2p_x$, $2p_y$ and $2p_z$ available to form covalent bonds with other atoms. The superposition of the 2s orbital with n (=1, 2 or 3) 2p orbitals is called sp^n hybridisation [12]. Hybridisation plays an essential role in covalent carbon bonds, and it is the reason why carbon can form a variety of materials such as diamond, graphite, graphene, carbon nanotubes and fullerenes. When carbon atoms combine to form graphene, sp^2 hybridisation or the superposition of the 2s orbital with two 2p orbitals (i.e. $2p_x$ and $2p_y$) occurs. The 2s orbital and two 2p orbitals form three hybrid sp^2 orbitals within one plane and have a mutual angle of 120°. These three sp^2 orbitals interact with the nearest neighbours of the carbon atoms to form in-plane σ bonds. The σ bonds are strong covalent bonds that bind the carbon atoms in the plane. The remaining $2p_z$ orbital forms a π bond, which is perpendicular to the atomic plane. The out-of-plane π bond is much weaker than the in-plane σ bond. The π bond plays an important role in the out-of-plane interactions, such as van der Waals interaction. Figure 1.2 illustrates the bonding mechanism between carbon atoms in graphene. Two adjacent hexagonal lattices are displayed, and the solid circles denote the nuclei of carbon atoms.

As shown in Figure 1.3, the unit cell of the graphene hexagonal crystalline lattice is made of two nonequivalent carbon atoms. Within the atomic plane, the distance between two nearest carbon atoms is 0.142 nm [13]. Graphene has two main types of edge terminations, zigzag and armchair, which are commonly used for referring to the directions of the graphene lattice. With a thickness of single atomic layer, graphene is the thinnest material known to mankind. Graphene can be regarded as the basic building unit of other carbon allotropes. For example, it can be wrapped up to form zero-dimensional fullerenes, rolled into one-dimensional carbon nanotubes or stacked up to form three-dimensional graphite.

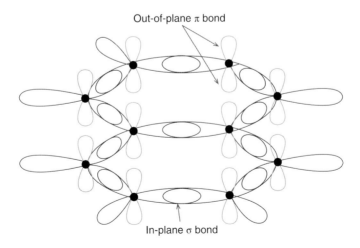

Figure 1.2   Bonding mechanism between carbon atoms in graphene

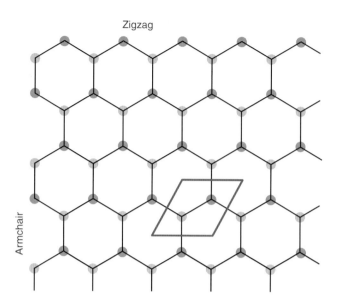

Figure 1.3   Unit cell and directionality of graphene

1.2.1.2   Synthesis

The initial approach used by Geim and Novoselov to produce graphene was simply mechanical exfoliation (i.e. repeated peeling) of highly oriented pyrolytic graphite [3]. In the recent years, both academia and industry have invested great effort in developing the techniques for mass production of graphene. In general, the methods for graphene synthesis can be classified into two categories, namely bottom-up method and top-down method.

Bottom-Up Method
The bottom-up methods for synthesising graphene mainly include chemical vapour deposition (CVD) and epitaxial growth. Abundant reports are available on the growth of graphene using CVD [14–30]. In this process, a substrate is exposed to thermally decomposed precursors, and the desired graphene layers are deposited on to the substrate surface at high temperatures. Various carbon-rich materials such as hydrocarbons can be used as the precursors. The substrate acts as a transition surface and it is usually made of metals such as nickel, copper, cobalt, ruthenium and platinum. On cooling the substrate, the solubility of carbon in the transition metal decreases, and the graphene layer is expected to precipitate from the surface. In 2009, Reina et al. [14] demonstrated the growth of continuous single- to few-layered graphene by CVD on polycrystalline nickel and transfer of graphene to a large variety of substrates using a wet etching method. They used diluted hydrocarbon gas flow as the precursor, and the growth was done at temperatures of 900 °C to 1000 °C and at ambient pressure. The resulting film exhibited a large fraction of single- and bilayered graphene regions with up to approximately 20 µm in lateral dimension. In 2010, Srivastava et al. [15] reported the growth of centimetre-sized, uniform and continuous single- and few-layered graphene by CVD on polycrystalline copper foils with liquid hexane precursor. Structural characterisations using the Raman spectroscopy, transmission electron microscopy and atomic force microscopy

suggested that the as-grown graphene has single to few layers over large areas and is highly continuous. This study demonstrated that these graphene layers can be easily transferred to any desired substrate without damage by dissolving the copper foil in diluted nitric acid.

Epitaxial growth is another substrate-based method for synthesising graphene, and it has been widely explored in the last few years [31–40]. In this method, the graphene layer is obtained from the high temperature reduction of silicon carbide substrates in an ultrahigh vacuum. As silicon is removed around 1000 °C, a carbon-rich layer is left behind, which later undergoes graphitisation to form graphene. Berger et al. [31] produced one- to three-layered graphene by epitaxial growth on the silicon-terminated (0001) surface of single-crystal 6H–SiC. In their work, the process started with surface preparation by H_2 etching. Next, the 6H–SiC samples were heated by electron bombardment in ultrahigh vacuum to approximately 1000 °C. Finally, the graphene layers were formed when the samples were heated to temperatures of 1250 °C to 1450 °C for up to 20 min. In 2006, Rollings et al. [32] used a similar epitaxial process and synthesised graphene down to one atom layer on the (0001) surface of 6H–SiC.

The synthesis of graphene by CVD and by epitaxial growth has several major advantages. Firstly, these methods produce graphene of high quality and high purity. Secondly, both methods are able to produce graphene in large sizes, in theory, up to an entire wafer. Thirdly, they are compatible with current semiconductor processes. By controlling the process parameters, graphene in desired morphology, shape and size is achievable. However, both methods suffer the disadvantages of high costs and complex processes.

Top-Down Method

The top-down methods for synthesising graphene mainly include mechanical, chemical and thermal exfoliation/reduction of graphite or graphite oxide. Mechanical exfoliation of graphite is the first recognised method for the synthesis of graphene. Graphite is actually graphene layers stacked together by weak van der Waals forces. The interlayer distance is 3.34 Å and the van der Waals binding energy is approximately 2 eV/nm². Mechanical cleaving of graphite into graphene of single atomic layer is achievable by applying an external normal force of about 300 nN/μm² [41]. Novoselov et al. [3] used a scotch tape to gradually peel off the graphite flakes and release them in acetone. The flakes were then transferred from the acetone solution to a silicon substrate. After cleaning with water and propanol, thin flakes of graphene including single-layered graphene were obtained. The mechanical exfoliation method can produce graphene of large size and high quality but in very limited quantities, which makes it only suitable for experimental purposes. In order to be applied in industry, this process needs to be further improved.

Chemical exfoliation of graphite is a process in which chemicals are intercalated with the graphite structure to isolate the graphene layers. Recent research studies reported that graphite can be chemically exfoliated into single- and few-layered graphene in the presence of per-fluorinated aromatic solvents [42], organosilane [43], *N*-methyl-pyrrolidone [44] and chloro-sulphonic acid [45]. The chemical exfoliation processes are usually assisted with sonication that facilitates the solubilisation and exfoliation of graphite. Chemical exfoliation has several advantages. Firstly, it produces graphene at a low cost. Secondly, the process is relatively simple and does not rely on complex infrastructure. Thirdly, the process is capable of depositing graphene on a wide variety of substrates and it can be extended to produce graphene-based composites. Most importantly, the process is scalable for high-volume manufacturing.

The reduction of graphite oxide is another promising method for graphene synthesis. This process starts from oxidising the raw material of graphite to graphite oxide. Through oxidisation, the interlayer distance between graphene layers increases from 3.34 Å to 7 Å because of the presence of oxygen moieties [46]. The interlayer van der Waals interactions in graphite oxide become much weaker in comparison to the original graphite structure. It is, therefore, much easier to exfoliate graphite oxide than it is to exfoliate graphite directly into graphene. Abundant reports have demonstrated that exfoliation and reduction of graphite oxide to graphene can be achieved through chemical and/or thermal processes [46–57]. The synthesis of graphene through reduction of graphite oxide has basically the same advantages as the chemical exfoliation method. Thus, the reduction of graphite oxide is another promising method for achieving mass production of graphene at a low cost.

To sum up, a number of methods have already been well developed for the synthesis of graphene, some of which are transferable to mass production at low cost. Nowadays, graphene has become a readily available material for use in various applications.

1.2.2 Properties of Graphene

Owing to its unique two-dimensional structure and sp^2 carbon bonding mechanism, graphene has been discovered to possess several extraordinary properties that endow it for a variety of potential applications.

1.2.2.1 Electrical Property

The charge carriers experience a very weak scattering while transporting in graphene. Recent experiments reported that the charge carrier mobility reaches more than $20,000 \, cm^2/V \, s$ at room temperature. This value is an order of magnitude higher than the most commonly used electronic material – silicon [58]. Graphene is one of the most promising replacements for silicon in electronic applications. It is believed that graphene-based field-effect transistors will exceed the frequency limits of conventional planar transistors, thereby paving the way for high-speed electronic devices. Intensive research studies have been carried out on this application of graphene and significant progress has been made. In 2010, Lin et al. [59] presented field-effect transistors fabricated on a 2-in. graphene wafer with a cut-off frequency as high as 100 GHz. Liao et al. [60] fabricated graphene-based field-effect transistors with a record of 300-GHz performance. In spite of the promising application of graphene in electronics, one of the major obstacles is that pristine graphene acts as a semi-metal with a zero electronic band gap at room temperature [61]. This feature of pristine graphene makes it unsuitable for the transistor digital circuits because of the relatively strong inter-band tunnelling in the field-effect transistor off state [62, 63]. Research efforts have been devoted to develop approaches to open the band gap of graphene and fabricate graphene-based field-effect transistor with a sufficiently large on/off ratio [64–66].

1.2.2.2 Mechanical Property

Graphene has been confirmed to be the strongest material ever discovered. In 2008, Lee et al. [67] measured the elastic properties and intrinsic breaking strength of a free-standing monolayer graphene by nanoindentation in an atomic force microscopy. They demonstrated that defect-free graphene possesses a Young's modulus of 1 TPa and an ultimate strength of 130 GPa. The experimental findings have been validated by theoretical and simulation works [68, 69]. The outstanding mechanical properties of graphene have attracted interest in a wide range of applications. For instance, the light but stiff graphene serves as a great reinforcement material for building nano-electro-mechanical systems [70, 71]. Many studies have been reported that graphene–polymer composites possess significantly enhanced Young's modulus and ultimate strength over their pristine polymer counterparts [72–76].

1.2.2.3   Optical Property

Another interesting property of graphene is its optical transparency, which is particularly important for optoelectronic applications. Experimental reports confirmed that the optical absorption of a monolayer graphene over the visible spectrum is 2.3%, and it linearly increases with an increase in the layer number [77–79]. This optical transparency of graphene in conjunction with its exceptional electrical and mechanical properties could lead to numerous novel photonic devices. The most direct applications of graphene are flexible electronic products, such as flexible screen displays, electronic papers and organic light-emitting diodes [80–82]. In these applications, the properties of graphene make it superior to the traditional material – indium tin oxide. Other potential applications of graphene include its use in photodetectors, solar cells, mode-locked lasers and optical modulators [83, 84].

1.2.3   Thermal Conductivity of Graphene

Conduction is one of the three ways of heat transfer, with the other two ways being convection and radiation. The basic principle of heat transfer by conduction is governed by Fourier's law. According to Fourier's law, the time rate of heat transfer through a material is directly proportional to the cross-sectional area and the negative temperature gradient in the heat transfer direction. Fourier's law may be expressed as

$$\frac{\partial Q}{\partial t} = -kA\frac{\partial T}{\partial x} \tag{1.1}$$

where Q is the heat, t the time, A is the cross-sectional area, $\partial T/\partial x$ is the temperature gradient along the heat transfer direction x. In Eq. (1.1), k is the thermal conductivity, which is a material-dependent constant. The thermal conductivity of a material is given by

$$k = \frac{1}{3}Cvl \tag{1.2}$$

where C is the specific heat, v is the average velocity of the thermal energy carriers and l is the mean free path of thermal energy carriers. The thermal energy carriers in non-metals are mainly phonons. In SI units, the thermal conductivity is measured in W/m K.

One of the superior properties of graphene is its extremely high thermal conductivity. In 2008, Balandin and his co-workers [85, 86] reported the first experimental investigation on the thermal conductivity of suspended single-layered graphene. The measurement was conducted in a non-contact manner using the confocal micro-Raman spectroscopy. Graphene has clear signatures in Raman spectra, and the G peak in graphene spectra manifest strong temperature dependence [87–91]. The temperature sensitivity of G peak allows one to monitor the local temperature change produced by the variation of the laser excitation power focused on a graphene layer. In a properly designed experiment, the local temperature rises as a function of the laser power and it can be utilised to extract the thermal conductivity. The single-layered graphene samples were suspended on a 3-μm- wide trench fabricated on a Si/SiO$_2$ substrate. Laser light of 488 nm in wavelength was chosen for this experiment, because it gives both efficient heating on the sample surface and clear Raman signatures for graphene. Using the aforementioned methodology, they discovered that the thermal conductivity value of a suspended single-layered graphene is up to 5300 W/m K at room temperature. The phonon mean free path in graphene was calculated to be approximately 775 nm. This experimental data was later confirmed by theoretical work. In 2009, Nika et al. [92, 93] conducted a theoretical study on the phonon thermal conductivity of a single-layered graphene by using the phonon dispersion obtained from the valence force field method, and treating the three-phonon Umklapp processes accounting for all phonon relaxation channels in the two-dimensional Brillouin zone of graphene. The uniqueness of graphene was reflected in the two-dimensional phonon density of states and restrictions on the phonon Umklapp scattering phase-space. They found that the thermal conductivity of a single-layered graphene at room temperature is in the range of 2000–5000 W/m K depending on the flake width, defect concentration and roughness of the edges, which is in good agreement with the experimental data.

Significant research studies have been pursued to further understand the thermal properties of graphene. In 2010, by using the same experimental technique based on confocal micro-Raman spectroscopy [85, 86], Ghosh et al. [94] showed that the thermal conductivity of graphene is dependent on the number of graphene layers. As the number of graphene layers increases from 2 to 4, the thermal conductivity changes from 2800 to 1300 W/m K. They explained the evolution of thermal transport from two-dimensional to bulk by the cross-plane coupling of phonons and the change in phonon Umklapp scattering. In a single-layered graphene, phonon Umklapp scattering is quenched and the thermal transport is limited mostly by the in-plane edge boundary scattering. While in a bilayered graphene, thermal conductivity decreases because the phonon Umklapp scattering increases as a result of cross-plane coupling. When the layer number increases to 4, the thermal conductivity is even lower because of stronger extrinsic effects resulted from the non-uniform thickness of samples. Also due to increased Umklapp scattering and extrinsic effects, several studies reported that the thermal conductivity of graphene supported by substrate is lower when compared to the suspended one [95–97]. In 2012, Chen et al. [98] reported their experimental study of the isotope effects on the thermal conductivity of graphene. They found that the thermal conductivity of graphene sheets composed of a 50:50 mixture of ^{12}C and ^{13}C is about half that of pure ^{12}C graphene. The reason is that isotopes, defects, impurities or vacancies reduce the point-defect-limited phonon lifetime, and thus reduce the phonon mean free path and the thermal conductivity.

In general, graphene outperforms all the known materials in heat conduction. The high thermal conductivity, together with its exceptional high specific surface area (2630 m^2/g), makes graphene a very promising material to be applied in thermal management applications [85, 86]. The reported value of greater than 5300 W/m K is more than a magnitude higher than any of the conventional filler materials used in nanocomposites. Therefore, graphene has been recognised as one of the most promising thermally conductive fillers for making high-performance nanocomposites with superior thermal conductivity.

1.3 Thermal Conductivity of Graphene–Polymer Nanocomposites

1.3.1 Measurement of Thermal Conductivity of Nanocomposites

Thermal conductivity of nanocomposites can be measured by either the steady-state methods or the transient methods.

1.3.1.1 Steady-State Method

The principle of the steady-state methods is to establish a steady temperature gradient over a known thickness of the sample and control the heat flow through the sample. The thermal conductivity may be determined from Fourier's law of heat conduction given in Eq. (1.1). The guarded hot plate method is the most representative steady-state method. In this method, the sample is first prepared as a disc with known thickness and area. Then it is placed between two parallel plates with constant but different temperatures, as a hot plate and a cold plate. Both plates have temperature sensors that measure the temperature gradient across the sample. When the equilibrium state is established, the thermal conductivity can be determined on the basis of the temperature gradient across the sample, the electrical power input, the thickness and the area of the sample. In order to minimise the effect of heat transfer through convection, the whole set-up is placed in a high vacuum environment. The effect of heat transfer through radiation is negligible. The details of the guarded hot plate method can be found in ASTM C177 [99] and ISO 8302 [100].

1.3.1.2 Transient Method

Among the transient methods, the most commonly used one is the laser flash method. This method does not measure thermal conductivity directly. Instead, it measures thermal diffusivity from which the thermal conductivity can be calculated through

$$k = \alpha C \rho \tag{1.3}$$

where α is the thermal diffusivity, C is the specific heat capacity and ρ is the density. In the laser flash method, the sample is a disc with a few millimetres in thickness. During the measurement, the front surface of the sample is irradiated with a pulse of energy from a laser and the subsequent temperature rise on the rear surface is recorded. The shape of the temperature–time curve on the rear surface of the sample can be used for extracting the thermal diffusivity of the sample. The details of laser flash method can be found in ASTM E1461 [101].

1.3.2 Modelling of Thermal Conductivity of Nanocomposites

Modelling is not only a powerful tool to predict the thermal conductivity of nanocomposites before synthesis, but it also interprets the experimental results during the development of new nanocomposites. Appropriately derived models are very useful for understanding the relationships between macroscopic thermal transport properties and their microstructures of nanocomposites. Thus, modelling of thermal conductivity of nanocomposites is extremely important for the design and development of new nanocomposites. Since nanocomposites are essentially composites of polymeric matrices with thermally conductive fillers, the theories and models developed for predicting the thermal conductivity of composite materials are readily applicable for nanocomposites. The determination of the effective thermal conductivity and other transport properties of composite materials has been a topic of considerable research interest, especially on theoretical development. After more than a century of research effort, several effective theoretical models have been established. This section gives a historical review of the most commonly used theoretical models for predicting the effective thermal conductivity of composite materials. Based on this review, the critical factors affecting the effective thermal conductivity of composite materials will be clearly identified. It should be emphasised that owing to the mathematical analogy between the subjects of thermal conduction, electrical conduction, electrostatics and magnetostatics, almost all results obtained in one area are readily applicable to the other.

1.3.2.1 Maxwell's Model

Theories for the transport properties of composite materials date back to the early investigations conducted by Maxwell [102]. The derivation of Maxwell's model starts from placing small spheres of resistance k_1 in a medium whose resistance is k_2. The distance between the small spheres is long enough, so that they do not interact with each other. For such a compound medium, Maxwell derived an analytical model for calculating its effective specific resistance based on k_1, k_2 and the volume ratio of all the small spheres to that of the whole compound medium. For thermal conductivity, Maxwell's model gives

$$\frac{k_e}{k_m} = 1 + 3f\frac{k_f - k_m}{2k_m + k_f - f(k_f - k_m)} \tag{1.4}$$

where k_e is the effective thermal conductivity of the compound medium, k_f the thermal conductivity of the small spheres, k_m the thermal conductivity of the medium and f is the ratio of the volume of all the small spheres to that of the whole compound medium. Equation (1.4) is also called the Maxwell–Garnett effective medium approximation. The validity of Maxwell's model has been confirmed by many subsequent theoretical and numerical research studies [103–106]. Maxwell' model has served as the foundation for the development of numerous theoretical models for studying the transport properties of composite materials in the past century. However, Maxwell's model has three main limitations. First, the interfacial thermal resistance between the sphere and medium was not considered in the model. Second, the volume ratio of spheres in the compound medium must be very low so that these spheres do not interfere with each other. Third, the model is only applicable for spherical fillers.

1.3.2.2 Hasselman and Johnson's Model

Based on the original Maxwell's theories, Hasselman and Johnson [107] derived another model for calculating the effective thermal conductivity of composite materials by taking the interfacial thermal resistance between filler and matrix into consideration. The effect of the interfacial thermal resistance is introduced by adding an interfacial conductance factor into the model. For composites filled with spherical fillers, Hasselman and Johnson described the effective thermal conductivity of the composites as

$$k_e = k_m \frac{\left[2\left(\frac{k_f}{k_m} - \frac{k_f}{rh} - 1\right)f + \frac{k_f}{k_m} + \frac{2k_f}{rh} + 2\right]}{\left[\left(1 - \frac{k_f}{k_m} + \frac{k_f}{rh}\right)f + \frac{k_f}{k_m} + \frac{2k_f}{rh} + 2\right]} \tag{1.5}$$

where k_e is the effective thermal conductivity of the composite, k_m the thermal conductivity of the matrix, k_f the thermal conductivity of the filler, f the volume fraction of fillers, r the radius of the spherical fillers and h is the interfacial thermal conductance between filler and matrix. Hasselman and Johnson made significant contributions in modelling the thermal conductivity of composite materials. These original contributions include the allowance for the effect of interfacial thermal resistance, the effect of filler size in the model (which was found to affect the effective thermal conductivity of composites) and the applicability of the models to analyse composites filled with non-spherical fillers (such as circular cylinders and flat plates). As for the limitations of the work by Hasselman and Johnson, they are as follows: (1) similar to Maxwell's model, the model does not consider the interference between fillers and so it can be only applied to dilute concentrations of fillers (i.e. for small volume fractions of fillers) and (2) for composites filled with circular cylinders and flat plates, the model is only applicable when the fillers are orientated perpendicular to heat flow.

1.3.2.3 Benveniste's Model

In 1986, Benveniste and Miloh [108] proposed a model to include the interfacial thermal resistance between filler and matrix in the calculation of the effective thermal conductivity of composites. It is one of the first two models that allowed for the effect of interface. The other one is Hasselman and Johnson's model discussed above which was published in the same year. These two aforementioned papers have common findings. In Benveniste and Miloh's work, they introduced a factor termed as 'skin effect' to account for the effect of interface between filler and matrix. For composites filled with spherical fillers, the model gives

$$\frac{k_e}{k_m} = 1 - 3f \frac{1 - \frac{k_f}{k_m} + \frac{k_f}{r\beta}}{2 + \frac{k_f}{k_m} + \frac{2k_f}{r\beta}} \tag{1.6}$$

where β is the interfacial thermal conductance between filler and matrix and all the other parameters are defined in Eq. (1.5). Similarly, Benveniste and Miloh discovered the filler size effect on the effective thermal conductivity of composites and included it in their model. Apart from spherical fillers, Benveniste and Miloh also presented the models for composites filled with prolate and oblate spheroidal fillers. Both aligned and randomly oriented distributions were considered. However, Benveniste and Miloh's model also neglected the interaction between the fillers, so it is only valid for small volume fractions of fillers. In order to overcome this limitation, Benveniste [109] developed another model for the effective thermal conductivity of composites filled with spherical fillers. In this 1978 work, two different micromechanical methods were used to take into account the interaction between the fillers. One method is the 'generalised self-consistent' scheme [110] while the other is the Mori–Tanaka theory [111]. These two methods are distinctively different in their approaches, but they yield the same analytical expression for the effective thermal conductivity of composites as follows:

$$\frac{k_e}{k_m} = \frac{2(1-f) + \frac{r\beta}{k_m}\left[1 + 2f + \frac{2k_m}{k_f}(1-f)\right]}{2 + f + \frac{r\beta}{k_m}\left[1 - f + \frac{k_m}{k_f}(2+f)\right]} \tag{1.7}$$

The parameters are the same as in Eq. (1.6). This model can be used to calculate the effective thermal conductivity of composites with spherical fillers at a volume fraction up to $f = \pi/6$. Benveniste and Miloh reported the filler shape effect in a later work in 1991 on the modelling of effective thermal conductivity of composites filled with coated short fibres [112].

1.3.2.4 Every's Model

Based on the model derived by Benvensite and Miloh in 1986, Every et al. [113] employed Bruggeman effective medium theory to account for the interaction between fillers and extended the modelling of effective thermal conductivity of composites to high volume fractions. Bruggeman theory assumes that the fields of neighbouring fillers can be taken into account by adding the dispersed fillers into the matrix incrementally. At each increment, the surrounding medium is taken to be the existing composite (i.e. the matrix and the previously added fillers). For composites filled with spherical fillers, Every's model gives

$$(1-f)^3 = \left(\frac{k_m}{k_e}\right)^{(1+2\alpha)/(1-\alpha)} \left[\frac{k_e - k_f(1-\alpha)}{k_m - k_f(1-\alpha)}\right]^{3/(1-\alpha)} \tag{1.8}$$

where $\alpha = a_k/a$, is a non-dimensional factor defined to account for the filler size effect, a is the radius of the fillers and a_k is the Kapitza radius defined as $a_k = R_k k_m$, where R_k is the interfacial thermal resistance between the filler and the matrix.

A number of special cases of Eq. (1.8) are of interest. By setting $\alpha \to 0$, that is the radius of the fillers are much larger than Kapitza radius, Eq. (1.8) becomes

$$(1-f)^3 = \frac{k_m}{k_e}\left(\frac{k_e - k_f}{k_m - k_f}\right)^3 \tag{1.9}$$

In the opposite limit, by setting $\alpha \to \infty$, Eq. (1.8) is reduced to

$$(1-f)^3 = \frac{k_e}{k_m} \tag{1.10}$$

When $k_f \gg k_m$, Eq. (1.8) is simplified as

$$\frac{k_e}{k_m} = \frac{1}{(1-f)^{3(1-\alpha)(1+2\alpha)}} \tag{1.11}$$

As pointed out in this work, both Maxwell model and Bruggeman model predict that the effective thermal conductivity of composites remains unchanged if the radius of the fillers is the same as the Kapitza radius. The contribution of the interfacial thermal resistance is then exactly balanced by the higher thermal conductivity of the fillers. If the fillers are much smaller than the Kapitza radius, the effective thermal conductivity of the composites is lowered by adding the fillers, even if the fillers themselves have a higher thermal conductivity than that of the matrix. Thus, the Kapitza radius and the interfacial thermal resistance have become physically important factors in the design of nanocomposites, where the objective is to increase the thermal conductivity by mixing two different constituents, that is thermally conductive fillers and polymer matrix. The modelling for composites filled with non-spherical fillers was not discussed in this work.

1.3.2.5 Nan's Model

In 1997, based on the Maxwell-based models, Nan et al. [114] developed a general effective medium approximation formulation for modelling the effective thermal conductivity of arbitrary particulate composites with the allowance of interfacial thermal resistance. Nan's model describes the effective thermal conductivity of the composites with equally sized ellipsoidal fillers as

$$k_{e11} = k_{e22} = k_m \frac{2 + f\left[\beta_{11}\left(1 - L_{11}\right)\left(1 + \langle\cos^2\theta\rangle\right) + \beta_{33}\left(1 - L_{33}\right)\left(1 - \langle\cos^2\theta\rangle\right)\right]}{2 - f\left[\beta_{11}L_{11}\left(1 + \langle\cos^2\theta\rangle\right) + \beta_{33}L_{33}\left(1 - \langle\cos^2\theta\rangle\right)\right]}$$

(1.12)

$$k_{e33} = k_m \frac{1 + f\left[\beta_{11}\left(1 - L_{11}\right)\left(1 - \langle\cos^2\theta\rangle\right) + \beta_{33}\left(1 - L_{33}\right)\langle\cos^2\theta\rangle\right]}{1 - f\left[\beta_{11}L_{11}\left(1 - \langle\cos^2\theta\rangle\right) + \beta_{33}L_{33}\langle\cos^2\theta\rangle\right]}$$

(1.13)

with

$$\langle\cos^2\theta\rangle = \frac{\int\rho(\theta)\cos^2\theta\sin\theta d\theta}{\int\rho(\theta)\sin\theta d\theta}$$

(1.14)

$$\beta_{ii} = \frac{k_{ii}^C - k_m}{k_m + L_{ii}\left(k_{ii}^C - k_m\right)}$$

(1.15)

$$k_{ii}^C = k_f/\left(1 + \gamma L_{ii}k_f/k_m\right)$$

(1.16)

$$L_{11} = \begin{cases} \frac{p^2}{2(p^2-1)} - \frac{p}{2(1-p^2)^{3/2}}\cosh^{-1}p, & \text{for } p > 1 \\ \frac{p^2}{2(p^2-1)} + \frac{p}{2(1-p^2)^{3/2}}\cos^{-1}p, & \text{for } p < 1 \end{cases}$$

(1.17)

$$L_{33} = 1 - 2L_{11}$$

(1.18)

$$\gamma = \begin{cases} (2 + 1/p)\,\alpha, & \text{for } p \geq 1 \\ (1 + 2p)\alpha, & \text{for } p \leq 1 \end{cases}$$

(1.19)

$$\alpha = \begin{cases} a_K/a_1, & \text{for } p \geq 1 \\ a_K/a_3, & \text{for } p \leq 1 \end{cases}$$

(1.20)

$$a_K = R_K k_m$$

(1.21)

where k_{eii} is the effective thermal conductivity of the composite along the ii axis, k_m the thermal conductivity of the matrix, k_f the thermal conductivity of the filler, f the volume fraction of the fillers, θ the angle between the composite material axis X_3 and the local filler symmetric axis X'_3, $\rho(\theta)$ the distribution function describing ellipsoidal filler orientation, k_{ii}^C is the equivalent thermal conductivity of a composite unit cell comprising an ellipsoidal fill and its surrounding interface layer along the X'_i symmetric axis. The radii of the ellipsoid along X'_1 or X'_3 axes are a_1 and a_3, respectively. The aspect ratio of the ellipsoid filler is given by $p = a_3/a_1, p > 1$ and $p < 1$ are for a prolate ($a_1 = a_2 < a_3$) and an oblate ($a_1 = a_2 > a_3$) ellipsoidal filler, respectively. A non-dimensional factor α is defined to account for the filler size effect. The Kapitza radius a_k is defined as $a_k = R_k k_m$, where R_k is the interfacial thermal resistance between filler and matrix.

So far, Nan's model has the most general formulation for modelling the effective thermal conductivity of composites. It contains the effects of thermal conductivity of the filler and the matrix, volume fraction, interfacial thermal resistance between the filler and matrix, filler size, filler shape and filler orientation distribution.

1.3.3 Progress and Challenge for Graphene–Polymer Nanocomposites

Recognising the promising application of graphene as thermally conductive fillers in nanocomposites, a large number of experimental attempts have been made by researchers worldwide to fabricate graphene–polymer nanocomposites and to characterise their thermal transport properties [115–133]. In 2012, Shahil et al. [118] conducted experiments to fabricate and characterise nanocomposites made of graphene fillers and epoxy resin matrix. The nanocomposite samples were prepared by mixing solutions of graphene with epoxy resin followed by moulding, degassing and curing in vacuum at a temperature up to 150 °C. The samples were prepared with different graphene loadings varying from 1 to 10 vol%. The sample thickness was around 1 to 1.5 mm. Thermal conductivity of the nanocomposites was measured using the laser flash method. With these setups, they demonstrated a thermal conductivity of 5.1 W/m K for graphene-filled epoxy nanocomposites at a filler loading fraction of 10 vol%, which corresponds to a thermal conductivity enhancement of 2300% over pristine epoxy (i.e. ~0.2 W/m K). By observing the decrease in thermal conductivity with respect to increasing environmental temperature, which is reminiscent of the phonon Umklapp scattering characteristic for crystalline materials, they suggested that the heat conduction in these nanocomposite samples is dominated by the graphene fillers rather than the epoxy matrix. Yu et al. [119] also conducted experiments to fabricate and characterise graphene–epoxy nanocomposites. In their work, the thermal conductivity of nanocomposites was measured using the guarded hot plate method. They reported that nanocomposites with about 25 vol% loading of graphene fillers in epoxy provide a thermal conductivity enhancement of 3000% and a thermal conductivity of 6.44 W/m K. In their experiments, the thermal conductivity enhancement results obtained with graphene fillers significantly outperformed the nanocomposite samples filled with other fillers such as carbon nanotubes, graphite and carbon black. In 2014, Guo et al. [120] used the ball milling method to prepare the graphene-reinforced epoxy nanocomposites. The expectation was that the mechanical ball milling is able to break the van der Waals interactions among the graphene layers and lead to an improved dispersion of graphene in polymer matrices over other mixing methods, such as sonication. With 25 wt% graphene fillers, they reported a thermal conductivity of 2.67 W/m K on their nanocomposites prepared by ball milling, and it outperforms the samples prepared by sonication in the same set of experiments. Debelak et al. [115] reported that their epoxy nanocomposites with 20 wt% graphene fillers yield a thermal conductivity of 4.3 W/m K, which gives about 2000% increase over pure epoxy. In 2011, Xiang et al. [121] prepared nanocomposites using graphene as fillers and paraffin as matrix. They reported that the thermal conductivity of graphene–paraffin nanocomposites with 5 vol% filler loading reaches 2.414 W/m K, which is about 10-fold higher than pure paraffin (i.e. ~0.25 W/m K). In 2014, Warzoha et al. [116] used 20 vol% graphene fillers and paraffin matrix to fabricate nanocomposites. It was reported that the graphene–paraffin nanocomposites has a thermal conductivity of more than 7 W/m K, which is 2800% enhancement over pristine paraffin. This result significantly outperforms the nanocomposite samples prepared in the same set of experiments using other fillers such as multiwalled carbon nanotubes, aluminium and TiO_2. In 2014, by using silicone as the matrix material, Yu et al. [122] produced nanocomposites with graphene fillers. With 4.25 vol% loading of fillers, the nanocomposites are found to have a thermal conductivity of 1.047 W/m K, indicating a 668% increase when compared to that of pure silicone (i.e. 0.13 W/m K). In 2011, Raza et al. [117] reported that the thermal conductivity of the graphene–silicone nanocomposites reaches 1.909 W/m K at a loading fraction of 20 wt%, which equals to a thermal conductivity enhancement of 1100%. So far, numerous experiments have also been carried out to characterise the thermal conductivity of the graphene-based polymer nanocomposites [121, 123–133]. Overall, the reported thermal conductivity of graphene–polymer nanocomposites falls in the range of 1 to 7 W/m K.

As identified by the theoretical models in Section 1.3.2, the effective thermal conductivity of composites can be affected by several key factors. In the case of graphene–polymer nanocomposites, most experimental reports have suggested that the key barrier in achieving superior thermal conductivity lies in the high interfacial thermal resistance between the graphene fillers and polymer matrices [118–126]. To quantitatively understand the impact of interfacial thermal resistance on the effective thermal conductivity of graphene–polymer nanocomposites, Nan's theoretical model may be employed [114]. In the case of graphene–polymer nanocomposites, Nan's model treats graphene as completely disoriented ellipsoidal fillers and describes the effective thermal conductivity of the nanocomposite K_e as

$$K_e = K_m \frac{3 + f \left[2\beta_{11} \left(1 - L_{11} \right) + \beta_{33} (1 - L_{33}) \right]}{3 - f \left[2\beta_{11} L_{11} + \beta_{33} L_{33} \right]} \tag{1.22}$$

The description of the notations used in this equation can be found in Section 1.3.2.5. For illustrative purposes, the parameters in the open literature are chosen for Eq. (1.22). Thermal conductivity of polymer matrix is 0.2 W/m K, thermal conductivity of graphene fillers is 5000 W/m K, volume fraction of fillers is 10%, and size of graphene fillers is 5 μm. Based on these parameters, the impact of interfacial thermal resistance on the effective thermal conductivity of the graphene–polymer nanocomposites, calculated by Nan's model, is plotted in Figure 1.4. It shows that the thermal conductivity of graphene–polymer nanocomposites increases significantly with reduced interfacial thermal resistance, especially when the interfacial thermal resistance is low. Therefore, the reduction of the interfacial thermal resistance between graphene and polymer is a key challenge for the development of graphene–polymer nanocomposites. In the following sections, reviews will be given on interfacial thermal resistance between graphene and polymer. Section 1.3.4 discusses the definition of interfacial

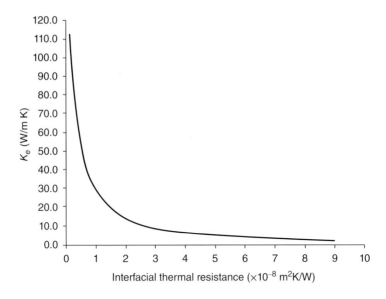

Figure 1.4   Impact of interfacial thermal resistance between filler and matrix on effective thermal conductivity of graphene–polymer nanocomposites

thermal resistance and the key factors affecting it. Section 1.3.5 reviews the approaches for reducing interfacial thermal resistance between graphene and polymer.

1.3.4   Interfacial Thermal Resistance

Interfacial thermal resistance, also known as Kapitza resistance, is a measure of the resistance to heat flow caused by an interface between two materials [134, 135]. It is due to the scattering of energy carriers (i.e. phonons) at the interface and the mismatch in vibrational spectra of different materials. In graphene, heat transfer is mainly conducted by the acoustic phonons and the contribution of electrons to thermal conductivity is negligible [85]. When a phonon attempts to cross the graphene–polymer interface, it will scatter at the interface due to the mismatch in the vibrational spectra of graphene and polymer. This phonon scattering is described as the interfacial thermal resistance between them in nanocomposites.

At the macroscopic scale, the interfacial thermal resistance R_K is defined in terms of the heat flux J, and the resultant temperature discontinuity ΔT, across the interface by

$$R_K = \Delta T / J \tag{1.23}$$

In SI units, the interfacial thermal resistance is measured in m^2K/W. The acoustic mismatch model and the diffuse mismatch model [136] are the two commonly used theoretical models for determining the interfacial thermal resistance and for understanding the factors affecting it.

1.3.4.1 Acoustic Mismatch Model

The acoustic mismatch model assumes that phonons are governed by continuum acoustics, that is, they behave like plane waves. It also assumes that the interface between two different materials is a perfect plane, and that the materials in which the phonons propagate are continua. The acoustic mismatch model assumes that no scattering occurs at the interface. Thus, when a phonon is incident on the interface, there are only four possible results. It can reflect, reflect and mode convert, refract, or refract and mode convert. The angle of reflection or refraction, as well as the probabilities of each, follows the Snell's law and is a consequence of the general principle of detailed balance. The exact probability of phonon transmission across an interface is complicated to derive, while an isotropic simplification is often made in calculating the phonon transmission probability using the acoustic mismatch model. For a phonon with normal incidence, its transmission probability from material 1 to material 2 may be given by

$$\alpha_{1\to 2} = \frac{4Z_2 Z_1}{(Z_1 + Z_2)^2} \tag{1.24}$$

with $Z_i = \rho_i v_i$ as the acoustic impedance of material i, ρ_i the density of material i, and v_i the phonon velocity in material i. Based on this estimation of phonon transmission probability, the interfacial thermal resistance between the matrix and filler in nanocomposites may be approximated as

$$R_{K,\mathrm{AMM}} = \frac{4}{Cv\alpha} \tag{1.25}$$

where C is the specific heat, v the Debye phonon velocity of the matrix and α is an appropriately averaged transmission probability across the matrix–filler interface [136, 137].

1.3.4.2 Diffusive Mismatch Model

The diffuse mismatch model differs from the acoustic mismatch model in the derivation of phonon transmission probability. Diffuse mismatch model assumes that all the phonons are diffusively scattered at the interface between two materials. The acoustic correlations at the interface are assumed to be completely destroyed by diffuse scattering. In other words, the scattered phonons lose all memory of polarisation and incident angle. Thus, the probability of phonon transmission across the interface is only determined by the mismatch in the vibrational density of states of the two materials [136, 137].

By using the diffuse mismatch model, the transmission probability of phonon with energy $\hbar\omega$ and mode j (longitudinal or transverse) from material i to $3 - i$ is given by

$$\alpha_i(\omega) = \frac{\sum_j v_{3-i,j} N_{3-i,j}(\omega, T)}{\sum_{i,j} v_{i,j} N_{i,j}(\omega, T)} \tag{1.26}$$

where $v_{i,j}$ is the phonon velocity in material i with mode j, and $N_{i,j}(\omega, T)$ is the density of phonons with energy $\hbar\omega$ in material i with mode j at temperature T. By using the Debye approximation for the phonon velocities and phonon densities of states, the interfacial thermal resistance may be written as

$$
R_{K,\text{DMM}} = \left[1.02 \times 10^{10} \frac{\left(\sum_j v_{i,j}^{-2} \right) \left(\sum_j c_{3-i,j}^{-2} \right)}{\sum_{i,j} v_{i,j}^{-2}} \right]^{-1} T^{-3} \tag{1.27}
$$

1.3.4.3 Vibrational Density of States

Vibrational (or phonon) density of states of a material describes its number of states per interval of energy at each energy level that are available to be occupied by phonons. The distribution of density of states is continuous. A high density of states at a specific energy level means that there are many states available for occupation. A zero density of states means that no states can be occupied at that energy level. In quantum mechanical systems, waves or wave-like particles can occupy states with wavelengths and propagation directions dictated by the system. Often only specific states are permitted. It can happen that many states are possible at a specific wavelength, and no states are available at the other wavelength. This distribution is characterised by the plots of the density of states.

Thermal properties such as heat capacity and thermal conductivity as well as many other material properties are strongly influenced by the vibrational density of states. As suggested by both the acoustic mismatch model and the diffuse mismatch model, the transmission probability of a phonon across the interface between two materials is strongly dependent on the existence of the common vibrational density of states in the two materials. In other words, the interfacial thermal resistance is determined by the mismatch in the vibrational density of states of the two materials [136, 137].

1.3.4.4 Molecular Dynamics Simulations for Interfacial Thermal Resistance

The interfacial thermal resistance at the interfaces among different materials have been studied by experiments [138–140] and numerical simulations [141–147]. Since experiments on thermal energy transport lack the control and sensitivity at the atomic scale, molecular dynamics simulations provide an accurate and close observation of the nanoscale thermal transport behaviours, including interfacial thermal resistance. Several molecular dynamics simulation works have been conducted on studying the graphene–polymer interfacial thermal resistance [148–150]. In 2012, Luo and Lloyd [148] studied the thermal transport behaviours across the interface between graphene and paraffin wax ($C_{30}H_{62}$) by using the molecular dynamics simulation method. In their work, firstly a composite system that consists of two amorphous polymer blocks was built with graphene in between, as shown in Figure 1.5. By applying and maintaining a heat source and heat sink at the two ends of the graphene–polymer composite system, a temperature gradient can be established across the system after steady state is reached. Because of the existence of interfacial thermal resistance, the temperature profile experiences a sudden temperature jump in the graphene–polymer interface (see Figure 1.5). The value of interfacial thermal resistance can be calculated by using Eq. (1.23). In the system under consideration, the interfacial thermal resistance between graphene and paraffin is found to be about 1.4–1.7×10^{-8} m^2 K/W. Luo and Lloyd [148] used vibrational density of states to interpret the results. The vibrational density of states is obtained by taking

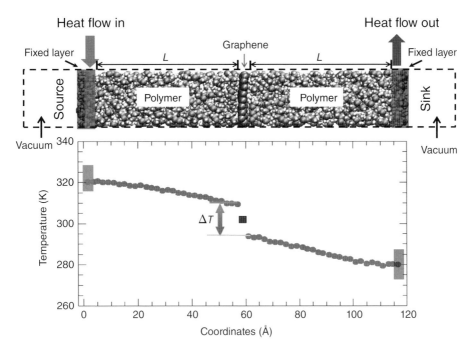

Figure 1.5 A graphene–polymer composite system and steady-state temperature gradient using molecular dynamics simulation [148]

the Fourier's transformation of the velocity autocorrelation functions of atoms in the system. Figure 1.6 shows the vibrational density of states of polymer atoms and graphene atoms in the composite system. The poor overlap in the vibrational density of states between graphene and polymer indicates that they do not couple each other; thus, the interfacial thermal resistance is high. In 2014, Liu et al. [149] studied the thermal transport behaviours at the graphene–octane interfaces by carrying out molecular dynamics simulations. They reported that the interfacial thermal resistance varies between 0.7 and 20×10^{-8} m^2 K/W depending on the modes of heat transfer in the composite system. Hu et al. [150] conducted molecular dynamics simulations to investigate the thermal transport behaviours across the graphene–phenol resin interfaces. They reported an interfacial thermal resistance of $0.9–10 \times 10^{-8}$ m^2 K/W depending on the different locations of heat source. These recent molecular dynamics simulation works have contributed to the good progress in quantifying the interfacial thermal resistance at graphene–polymer interfaces.

1.3.5 Approaches for Reduction of Interfacial Thermal Resistance

Recognising the critical impact of interfacial thermal resistance on the thermal conductivity of graphene–polymer nanocomposites, a number of researchers have started working on possible methods to reduce it. Because high graphene–polymer interfacial thermal resistance is due to the poor coupling in vibrational density of states of graphene and polymer, the fundamental principle for reducing the interfacial thermal resistance is to adjust the vibrational density of

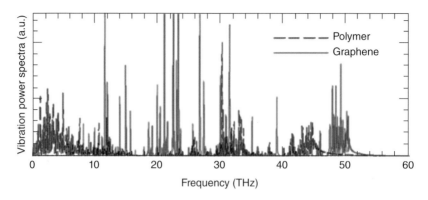

Figure 1.6 Vibrational density of states of polymer atoms and graphene atoms [148]

states for better matching. Up to now, researchers have focused on two methods: covalent functionalisation and non-covalent functionalisation in graphene.

1.3.5.1 Covalent Functionalisation

Covalent functionalisation has been recognised as an effective method for promoting interfacial acoustic coupling between carbon nanostructures and polymers. Earlier research studies have proven that it is able to enhance the interfacial thermal transport across carbon nanotube–polymer interfaces [151–155]. Recently, researchers started investigating the application of covalent functionalisation on graphene. A few experiments have been conducted to measure the thermal conductivity of nanocomposites filled with covalently functionalised graphene fillers [156–166]. In the experiments, the functional groups and polymers used for sample preparation vary significantly. Ganguli et al. [156] prepared graphene fillers covalently functionalised with silane via chemical reaction. By using epoxy as the matrix, two types of nanocomposites were prepared with the silane functionalised graphene fillers and unmodified graphene fillers, respectively. They found that the thermal conductivity of nanocomposites filled with functionalised graphene fillers is up to 50% higher than the nanocomposites filled with unmodified graphene fillers. These results demonstrated that covalent functionalisation provides a significant improvement on the overall thermal conductivity of the graphene–polymer nanocomposites. In 2014, Gu et al. [157] prepared nanocomposites using polyphenylene sulphide matrix and graphene fillers covalently functionalised with titanate (NDZ-105: $C_{54}H_{106}O_7Ti$). With a 29.3 vol% functionalised graphene fillers, the resultant thermal conductivity of the nanocomposites reaches 4.414 W/m K, which is 19 times higher than that of the pure polyphenylene sulphide matrix (i.e. 0.226 W/m K). This improvement of the thermal conductivity is partially due to the functionalisation of graphene that minimises the interfacial thermal resistance. Several other experiments have also reported thermal conductivity measurements on various nanocomposites filled with covalently functionalised graphene-based fillers [158–166]. For example, Chatterjee et al. [158] prepared nanocomposites using epoxy matrix and graphene fillers covalently functionalised with amine. Fang et al. [159] synthesised nanocomposites using polystyrene matrix and graphene fillers covalently functionalised with polystyrene. In these experiments, the filler volume fractions in the nanocomposite samples are relatively low (i.e. 0.1–2.5 wt%), and so the resultant thermal conductivity improvement is not significant (i.e. in the range of 15%–200%).

Very recently, two simulation studies on molecular dynamics have been carried out to investigate the effect of covalent functionalisation on enhancing thermal transport across the graphene–polymer interfaces. In 2014, Wang et al. [167] conducted molecular dynamics simulations to study the thermal transport across the interface between covalently functionalised graphene and polyethylene. In their simulations, a heat source was placed at the centre of the polymer model while two heat sinks were placed at the two ends to generate heat flux. The graphene was covalently functionalised with hydrocarbon chains ($C_{15}H_{31}$) at different densities and the samples were simulated in the polymer system individually. After equilibrium, the interfacial thermal resistance values can be obtained for different models by using Eq. (1.23). As a result, they demonstrated that the thermal transport across graphene–polyethylene interface can be enhanced by 300% when the density of covalent functionalisation with $C_{15}H_{31}$ hydrocarbon chains reaches $0.015\,Å^{-2}$. Konatham et al. [168] performed molecular dynamics simulations to study the thermal transport at the graphene–octane interface. By applying covalent functionalisation of hydrocarbon chains on the edges of graphene sheets, the graphene–octane interfacial thermal resistance was reduced.

1.3.5.2 Non-covalent Functionalisation

Very recently, non-covalent functionalisation has been used to reduce the interfacial thermal resistance between the graphene–polymer interfaces [169–171]. The principle behind this method is that the chemical structure of the functional molecules can be tuned to couple the vibrational density of states of graphene with that of the polymer matrix. One of the key advantages for non-covalent functionalisation is that it does not introduce defects or add atoms to the graphene basal plane. Thus, it does not disrupt the intrinsic thermal conduction property of pristine graphene. Two experimental studies have been carried out by using this new method to enhance the thermal conductivity of graphene–polymer nanocomposites. Teng et al. [169] used poly(glycidyl methacrylate) containing localised pyrene groups to functionalise graphene non-covalently. The non-covalently functionalised graphene is then dispersed in epoxy matrix to synthesise nanocomposites. With 4 wt% filler loading, it is reported that the non-covalently functionalised graphene–polymer nanocomposites show about 20% higher thermal conductivity than the pristine one. In 2013, Song et al. [170] performed non-covalent functionalisation on graphene with 1-pyrenebutyric acid. The non-covalently functionalised graphene was then dispersed in an epoxy matrix. It is found that the non-covalently functionalised graphene–polymer nanocomposites reach 1.53 W/m K at 10 wt% filler loading. In 2013, Lin et al. [171] conducted molecular dynamics simulations to study the thermal transport across the non-covalently functionalised graphene–octane interface. Alkylpyrene was used as the functional molecule and the length of linear alkyl chain varies C_4H_9, C_8H_{17} and $C_{12}H_{25}$. The functional molecules were placed at the graphene–octane interfaces without covalent bonding. The heat source was placed at the centre and heat sinks were placed at the two ends of the polymer system. The interfacial thermal resistance was obtained for different simulation cases by using Eq. (1.23). The results showed that the C_8-pyrene molecule can enhance the interfacial thermal transport by about 22% when compared to that without functional molecule. The molecular dynamics simulations have demonstrated that non-covalent functionalisation is promising in reducing the graphene–octane interfacial thermal resistance.

1.4   Concluding Remarks

Thermal conductivity of graphene has been investigated by both experimental and theoretical studies. It has been established that graphene is hitherto the best thermal conductor ever known to mankind. Many researchers have devoted their efforts to produce graphene–polymer nanocomposites with superior thermal conductivity, and some progress has been achieved. Currently, interfacial thermal resistance at the graphene–polymer interfaces is the key barrier in further improving the thermal conductivity of graphene–polymer nanocomposites. Recent research studies have shown that covalent and non-covalent functionalisation techniques are promising in reducing the interfacial thermal resistance. In order to achieve graphene–polymer nanocomposites with superior thermal conductivity, further in-depth research studies are needed to explore the mechanisms of thermal transport across the graphene–polymer interfaces.

References

[1] Wallace, P.R. (1947) The band theory of graphite. *Physical Review*, **71**, 622–634.

[2] Mermin, N. (1968) Crystalline order in two dimensions. *Physical Review*, **176**, 250–254.

[3] Novoselov, K.S., Geim, A.K., Morozov, S.V. et al. (2004) Electric field effect in atomically thin carbon films. *Science*, **306**, 666–669.

[4] Novoselov, K.S., Geim, A.K., Morozov, S.V. et al. (2005) Two-dimensional gas of massless Dirac fermions in graphene. *Nature*, **438**, 197–200.

[5] Geim, A.K. and Novoselov, K.S. (2007) The rise of graphene. *Nature Materials*, **6**, 183–191.

[6] Li, D. and Kaner, R.B. (2008) Materials science. Graphene-based materials. *Science*, **320**, 1170–1171.

[7] Schedin, F., Geim, A.K., Morozov, S.V. et al. (2007) Detection of individual gas molecules adsorbed on graphene. *Nature Materials*, **6**, 652–655.

[8] Elias, D.C., Nair, R.R., Mohiuddin, T.M. et al. (2009) Control of graphene's properties by reversible hydrogenation: evidence for graphane. *Science*, **323**, 610–613.

[9] Meyer, J.C., Geim, A.K., Katsnelson, M.I. et al. (2007) The structure of suspended graphene sheets. *Nature*, **446**, 60–63.

[10] Stolyarova, E., Rim, K.T., Ryu, S. et al. (2007) High-resolution scanning tunneling microscopy imaging of mesoscopic graphene sheets on an insulating surface. *Proceedings of the National Academy of Sciences of the United States of America*, **104**, 9209–9212.

[11] Brown, T.L., Lemay, H.E., Bursten, B.E. et al. (2012) *Chemistry: The Central Science*, 12 edn, Prentice Hall, Boston.

[12] Dresselhaus, M.S., Dresselhaus, G., Sugihara, K. et al. (1988) *Graphite Fibers and Filaments*, Springer-Verlag.

[13] Enoki, T. and Ando, T. (2013) *Physics and Chemistry of Graphene: Graphene to Nanographene*, Taylor & Francis, Boca Raton.

[14] Reina, A., Jia, X., Ho, J. et al. (2009) Large area, few-layer graphene films on arbitrary substrates by chemical vapour deposition. *Nano Letters*, **9**, 30–35.

[15] Srivastava, A., Galande, C., Ci, L. et al. (2010) Novel liquid precursor-based facile synthesis of large-area continuous, single, and few-layer graphene films. *Chemistry of Materials*, **22**, 3457–3461.

[16] Wang, J.J., Zhu, M.Y., Outlaw, R.A. et al. (2004) Synthesis of carbon nanosheets by inductively coupled radio-frequency plasma enhanced chemical vapor deposition. *Carbon*, **42**, 2867–2872.

[17] Wang, X.B., You, H.J., Liu, F.M. et al. (2009) Large-scale synthesis of few-layered graphene using CVD. *Chemical Vapor Deposition*, **15**, 53–56.

[18] Wang, Y., Chen, X.H., Zhong, Y.L. et al. (2009) Large area, continuous, few-layered graphene as anodes in organic photovoltaic devices. *Applied Physics Letters*, **95**, 063302.

[19] Yuan, G.D., Zhang, W.J., Yang, Y. et al. (2009) Graphene sheets via microwave chemical vapor deposition. *Chemical Physics Letters*, **467**, 361–364.

[20] Zhang, Y., Gomez, L., Ishikawa, F.N. et al. (2010) Comparison of graphene growth on single-crystalline and polycrystalline Ni by chemical vapour deposition. *Journal of Physical Chemistry Letters*, **1**, 3101–3107.

[21] Zhu, M.Y., Wang, J.J., Holloway, B.C. et al. (2007) A mechanism for carbon nanosheet formation. *Carbon*, **45**, 2229–2234.

[22] Song, H.J., Son, M., Park, C. et al. (2012) Large scale metal-free synthesis of graphene on sapphire and transfer-free device fabrication. *Nanoscale*, **4**, 3050–3054.

[23] Chae, S.J., Gunes, F., Kim, K.K. et al. (2009) Synthesis of large-area graphene layers on poly-nickel substrate by chemical vapour deposition: wrinkle formation. *Advanced Materials*, **21**, 2328–2333.

[24] Dervishi, E., Li, Z., Watanabe, F. et al. (2009) Large-scale graphene production by RF-cCVD method. *Chemical Communications*, 4061–4063.

[25] Di, C.A., Wei, D.C., Yu, G. et al. (2008) Patterned graphene as source/drain electrodes for bottom-contact organic field-effect transistors. *Advanced Materials*, **20**, 3289–3293.

[26] Kim, K.S., Zhao, Y., Jang, H. et al. (2009) Large-scale pattern growth of graphene films for stretchable transparent electrodes. *Nature*, **457**, 706–710.

[27] Li, X., Cai, W., An, J. et al. (2009) Large-area synthesis of high-quality and uniform graphene films on copper foils. *Science*, **324**, 1312–1314.

[28] Obraztsov, A.N. (2009) Chemical vapour deposition: Making graphene on a large scale. *Nature Nanotechnology*, **4**, 212–213.

[29] Obraztsov, A.N., Obraztsova, E.A., Tyurnina, A.V. and Zolotukhin, A.A. (2007) Chemical vapour deposition of thin graphite films of nanometer thickness. *Carbon*, **45**, 2017–2021.

[30] Somani, P.R., Somani, S.P. and Umeno, M. (2006) Planer nano-graphenes from camphor by CVD. *Chemical Physics Letters*, **430**, 56–59.

[31] Berger, C., Song, Z.M., Li, T.B. et al. (2004) Ultrathin epitaxial graphite: 2D electron gas properties and a route toward graphene-based nanoelectronics. *Journal of Physical Chemistry B*, **108**, 19912–19916.

[32] Rollings, E., Gweon, G.H., Zhou, S.Y. et al. (2006) Synthesis and characterization of atomically thin graphite films on a silicon carbide substrate. *Journal of Physics and Chemistry of Solids*, **67**, 2172–2177.

[33] Seyller, T., Bostwick, A., Emtsev, K.V. et al. (2008) Epitaxial graphene: a new material. *Physica Status Solidi B*, **245**, 1436–1446.

[34] Sprinkle, M., Soukiassian, P., De Heer, W.A. et al. (2009) Epitaxial graphene: the material for graphene electronics. *Physica Status Solidi RRL: Rapid Research Letters*, **3**, A91–A94.

[35] Berger, C., Song, Z.M., Li, X.B. et al. (2007) Magnetotransport in high mobility epitaxial graphene. *Physica Status Solidi A: Applications and Materials Science*, **204**, 1746–1750.

[36] De Heer, W.A., Berger, C., Wu, X. et al. (2007) Epitaxial graphene. *Solid State Communications*, **143**, 92–100.

[37] Kedzierski, J., Hsu, P.L., Healey, P. et al. (2008) Epitaxial graphene transistors on SIC substrates. *IEEE Transactions on Electron Devices*, **55**, 2078–2085.

[38] Mattausch, A. and Pankratov, O. (2008) Density functional study of graphene overlayers on SiC. *Physica Status Solidi B*, **245**, 1425–1435.

[39] Ni, Z.H., Chen, W., Fan, X.F. et al. (2008) Raman spectroscopy of epitaxial graphene on a SiC substrate. *Physical Review B*, **77**, 115416.

[40] Emtsev, K.V., Bostwick, A., Horn, K. et al. (2009) Towards wafer-size graphene layers by atmospheric pressure graphitization of silicon carbide. *Nature Materials*, **8**, 203–207.

[41] Zhang, Y., Small, J.P., Pontius, W.V. and Kim, P. (2005) Fabrication and electric-field-dependent transport measurements of mesoscopic graphite devices. *Applied Physics Letters*, **86**, 073104.

[42] Bourlinos, A.B., Georgakilas, V., Zboril, R. et al. (2009) Liquid-phase exfoliation of graphite towards solubilized graphenes. *Small*, **5**, 1841–1845.

[43] Nuvoli, D., Alzari, V., Sanna, R. et al. (2012) The production of concentrated dispersions of few-layer graphene by the direct exfoliation of graphite in organosilanes. *Nanoscale Research Letters*, **7**, 674.

[44] Hernandez, Y., Nicolosi, V., Lotya, M. et al. (2008) High-yield production of graphene by liquid-phase exfoliation of graphite. *Nature Nanotechnology*, **3**, 563–568.

[45] Behabtu, N., Lomeda, J.R., Green, M.J. et al. (2010) Spontaneous high-concentration dispersions and liquid crystals of graphene. *Nature Nanotechnology*, **5**, 406–411.

[46] Jeong, H.K., Lee, Y.P., Lahaye, R.J. et al. (2008) Evidence of graphitic AB stacking order of graphite oxides. *Journal of the American Chemical Society*, **130**, 1362–1366.

[47] Stankovich, S., Dikin, D.A., Piner, R.D. et al. (2007) Synthesis of graphene-based nanosheets via chemical reduction of exfoliated graphite oxide. *Carbon*, **45**, 1558–1565.

[48] Tung, V.C., Allen, M.J., Yang, Y. and Kaner, R.B. (2009) High-throughput solution processing of large-scale graphene. *Nature Nanotechnology*, **4**, 25–29.

[49] Wu, Z.S., Ren, W.C., Gao, L.B. et al. (2009) Synthesis of high-quality graphene with a pre-determined number of layers. *Carbon*, **47**, 493–499.

[50] Gilje, S., Han, S., Wang, M. et al. (2007) A chemical route to graphene for device applications. *Nano Letters*, **7**, 3394–3398.

[51] Nethravathi, C. and Rajamathi, M. (2008) Chemically modified graphene sheets produced by the solvothermal reduction of colloidal dispersions of graphite oxide. *Carbon*, **46**, 1994–1998.

[52] Park, S. and Ruoff, R.S. (2009) Chemical methods for the production of graphenes. *Nature Nanotechnology*, **4**, 217–224.

[53] Si, Y. and Samulski, E.T. (2008) Synthesis of water soluble graphene. *Nano Letters*, **8**, 1679–1682.

[54] Wang, G.X., Shen, X.P., Wang, B. et al. (2009) Synthesis and characterisation of hydrophilic and organophilic graphene nanosheets. *Carbon*, **47**, 1359–1364.

[55] Wang, H., Robinson, J.T., Li, X. and Dai, H. (2009) Solvothermal reduction of chemically exfoliated graphene sheets. *Journal of the American Chemical Society*, **131**, 9910–9911.

[56] Zhou, Y., Bao, Q., Tang, L.A.L. et al. (2009) Hydrothermal dehydration for the "Green" reduction of exfoliated graphene oxide to graphene and demonstration of tunable optical limiting properties. *Chemistry of Materials*, **21**, 2950–2956.

[57] Zhu, Y., Stoller, M.D., Cai, W. et al. (2010) Exfoliation of graphite oxide in propylene carbonate and thermal reduction of the resulting graphene oxide platelets. *ACS Nano*, **4**, 1227–1233.

[58] Bolotin, K.I., Sikes, K.J., Jiang, Z. et al. (2008) Ultrahigh electron mobility in suspended graphene. *Solid State Communications*, **146**, 351–355.

[59] Lin, Y.M., Dimitrakopoulos, C., Jenkins, K.A. et al. (2010) 100-GHz transistors from wafer-scale epitaxial graphene. *Science*, **327**, 662.

[60] Liao, L., Lin, Y.C., Bao, M. et al. (2010) High-speed graphene transistors with a self-aligned nanowire gate. *Nature*, **467**, 305–308.

[61] Zhang, Y., Tan, Y.W., Stormer, H.L. and Kim, P. (2005) Experimental observation of the quantum Hall effect and Berry's phase in graphene. *Nature*, **438**, 201–204.

[62] Ryzhii, V., Ryzhii, M. and Otsuji, T. (2008) Tunneling current-voltage characteristics of graphene field-effect transistor. *Applied Physics Express*, **1**, 013001.

[63] Ryzhii, V., Ryzhii, M., Satou, A. and Otsuji, T. (2008) Current-voltage characteristics of a graphene-nanoribbon field-effect transistor. *Journal of Applied Physics*, **103**, 094510.

[64] Obradovic, B., Kotlyar, R., Heinz, F. et al. (2006) Analysis of graphene nanoribbons as a channel material for field-effect transistors. *Applied Physics Letters*, **88**, 142102.

[65] Bai, J., Zhong, X., Jiang, S. et al. (2010) Graphene nanomesh. *Nature Nanotechnology*, **5**, 190–194.

[66] Ohta, T., Bostwick, A., Seyller, T. et al. (2006) Controlling the electronic structure of bilayer graphene. *Science*, **313**, 951–954.

[67] Lee, C., Wei, X., Kysar, J.W. and Hone, J. (2008) Measurement of the elastic properties and intrinsic strength of monolayer graphene. *Science*, **321**, 385–388.

[68] Scarpa, F., Adhikari, S. and Srikantha Phani, A. (2009) Effective elastic mechanical properties of single layer graphene sheets. *Nanotechnology*, **20**, 065709.

[69] Gao, Y.W. and Hao, P. (2009) Mechanical properties of monolayer graphene under tensile and compressive loading. *Physica E: Low-Dimensional Systems & Nanostructures*, **41**, 1561–1566.

[70] Poetschke, M., Rocha, C.G., Torres, L.E.F.F. et al. (2010) Modeling graphene-based nanoelectromechanical devices. *Physical Review B*, **81**, 193404.

[71] Isacsson, A. (2011) Nanomechanical displacement detection using coherent transport in graphene nanoribbon resonators. *Physical Review B*, **84**, 125452.

[72] Rafiee, M.A., Rafiee, J., Wang, Z. et al. (2009) Enhanced mechanical properties of nanocomposites at low graphene content. *ACS Nano*, **3**, 3884–3890.

[73] Das, B., Eswar Prasad, K., Ramamurty, U. and Rao, C.N. (2009) Nano-indentation studies on polymer matrix composites reinforced by few-layer graphene. *Nanotechnology*, **20**, 125705.

[74] Ramanathan, T., Abdala, A.A., Stankovich, S. et al. (2008) Functionalized graphene sheets for polymer nanocomposites. *Nature Nanotechnology*, **3**, 327–331.

[75] Gong, K., Pan, Z., Korayem, A.H. et al. (2014) Reinforcing effects of graphene oxide on portland cement paste. *Journal of Materials in Civil Engineering*, **27**, A4014010.

[76] Chuah, S., Pan, Z., Sanjayan, J.G. et al. (2014) Nano reinforced cement and concrete composites and new perspective from graphene oxide. *Construction and Building Materials*, **73**, 113–124.

[77] Nair, R.R., Blake, P., Grigorenko, A.N. et al. (2008) Fine structure constant defines visual transparency of graphene. *Science*, **320**, 1308.

[78] Roddaro, S., Pingue, P., Piazza, V. et al. (2007) The optical visibility of graphene: interference colors of ultrathin graphite on SiO(2). *Nano Letters*, **7**, 2707–2710.

[79] Bonaccorso, F., Sun, Z., Hasan, T. and Ferrari, A.C. (2010) Graphene photonics and optoelectronics. *Nature Photonics*, **4**, 611–622.

[80] Bae, S., Kim, H., Lee, Y. et al. (2010) Roll-to-roll production of 30-inch graphene films for transparent electrodes. *Nature Nanotechnology*, **5**, 574–578.

[81] Li, X., Zhu, Y., Cai, W. et al. (2009) Transfer of large-area graphene films for high-performance transparent conductive electrodes. *Nano Letters*, **9**, 4359–4363.

[82] Eda, G., Fanchini, G. and Chhowalla, M. (2008) Large-area ultrathin films of reduced graphene oxide as a transparent and flexible electronic material. *Nature Nanotechnology*, **3**, 270–274.

[83] Novoselov, K.S., Fal'ko, V.I., Colombo, L. et al. (2012) A roadmap for graphene. *Nature*, **490**, 192–200.

[84] Li, X., Zhang, G., Bai, X. et al. (2008) Highly conducting graphene sheets and Langmuir-Blodgett films. *Nature Nanotechnology*, **3**, 538–542.

[85] Balandin, A.A., Ghosh, S., Bao, W. et al. (2008) Superior thermal conductivity of single-layer graphene. *Nano Letters*, **8**, 902–907.

[86] Ghosh, S., Calizo, I., Teweldebrhan, D. et al. (2008) Extremely high thermal conductivity of graphene: Prospects for thermal management applications in nanoelectronic circuits. *Applied Physics Letters*, **92**, 151911.

[87] Ferrari, A.C., Meyer, J.C., Scardaci, V. et al. (2006) Raman spectrum of graphene and graphene layers. *Physical Review Letters*, **97**, 187401.

[88] Gupta, A., Chen, G., Joshi, P. et al. (2006) Raman scattering from high-frequency phonons in supported n-graphene layer films. *Nano Letters*, **6**, 2667–2673.

[89] Calizo, I., Bao, W., Miao, F. et al. (2007) The effect of substrates on the Raman spectrum of graphene: graphene-on-sapphire and graphene-on-glass. *Applied Physics Letters*, **91**, 201904.

[90] Calizo, I., Balandin, A.A., Bao, W. et al. (2007) Temperature dependence of the Raman spectra of graphene and graphene multilayers. *Nano Letters*, **7**, 2645–2649.

[91] Calizo, I., Miao, F., Bao, W. et al. (2007) Variable temperature Raman microscopy as a nanometrology tool for graphene layers and graphene-based devices. *Applied Physics Letters*, **91**, 071913.

[92] Nika, D.L., Pokatilov, E.P., Askerov, A.S. and Balandin, A.A. (2009) Phonon thermal conduction in graphene: Role of Umklapp and edge roughness scattering. *Physical Review B*, **79**, 155413.

[93] Nika, D.L., Ghosh, S., Pokatilov, E.P. and Balandin, A.A. (2009) Lattice thermal conductivity of graphene flakes: Comparison with bulk graphite. *Applied Physics Letters*, **94**, 203103.

[94] Ghosh, S., Bao, W., Nika, D.L. et al. (2010) Dimensional crossover of thermal transport in few-layer graphene. *Nature Materials*, **9**, 555–558.

[95] Cai, W., Moore, A.L., Zhu, Y. et al. (2010) Thermal transport in suspended and supported monolayer graphene grown by chemical vapor deposition. *Nano Letters*, **10**, 1645–1651.

[96] Jauregui, L.A., Yue, Y.N., Sidorov, A.N. et al. (2010) Thermal transport in graphene nanostructures: experiments and simulations. *Graphene, Ge/Iii-V, and Emerging Materials for Post-Cmos Applications 2*, **28**, 73–83.

[97] Chen, S., Moore, A.L., Cai, W. et al. (2011) Raman measurements of thermal transport in suspended monolayer graphene of variable sizes in vacuum and gaseous environments. *ACS Nano*, **5**, 321–328.

[98] Chen, S., Wu, Q., Mishra, C. et al. (2012) Thermal conductivity of isotopically modified graphene. *Nature Materials*, **11**, 203–207.

[99] ASTM (2013) *C177, Standard Test Method for Steady-State Heat Flux Measurements and Thermal Transmission Properties by Means of the Guarded-Hot-Plate Apparatus*, ASTM International.

[100] ISO 8302 (1991) *Thermal Insulation - Determination of Steady-State Thermal Resistance and Related Properties - Guarded Hot Plate Apparatus*, International Organization for Standardization.

[101] ASTM E1461 (2013) *Standard Test Method for Thermal Diffusivity by the Flash Method*, ASTM International.

[102] Maxwell, J.C. (1954) *A Treatise on Electricity and Magnetism*, 3rd edn, Dover, New York.

[103] Fricke, H. (1954) The Maxwell Wagner dispersion in a suspension of ellipsoids. *Journal of Physical Chemistry*, **57**, 934–937.

[104] Hashin, Z. (1968) Assessment of the self consistent scheme approximation: conductivity of particulate composites. *Journal of Composite Materials*, **2**, 284–300.

[105] Davis, R.H. (1986) The effective thermal-conductivity of a composite-material with spherical inclusions. *International Journal of Thermophysics*, **7**, 609–620.

[106] Davis, L.C. and Artz, B.E. (1995) Thermal conductivity of metal-matrix composites. *Journal of Applied Physics*, **77**, 4954.

[107] Hasselman, D.P.H. and Johnson, L.F. (1987) Effective thermal-conductivity of composites with interfacial thermal barrier resistance. *Journal of Composite Materials*, **21**, 508–515.

[108] Benveniste, Y. and Miloh, T. (1986) The effective conductivity of composites with imperfect thermal contact at constituent interfaces. *International Journal of Engineering Science*, **24**, 1537–1552.

[109] Benveniste, Y. (1987) Effective thermal-conductivity of composites with a thermal contact resistance between the constituents - nondilute case. *Journal of Applied Physics*, **61**, 2840–2843.

[110] Kerner, E.H. (1956) The electrical conductivity of composite media. *Proceedings of the Physical Society of London, Section B*, **69**, 802–807.

[111] Mori, T. and Tanaka, K. (1973) Average stress in matrix and average elastic energy of materials with misfitting inclusions. *Acta Metallurgica*, **21**, 571–574.

[112] Benveniste, Y. and Miloh, T. (1991) On the effective thermal-conductivity of coated short-fiber composites. *Journal of Applied Physics*, **69**, 1337–1344.

[113] Every, A.G., Tzou, Y., Hasselman, D.P.H. and Raj, R. (1992) The effect of particle-size on the thermal-conductivity of ZnS/diamond composites. *Acta Metallurgica Et Materialia*, **40**, 123–129.

[114] Nan, C.W., Birringer, R., Clarke, D.R. and Gleiter, H. (1997) Effective thermal conductivity of particulate composites with interfacial thermal resistance. *Journal of Applied Physics*, **81**, 6692–6699.

[115] Debelak, B. and Lafdi, K. (2007) Use of exfoliated graphite filler to enhance polymer physical properties. *Carbon*, **45**, 1727–1734.

[116] Warzoha, R.J. and Fleischer, A.S. (2014) Improved heat recovery from paraffin-based phase change materials due to the presence of percolating graphene networks. *International Journal of Heat and Mass Transfer*, **79**, 314–323.

[117] Raza, M.A., Westwood, A., Brown, A. et al. (2011) Characterisation of graphite nanoplatelets and the physical properties of graphite nanoplatelet/silicone composites for thermal interface applications. *Carbon*, **49**, 4269–4279.

[118] Shahil, K.M.F. and Balandin, A.A. (2012) Graphene-multilayer graphene nanocomposites as highly efficient thermal interface materials. *Nano Letters*, **12**, 861–867.

[119] Yu, A.P., Ramesh, P., Itkis, M.E. et al. (2007) Graphite nanoplatelet-epoxy composite thermal interface materials. *Journal of Physical Chemistry C*, **111**, 7565–7569.

[120] Guo, W. and Chen, G. (2014) Fabrication of graphene/epoxy resin composites with much enhanced thermal conductivity via ball milling technique. *Journal of Applied Polymer Science*, **131**, 40565.

[121] Xiang, J.L. and Drzal, L.T. (2011) Investigation of exfoliated graphite nanoplatelets (xGnP) in improving thermal conductivity of paraffin wax-based phase change material. *Solar Energy Materials and Solar Cells*, **95**, 1811–1818.

[122] Yu, W., Xie, H.Q., Chen, L.F. et al. (2014) Graphene based silicone thermal greases. *Physics Letters A*, **378**, 207–211.

[123] Chu, K., Li, W.S. and Dong, H. (2013) Role of graphene waviness on the thermal conductivity of graphene composites. *Applied Physics A: Materials Science and Processing*, **111**, 221–225.

[124] Yu, A.P., Ramesh, P., Sun, X.B. et al. (2008) Enhanced thermal conductivity in a hybrid graphite nanoplatelet-carbon nanotube filler for epoxy composites. *Advanced Materials*, **20**, 4740–4744.

[125] Yang, S.Y., Lin, W.N., Huang, Y.L. et al. (2011) Synergetic effects of graphene platelets and carbon nanotubes on the mechanical and thermal properties of epoxy composites. *Carbon*, **49**, 793–803.

[126] Sun, X., Ramesh, P., Itkis, M.E. et al. (2010) Dependence of the thermal conductivity of two-dimensional graphite nanoplatelet-based composites on the nanoparticle size distribution. *Journal of Physics: Condensed Matter*, **22**, 334216.

[127] Ghose, S., Working, D.C., Connell, J.W. et al. (2006) Thermal conductivity of UltemTM/carbon nanofiller blends. *High Performance Polymers*, **18**, 961–977.

[128] Yu, L., Park, J.S., Lim, Y.S. et al. (2013) Carbon hybrid fillers composed of carbon nanotubes directly grown on graphene nanoplatelets for effective thermal conductivity in epoxy composites. *Nanotechnology*, **24**, 155604.

[129] Yan, H., Tang, Y., Long, W. and Li, Y. (2014) Enhanced thermal conductivity in polymer composites with aligned graphene nanosheets. *Journal of Materials Science*, **49**, 5256–5264.

[130] Wang, S.R., Tambraparni, M., Qiu, J.J. et al. (2009) Thermal expansion of graphene composites. *Macromolecules*, **42**, 5251–5255.

[131] Huang, X., Zhi, C. and Jiang, P. (2012) Toward effective synergetic effects from graphene nanoplatelets and carbon nanotubes on thermal conductivity of ultrahigh volume fraction nanocarbon epoxy composites. *Journal of Physical Chemistry C*, **116**, 23812–23820.

[132] Wang, W., Wang, C., Wang, T. et al. (2014) Enhancing the thermal conductivity of n-eicosane/silica phase change materials by reduced graphene oxide. *Materials Chemistry and Physics*, **147**, 701–706.

[133] Li, M. (2013) A nano-graphite/paraffin phase change material with high thermal conductivity. *Applied Energy*, **106**, 25–30.

[134] Kapitza, P.L. (1941) The study of heat transfer in helium II. *Journal of Physics (Ussr)*, **4**, 181–210.

[135] Pollack, G.L. (1969) Kapitza resistance. *Reviews of Modern Physics*, **41**, 48–81.

[136] Swartz, E. and Pohl, R. (1989) Thermal boundary resistance. *Reviews of Modern Physics*, **61**, 605–668.

[137] Fisher, T.S. (2014) *Thermal Energy at the Nanoscale*, World Scientific, Singapore.

[138] Xu, Y.B., Wang, H.T., Tanaka, Y. et al. (2007) Measurement of interfacial thermal resistance by periodic heating and a thermo-reflectance technique. *Materials Transactions*, **48**, 148–150.

[139] Xu, Y.B., Kato, R. and Goto, M. (2010) Effect of microstructure on Au/sapphire interfacial thermal resistance. *Journal of Applied Physics*, **108**, 104317.

[140] Chien, H.C., Yao, D.J. and Hsu, C.T. (2008) Measurement and evaluation of the interfacial thermal resistance between a metal and a dielectric. *Applied Physics Letters*, **93**, 231910.

[141] Phillpot, S.R., Schelling, P.K. and Keblinski, P.K. (2005) Interfacial thermal conductivity: Insights from atomic level simulation. *Journal of Materials Science*, **40**, 3143–3148.

[142] Schelling, P.K., Phillpot, S.R. and Keblinski, P. (2004) Kapitza conductance and phonon scattering at grain boundaries by simulation. *Journal of Applied Physics*, **95**, 6082–6091.

[143] Hu, M., Shenogin, S. and Keblinski, P. (2007) Molecular dynamics simulation of interfacial thermal conductance between silicon and amorphous polyethylene. *Applied Physics Letters*, **91**, 241910.

[144] Yang, J.K., Chen, Y.F. and Yan, J.P. (2003) Molecular dynamics simulation of thermal conductivities of superlattice nanowires. *Science in China Series E: Technological Sciences*, **46**, 278–286.

[145] Hegedus, P.J. and Abramson, A.R. (2006) A molecular dynamics study of interfacial thermal transport in heterogeneous systems. *International Journal of Heat and Mass Transfer*, **49**, 4921–4931.

[146] Shenogin, S., Xue, L.P., Ozisik, R. et al. (2004) Role of thermal boundary resistance on the heat flow in carbon-nanotube composites. *Journal of Applied Physics*, **95**, 8136–8144.

[147] Chen, J., Zhang, G. and Li, B.W. (2012) Thermal contact resistance across nanoscale silicon dioxide and silicon interface. *Journal of Applied Physics*, **112**, 064319.

[148] Luo, T.F. and Lloyd, J.R. (2012) Enhancement of thermal energy transport across graphene/graphite and polymer interfaces: a molecular dynamics study. *Advanced Functional Materials*, **22**, 2495–2502.

[149] Liu, Y., Huang, J.S., Yang, B. et al. (2014) Duality of the interfacial thermal conductance in graphene-based nanocomposites. *Carbon*, **75**, 169–177.

[150] Hu, L., Desai, T. and Keblinski, P. (2011) Thermal transport in graphene-based nanocomposite. *Journal of Applied Physics*, **110**, 033517.

[151] Clancy, T.C. and Gates, T.S. (2006) Modeling of interfacial modification effects on thermal conductivity of carbon nanotube composites. *Polymer*, **47**, 5990–5996.

[152] Shenogin, S., Bodapati, A., Xue, L. et al. (2004) Effect of chemical functionalization on thermal transport of carbon nanotube composites. *Applied Physics Letters*, **85**, 2229–2231.

[153] Yang, S.Y., Ma, C.C.M., Teng, C.C. et al. (2010) Effect of functionalized carbon nanotubes on the thermal conductivity of epoxy composites. *Carbon*, **48**, 592–603.

[154] Pongsa, U. and Somwangthanaroj, A. (2013) Effective thermal conductivity of 3,5-diaminobenzoyl-functionalized multiwalled carbon nanotubes/epoxy composites. *Journal of Applied Polymer Science*, **130**, 3184–3196.

[155] Yang, K., Gu, M.Y., Guo, Y.P. et al. (2009) Effects of carbon nanotube functionalization on the mechanical and thermal properties of epoxy composites. *Carbon*, **47**, 1723–1737.

[156] Ganguli, S., Roy, A.K. and Anderson, D.P. (2008) Improved thermal conductivity for chemically functionalized exfoliated graphite/epoxy composites. *Carbon*, **46**, 806–817.

[157] Gu, J., Xie, C., Li, H. et al. (2014) Thermal percolation behavior of graphene nanoplatelets/polyphenylene sulfide thermal conductivity composites. *Polymer Composites*, **35**, 1087–1092.

[158] Chatterjee, S., Wang, J.W., Kuo, W.S. et al. (2012) Mechanical reinforcement and thermal conductivity in expanded graphene nanoplatelets reinforced epoxy composites. *Chemical Physics Letters*, **531**, 6–10.

[159] Fang, M., Wang, K.G., Lu, H.B. et al. (2010) Single-layer graphene nanosheets with controlled grafting of polymer chains. *Journal of Materials Chemistry*, **20**, 1982–1992.

[160] Ding, P., Su, S., Song, N. et al. (2014) Influence on thermal conductivity of polyamide-6 covalently-grafted graphene nanocomposites: Varied grafting-structures by controllable macromolecular length. *RSC Advances*, **4**, 18782–18791.

[161] Qian, R., Yu, J., Wu, C. et al. (2013) Alumina-coated graphene sheet hybrids for electrically insulating polymer composites with high thermal conductivity. *RSC Advances*, **3**, 17373–17379.

[162] Martin-Gallego, M., Verdejo, R., Khayet, M. et al. (2011) Thermal conductivity of carbon nanotubes and graphene in epoxy nanofluids and nanocomposites. *Nanoscale Research Letters*, **6**, 1–10.

[163] Li, Z., Wang, D., Zhang, M. and Zhao, L. (2014) Enhancement of the thermal conductivity of polymer composites with Ag-graphene hybrids as fillers. *Physica Status Solidi A*, **211**, 2142–2149.

[164] Pu, X., Zhang, H.B., Li, X.F. et al. (2014) Thermally conductive and electrically insulating epoxy nanocomposites with silica-coated graphene. *RSC Advances*, **4**, 15297–15303.

[165] Hung, M.T., Choi, O., Ju, Y.S. and Hahn, H.T. (2006) Heat conduction in graphite-nanoplatelet-reinforced polymer nanocomposites. *Applied Physics Letters*, **89**, 023117.

[166] Heo, C. and Chang, J.H. (2013) Polyimide nanocomposites based on functionalized graphene sheets: Morphologies, thermal properties, and electrical and thermal conductivities. *Solid State Sciences*, **24**, 6–14.

[167] Wang, M., Galpaya, D., Lai, Z.B. et al. (2014) Surface functionalization on the thermal conductivity of graphene-polymer nanocomposites. *International Journal of Smart and Nano Materials*, **5**, 123–132.

[168] Konatham, D. and Striolo, A. (2009) Thermal boundary resistance at the graphene-oil interface. *Applied Physics Letters*, **95**, 163105.

[169] Teng, C.C., Ma, C.C.M., Lu, C.H. et al. (2011) Thermal conductivity and structure of non-covalent functionalized graphene/epoxy composites. *Carbon*, **49**, 5107–5116.

[170] Song, S.H., Park, K.H., Kim, B.H. et al. (2013) Enhanced thermal conductivity of epoxy-graphene composites by using non-oxidized graphene flakes with non-covalent functionalization. *Advanced Materials*, **25**, 732–737.

[171] Lin, S. and Buehler, M.J. (2013) The effect of non-covalent functionalization on the thermal conductance of graphene/organic interfaces. *Nanotechnology*, **24**, 165702.

2

Mechanics of CNT Network Materials

Mesut Kirca[1] and Albert C. To[2]

[1]*Department of Mechanical Engineering, Istanbul Technical University, Istanbul, Turkey*
[2]*Department of Mechanical Engineering and Materials Science, University of Pittsburgh, Pittsburgh, PA, USA*

2.1 Introduction

In recent years, carbon nanotube (CNT) has been one of the most prominent nanomaterials ever to be investigated due to its extraordinary mechanical, thermal, and electrical properties. Recently, the application of CNTs has been extended to CNT networks in which the CNTs are joined together in a two- or three-dimensional space. In this regard, CNT networks have enabled nanotubes to be used as building blocks for synthesizing novel advanced materials, thus taking full advantage of the superior properties of individual CNTs. In this chapter, a thorough literature review including the most recent theoretical and experimental studies is presented, with a focus on the mechanical characteristics of CNT network materials. As a supporting material, some recent studies of authors will also be introduced to provide a deeper understanding in the mechanical behavior of CNT network materials.

Since their discovery in 1991, research has remained intense on CNTs through experimental, theoretical, and computational investigations, which demonstrated their remarkable mechanical, thermal, optical, and electrical properties. For instance, it was demonstrated that the thermal conductivity of a single-walled carbon nanotube (SWCNT) at room temperature is as high as 3500 W/mK, which is larger than the thermal conductivity of diamond. Regarding their mechanical behavior, most of the initial studies revealed that elastic modulus and tensile strength values of individual SWCNTs are approximately 1 TPa and 50 GPa, respectively, which are significantly higher than the values of conventional structural materials such as steel or industrial fibers. Following a maturation period that builds up more information on the physical and chemical characteristics of individual CNTs, in the last decade, network-type materials consisting of CNTs have emerged extending spatial functionalities of individual CNTs to the microscales and macroscales. Through the synthesis of CNT network materials,

Advanced Computational Nanomechanics, First Edition. Edited by Nuno Silvestre.
© 2016 John Wiley & Sons, Ltd. Published 2016 by John Wiley & Sons, Ltd.

limitations on capitalizing the superior properties of individual CNTs are being overcome, and the scope of applications are being widened. However, mostly based on the experimental efforts, researchers have noted that the mechanical properties of CNT-based network materials can differ from the individual CNTs significantly depending on the network morphology and characteristics of the CNT interactions within the network.

2.1.1   Types of CNT Network Materials

CNT network materials refer to porous nanostructures formed by ordered or randomly organized CNTs, which can be SWCNTs or multiwalled carbon nanotubes (MWCNTs). Opposing to CNT arrays in which CNT units are aligned in one direction, CNTs throughout the CNT networks are interconnected with each other. The type of interconnection that dictates the physical properties of the material could be noncovalent (i.e., van der Waals, physical entanglement) or covalent depending on the synthesis method. Because both SWCNTs and MWCNTs inherently have a nearly inert surface, covalent bonding between CNTs requires the establishment of special processes such as chemical surface functionalization, doping, surface acid treatment, and electron beam irradiation combined with high temperature. CNT network materials can be simply categorized into two main groups based on the dimensionality. In this regard, two-dimensional (2D) CNT network materials, which are generally termed as CNT thin films, sheets or buckypaper, attempt to extrapolate the unique electrical, optical, and mechanical properties of one-dimensional (1D) individual CNTs to 2D macroscopic scale. In the CNT network films consisting of randomly oriented CNT units, CNT bundles intersecting and joining with other bundles with approximate diameter of 50 nm are formed by the aggregation of CNTs [1]. In the literature, 2D CNT network materials, namely CNT thin films or CNT aerogel sheets, have received considerable attention [2–5] among researchers especially for their applications in electronics such as sensors [6–9] or transparent electrodes [10–12]. Owing to their exceptional mechanical flexibility and stretchability in addition to optical transparence, 2D conductive CNT films have also been employed as super-elastic actuators [13]. Along the same line, by three-dimensional (3D) CNT network materials, which are commonly called CNT foams or CNT sponges, the scope of the extrapolation is extended to 3D scale.

Classification of CNT network materials can also be made according to the type of interaction between CNT units in the network such as noncovalent and covalent networks. Noncovalent CNT networks are sparse materials through which the only interaction between CNT units is via van der Waals forces and physical contacts by entanglements. Owing to the weak interaction between CNT units, extraordinary mechanical, electrical, and thermal characteristics of individual CNTs cannot be carried over to the bulk form of CNT networks, which in turn mitigate the performance of CNT networks and thus restrict the variety of their applications. Motivated by this, in recent years, several methods have been developed to synthesize covalently bonded 3D CNT networks in which chemical bonding between CNTs are created to increase overall mechanical strength and to decrease thermal and electrical contact resistance between CNTs. Details of the covalent bonding schemes will be evaluated in Section 2.1.2.

Owing to their outstanding mechanical properties, CNT-based network materials including buckypaper, sponges, fibers of yarns, foams, and aerogels are of great interest, and there-fore they have significant potential to be utilized in a wide range of applications such as

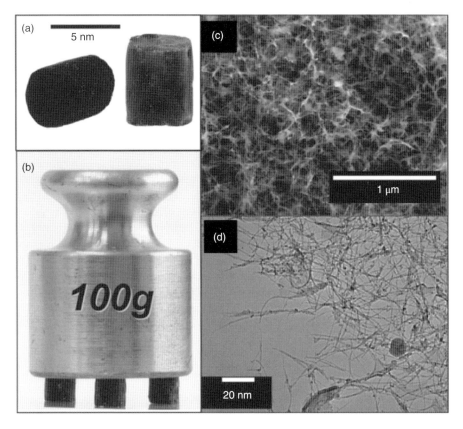

Figure 2.1 Images of aerogels. (a) Macroscopic pieces of 7.5 mg/mL CNT aerogels. Pristine CNT aerogel (i) appears black, whereas the aerogel reinforced in a 1 wt% PVA bath (ii) is slightly gray. (b) Three PVA-reinforced aerogel pillars (total mass = 13.0 mg) supporting 100 g, or *ca.* 8000 times their weight. (c) This scanning electron microscopy (SEM) image of a critical-point-dried aerogel reinforced in a 0.5 wt% PVA solution (CNT content = 10 mg/mL) reveals an open, porous structure. (d) This high-magnification transmission electron microscopy (TEM) image of an unreinforced aerogel reveals small-diameter CNTs arranged in a classic filamentous network. *Source*: Bryning et al [15]. Reproduced with permission of Wiley-VCH

multifunctional composite, electrochemical energy storage, and particulate filters [14]. For instance, a recently developed CNT aerogel [15] (see Figure 2.1(a)–(d)) material having a random network of self-intersecting CNTs at the macroscopic scale is one of the most important example materials that displays an ultra-high stiffness-to-weight ratio with thermally and electrically conductive properties. Rather than having a reinforcement role in CNT-based composites, this material enables the CNT network to be the bulk material on its own.

2.1.2 Synthesis of CNT Network Materials

The main objective of synthesizing CNT network materials is to extend the excellent mechanical, thermal, and electrical properties of individual CNTs to macroscopic applications.

However, bulk properties of CNT network materials are highly dependent on the spatial arrangement of the CNTs as well as on the physical and chemical interactions between the CNTs. In this perspective, the synthesis process that dictates the network morphology and intra-network interactions has direct effects on the properties of CNT network materials. Fabrication of CNT network materials can be generally divided into two groups. In the first group of synthesis techniques, CNT networks are synthesized by direct growth of CNTs on substrates by using dry metal catalysts accompanied by gaseous precursors. Chemical vapor deposition (CVD) is the most popular direct growth technique for manufacturing SWCNT networks consisting of randomly oriented or aligned CNTs. In the general procedure of CVD, a metallic catalyst (e.g., Fe/Co/Mo) is utilized for the growth of CNTs to grow on a substrate (e.g., SiO_2/Si substrate). The CVD method enables researchers to produce long nanotubes with controllable parameters such as nanotube density, alignment, and electrical (i.e., semiconducting/metallic) properties [16–18]. Alignment of the CNT networks can be tuned by characteristics of substrate and several growth parameters such as heating process [19]. In addition to conventional Si, several types of substrates can be used such as mesoporous silica [20], nanoporous silica [21], and alumina [22]. For example, Kaur et al. presented a manufacturing method for chemically and mechanically stable 3D CNT networks by growing MWCNTs using colloidal silica templates through the CVD process [23]. Based on that method, empty volume between the closely packed monolayered or multilayered colloidal spheres are filled by CNTs through a CVD step in which CNTs are grown. Following the thermal CVD process, silica spheres are selectively removed by etching to yield the 3D CNT foams.

The other group of methods, namely suspension-based techniques, employs the deposition of CNTs that are dispersed in a solvent. Owing to its simplicity, many deposition techniques have been developed based on this strategy, including spray coating, spin coating, and ink-jet printing.

In addition to CNT networks grown on substrates, stand-alone CNT networks that are independent of any substrate structures can be fabricated by similar methods except that a temporary substrate is used during the synthesis. In this regard, common methods such as drop casting [24], spin coating [25], spray coating [26], vacuum filtration [27–29], and direct deposition [30, 31] are utilized to fabricate stand-alone CNT networks by removing the temporary substrates. Ultra-long CNTs (*ca.* 100 μm) with high purity are desirable in stand-alone CNT networks in order to maintain self-supported structure through stronger intertube interactions. For instance, individual CNTs with approximately 100 μm long have been used by Liu et al. to synthesize free-standing CNT network films [26]. In another study, Shi et al. manufactured free-standing ultra-thin (i.e., 20-nm thick) CNT network films by employing approximately 15–25 μm long SWCNTs through the vacuum filtration method [27].

As another example of stand-alone CNT network materials, Bryning et al. presented a new class of aerogels called the "CNT aerogels" consisting percolating networks of CNTs [15]. They demonstrated that an ensemble of CNTs with small diameters including single and a few MWCNTs can be used as a precursor to obtain CNT aerogels which are electrically conductive and ultra-light in weight. In their study, CNT aerogels are fabricated from aqueous-gel precursors composed of CNTs in water with a surfactant through critical-point-drying (CPD) and freeze-drying methods. Individually, CNTs are highly stiff [32] and effectively electrically conducting nanostructures. Therefore, an ensemble of CNTs forms an attractive precursor for the generation of aerogels. Due to insufficient mechanical strength of pristine CNT aerogels, reinforcement is achieved by polyvinyl alcohol (PVA). Although the mechanical strength is improved, PVA degrades the conductivity and elevates the density of the aerogel. As shown in Figure 2.1(b), three pieces of PVA-reinforced CNT aerogel pillars with a total mass of 13 mg can support 100 g weight, which is 8000 times their weight.

For the purpose of enhancing the physical properties of CNT networks, new methods have been developed to increase the degree of interaction between the individual CNTs through cross-link nodes. One of the techniques used to generate cross-links between CNTs is electron or ion beam irradiation, which can be used for both SWCNTs and MWCNTs as a post-growth processing. For example, exposure of electron beam to the atoms on the CNT surface yields structural defects that lead to the formation of covalent linking via dangling bonds. For this strategy there is no need for the utilization of any chemical reactions, thus application of this method is relatively simpler. Furthermore, by irradiation chemical bonds, which are formed between CNTs, are much stronger than van der Waals forces that exist within the freestanding CNT networks. However, apart from those advantages, irradiation has some drawbacks such that sp^2 bonding of nanotubes is destroyed, which in turn affects the intrinsic properties of CNTs. In addition to this, by electron or ion beam irradiation, cross-links can be generated at only local sites where the irradiation can reach below the surface layers of the CNT network [33–36]. Moreover, excessive application of electron or ion irradiation around the junction region causes deterioration of CNT structure, which as a consequence degrades the performance of the CNTs. Therefore, those post-growth techniques are more suitable for the manufacturing of small-scale devices or structures in contrast to manufacturing of macroscale network materials [37].

Inefficiency of the post-growth methods for the CNT network materials stimulated the development of second group processing techniques, which follow a direct growth on a predefined template or drive the growth or annealing process by additive agents [37]. For example, Fu et al. developed a new approach for the mass production of covalently linked CNT network materials. According to that procedure, CNTs are grown on nickel template, which is in the interconnected tubular form, by CVD technique. After application of CVD by growing graphene layers on the cylindrical surfaces of Ni template, Ni core is sacrificed by etching to accomplish interconnected CNT networks. Figure 2.2(a)–(f) provides a summary of their manufacturing process starting with the electrospinning of polyvinyl butyral (PVB) that contains nickel nitrate (i.e., $Ni(NO_3)_2$). Sprayed PVB fibers containing nickel nitrate merge with each other to form an interconnected network. Graphene layers are grown on the interconnected network after a pure Ni fiber network is obtained by firstly burning off the PVB followed by the decomposition of nickel nitrate in O_2. Finally, following the removal of the Ni core by chemical etching, covalently bonded 3D interconnected CNT network is derived.

Another strategy to produce covalently bonded 3D CNT networks is to introduce heteroatoms such as boron and nitrogen [38, 39]. Both theoretical and experimental studies have indicated that the presence of heteroatoms may stimulate significant morphological changes through the CNT networks such as stable bend and junction formations [39–47]. For example, recently Hashim et al. have reported the manufacturing of 3D CNT network solids consisting of MWCNTs with covalent nanojunctions formed by boron doping through an aerosol-assisted chemical vapor deposition (AACVD) process. Resultant stable covalent multijunctions (e.g., Y-junction, four-way junction) and "elbow" formations, which provide remarkable mechanical durability and flexibility, are attributed to boron dopant. For example,

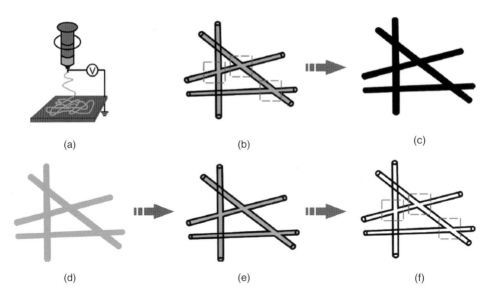

Figure 2.2 Synthesis flow chart of the covalently bonded 3D CNT networks. (a) Electrospinning of $Ni(NO_3)_2$ contained PVB nanofiber networks. (b) Interconnected electrospun $Ni(NO_3)_2$/PVB nanofibers in the network. The fused interconnected joints are marked by dashed frames. (c) NiO_x nanofiber network after the burning of PVB and decomposition of $Ni(NO_3)_2$ in O_2. (d) Pure Ni fiber network reduced from NiO_x by H_2, acting as graphene-tube growth template. (e) Graphitic layer growth on the interconnected 3D Ni template. (f) Chemical etching of Ni core in $FeCl_3$ solution so that covalently bonded 3D CNT network is obtained. The interconnection joints are marked by dashed frames. *Source*: Fu et al. [37]. Reproduced with permission of Wiley-VCH

as a result of utilizing several mapping methods such as high-angle annular-dark-field (HAADF) imaging and electron energy loss spectroscopy (EELS) line-scans, the highest boron concentrations are noticed to be around "elbow-like" junctions signifying the role of boron in the formation of negative curvature areas. Moreover, Hashim et al. also demonstrated that boron-doped CNT network solids combine many different properties as a multifunctional material such as high structural damping, electrical conductivity, low relative density, super-hydrophobicity, strong ferromagnetism, thermal stability, and oleophilic behavior [39].

In a more recent study, Shan et al. reported the production of nitrogen-doped macroscopic 3D MWCNT sponges (i.e., N-MWCNT sponges) with covalent junctions by CVD technique (see Figure 2.3(a)). As shown in Figure 2.3(b) with a scanning electron microscopy (SEM) image, N-MWCNT sponges are composed of individual CNTs, which are randomly oriented and entangled. Similar to boron-doped MWCNT networks, formation of covalent junctions and "elbow-like" is observed as shown in Figure 2.3(c) and (d), and controlled experiments show that those morphological formations are referred to doping with sulfur and nitrogen during CVD [38]. More importantly, it is also reported that mechanical and thermal properties can be tuned by the N-MWCNT diameter, which can be easily controlled by the concentration of sulfur. Detailed characterization of junctions by transmission electron microscopy (TEM) as depicted in Figure 2.3(e) and (f), junction between nitrogen-doped MWCNTs, consists of graphitic layers that ensure the fusion between the CNTs.

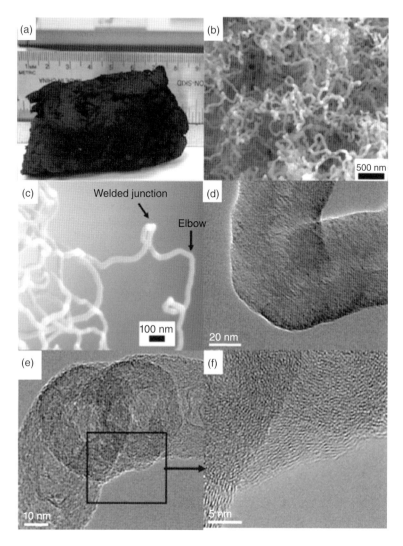

Figure 2.3 (a) Photo of N-MWCNT sponge. (b and c) SEM images of N-MWCNT sponges. (d) TEM image of an "elbow" junction in N-MWCNT sponge. (e) TEM and (f) HRTEM images of "welded" junction between two N-MWCNTs. *Source*: Shan et al. [38]. Reproduced with permission of the American Chemical Society

2.1.3 Applications

Unusual properties of individual CNTs have stimulated an extensive research for their potential usage in various applications of solar cells, hydrogen storage devices, avionic structures, energy conversion devices, logical sensors, and thermal interface materials [2, 4, 48–58]. However, many of those applications require bulk form of CNT materials such as thin films, foams, or composites consisting of random or ordered CNT networks. Due to their mechanical flexibility, good electrical conductivity, lightweight, and highly porous structure,

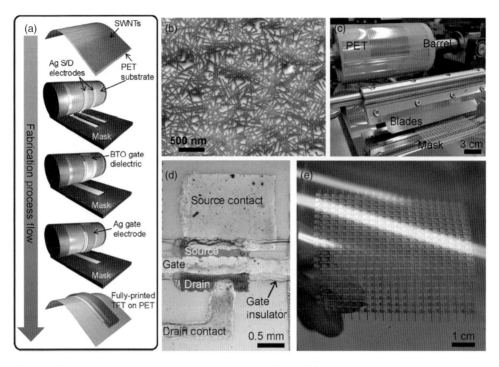

Figure 2.4 Fully printed SWCNT TFTs on mechanically flexible PET substrates. (a) Schematic diagram showing the printing process scheme. (b) Representative SEM image of an SWCNT network deposited on a PET substrate. (c) Image of the inverse photogravure printer used in this work. (d and e) Optical micrograph of a single TFT and optical image of a fully printed 20×20 device array on a PET substrate, respectively. *Source*: Lau et al. [62]. Reproduced with permission of the American Chemical Society

CNT networks are initially considered as attractive materials in the development of different electronic devices. In this context, because of their strong electrical conductivity with high light transmittance, CNT networks are proposed as alternatives to silicon-based macroelectronic devices [59]. Initial studies on 2D CNT networks deposited onto flexible and polymeric substrates have focused on their electronic and sensing properties [60, 61]. For example, CNT thin films have been extensively studied as thin-film transistors (TFTs) by increasing the performance of flexible electronics for a wide range of applications including lightweight and unbreakable displays [48, 62], biological sensors [48], and flexible antennae [4, 56]. Figure 2.4(a)–(e) depicts a macrosample of CNT network TFT with SEM and optical images and summarizes the manufacturing process realized on poly(ethylene terephthalate) (PET).

Furthermore, CNT sponges have been utilized in super-capacitor devices as a conductive substrate for composite electrode development, which boosts the areal capacitance together with outstanding mechanical and chemical stability [63]. CNT networks are also reviewed as ideal candidates to replace indium tin oxide (ITO), which is commonly used as high-conductivity transparent electrodes for flexible electronic applications [64]. ITO is one of the most used materials for transparent electrodes because of their good transparency and low sheet resistance; however, their mechanical behavior is brittle, which causes material failure at low strains. At this point, CNT networks have been validated to maintain electrical performance under severe bending deformation in addition to impressive electronic potential [65].

CNT sponges are also investigated for their oil and organic reagent absorption applications owing to their significant features such as super-hydrophobic nature (see Figure 2.5(a) and (b)), large specific surface area (SSA), and oleophilic characteristics, which make them ideal instruments for absorbing oil and for purifying water polluted by oil spill disasters. Both doped and undoped CNT sponges have been examined for their oil absorption characteristics (see Figure 2.5(c)) [39, 66]. As an example, in one of those studies [39], CNT sponges doped with boron have been presented as a reusable oil absorbent material in seawater. It has also shown that after CNT sponge is saturated by oil, the oil content can be removed by burning without destroying the CNT sponge, and by this way the sponge can be used repeatedly as shown in Figure 2.5(d)–(g). In another study focusing on the recyclability of CNT sponges for oil absorption, it has been shown that 98% of absorbed oil can be regained just by squeezing or can be burnt directly to obtain heat energy [66]. Although CNT sponges are not cost efficient as compared to the widely used commercial sorbents such as exfoliated graphite, cotton grass fiber, or polypropylene, the adsorption capacity of CNT sponges (i.e., 100 g/g) is much higher than its commercial rivals (i.e., 10–30 g/g).

In a different study, 2D CNT network films synthesized by airbrushing spraying of double-walled and multiwalled alumina substrates have been investigated for their usage as gas sensors for NO_2 detection [67]. Examination of several CNT network thin films has demonstrated significant performance despite the existence of interfering gasses even at low NO_2 concentrations. In addition, response capability of CNT networks has been shown to be enhanced by applying successive heat treatments. Sayago et al. also proposed that sensitivity of CNT networks as sensors can be controlled by adjusting CNT deposition rate and CNT transport properties or by selecting the CNT units properly [67].

Bottom-up controlled production of such CNT networks may enable them to be employed in applications such as sensors, filters, composites, and electromechanical actuators [68]. However, CNT networks also exhibit charming mechanical behavior of individual CNTs to macromechanical applications such as nanocomposites. Similar to CNTs that are used as reinforcing units in nanocomposites, CNT networks can also be employed in nanocomposites for both reinforcement and damage detection purposes [23]. For instance, Zhang et al. investigated real-time damage-sensing capacity of CNT networks in fiber-reinforced composite materials [69]. According to them, well-dispersed homogeneous CNT network layers are formed by airbrushing onto the glass fiber-woven fabric and maintaining CNT localization over the damage-prone matrix. After forming a conductive CNT network within the composite specimen, *in situ* damage-sensing tests were established by monitoring the load–displacement and volumetric electrical resistance of the test specimen simultaneously. By comparing the load–displacement and volumetric resistance change, significant correlation between the mechanical load variations or crack propagation and electrical resistance has been confirmed in that internal damage condition could be determined by measuring electrical resistance of the percolated conductive CNT network. Along this direction, online structural health-monitoring systems can be developed by using CNT networks embedded within a structural composite system without any need to perform out-of-service inspections, which may serve as a tremendous cost-effective solution especially for aviation industry.

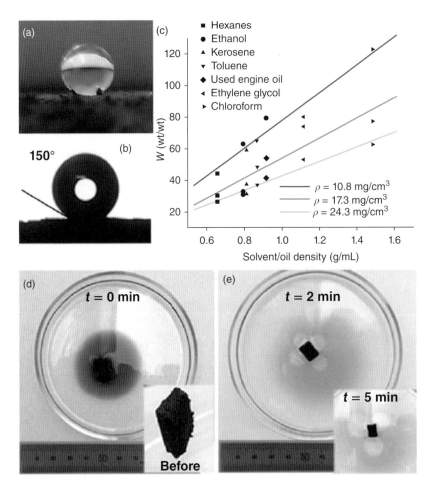

Figure 2.5 Oil absorption application. (a) Photograph of super-hydrophobic surface; (b) an approximately 150° contact angle measurement with a 2-mm-diameter water droplet resting on the sponge surface; (c) weight-to-weight (wt/wt) absorption capacity plot for common solvents measured with samples of density 24.3 mg/cm^3 (green), 17.3 mg/cm^3 (red), and 10.8 mg/cm^3 (blue); (d)–(g) demonstration using CB$_x$MWCNT sponge to clean up used engine oil spill (0.26 mL) in seawater; sample ($m \approx 4.8$ mg, $\rho \approx 17$ mg/cm^3); (d) photograph shows the sponge dropped into the oil at $t = 50$ min, inset shows sponge before use; (e) sponge sample absorbing the oil at $t = 2$ min, inset shows $t = 5$ min; (f) by burning or squeezing (inset) one can salvage the oil; (g) the sponge can then be reused repetitively. By using a magnet one can track or move the oil; inset shows sponge after burning and before reuse. *Source*: Hashim et al. [39]. Reproduced with permission of Nature Publishing Group

As already mentioned, CNT network materials can also be used for the reinforcement of composite materials. In this context, Tai et al. utilized CNT materials in both dispersed and network forms for reinforcing CNT/phenolic composites [70]. According to the tensile experiments established for the reinforced composites, it is concluded that the CNT networks provide higher enhancement on the mechanical properties than on dispersed CNTs because of the entanglement of longer and less agglomerated CNTs in networks. Furthermore, it is

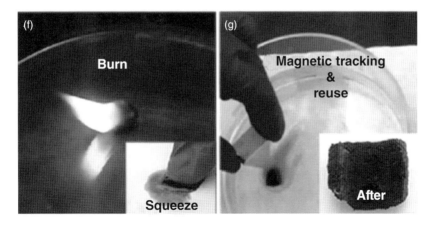

Figure 2.5 *Continued*

reported that by inclusion of 3 wt% CNT network, tensile strength and elastic modulus can be increased approximately by 97.0% and 49.8%, respectively [70].

In addition to serving as a reinforcing agent within another material, CNT networks can also be employed as stand-alone materials providing high-energy dissipation. Supporting this idea, Zeng et al. presented CNT sponge–array composites (see Figure 2.6(a)–(f)) consisting of stacked layers of CNT sponges and CNT arrays in tandem arrangement and showed their stability in tension and compression [71]. Following this study, Gui et al. showed improved energy dissipation in CNT sponge–array composites that can be manufactured in different configurations such as series, parallel, and sandwich architectures, which in turn inspires potential applications in energy absorption and cushioning [66].

2.2 Experimental Studies on Mechanical Characterization of CNT Network Materials

Most studies investigating the properties of CNT networks are carried out experimentally following a corresponding manufacturing phase. For instance, the thermal stability of CNT networks produced by the dielectrophoresis method on microelectrodes is studied experimentally to predict their performance in electrical applications [2]. Similarly, Wang et al. [72] investigated the electrical and mechanical properties of CNT networks with measurements taken *in situ* inside the TEM. Furthermore, because covalently bonded CNT networks require special treatments that need to be developed, initial CNT networks reported in the literature consists of entangled CNTs without any chemical-bonded linkages. Due to their weaker stand-alone mechanical properties, from the mechanistic point of view, noncovalent CNT networks are generally considered as the reinforcement units within nanocomposites [73] as a further enhancement to the composites reinforced by dispersed CNTs. In addition, Thostenson and Chou studied the damage and strain-sensing capabilities of CNT networks utilized in glass fiber–epoxy composites and have shown through tensile testing that conductive percolating CNT networks can detect the initiation and progression of damage [74]. Several other studies [75–77] exist in the literature on the electrical, thermal, or mechanical properties

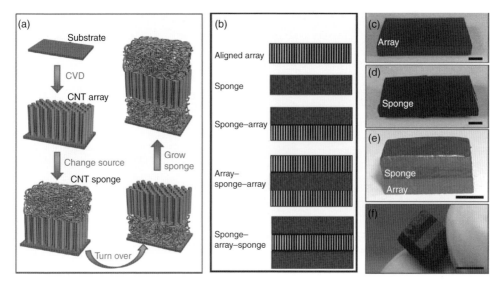

Figure 2.6 Illustration of CNT sponge–array composite structures and growth process: (a) CVD process for synthesizing a sponge–array–sponge triple-layered composite, involving the growth of an aligned array (first layer) on a quartz substrate, the growth of a sponge layer on top of the array by switching the carbon source, and the growth of another sponge layer on the other side of the array. (b) Illustration of single-layered and composite structures that can be controllably produced, including individual array and sponge, double-layered and triple-layered structures. Pictures of as-synthesized (c) aligned CNT array, (d) CNT sponge, (e) sponge-on-array double-layered structure, and (f) sponge–array–sponge triple-layered structures. Scale bars in (c)–(f): 5 mm. *Source*: Zeng et al. [71]. Reproduced with permission of John Wiley & Sons Ltd

of CNT networks carrying out experimental investigations, which, as the other aforementioned studies, inherently require equipment of manufacturing and testing. In this section, experimental studies are reviewed by considering the noncovalent and covalent CNT networks separately.

2.2.1 Non-covalent CNT Network Materials

Among the studies investigating the mechanical characteristics of noncovalent CNT networks, Slobodian and Saha performed cyclic compression tests to trace stress–strain hysteresis loops of nonbonded CNT networks composed of randomly oriented and entangled MWCNTs [78]. In this aspect, manufactured MWCNT network specimens with length 10 mm, width 8 mm, and thickness range of 0.02–0.4 mm are successively subjected to several compression and relaxation cycles, which leads to the accumulation of residual strain after the first compression cycle. As the specimens are experienced with compressive loading, deformation mechanisms including buckling of CNTs as well as frictional slippage at the contact points result in irreversible reorganization of the network. Figure 2.7(a) shows that compressive stress increases with growing rate due to densification of the structure. Following several loading and unloading cycles, it is also noticed from Figure 2.7(b) that residual strain after

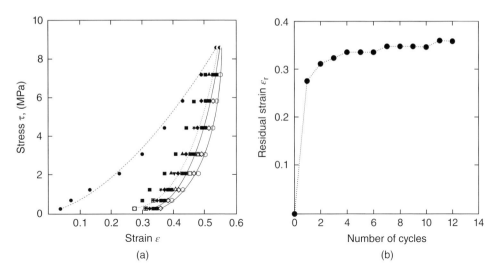

Figure 2.7 (a) Absolute stress–strain loops in cyclic compression test for MWCNT network subjected to 12 compression/expansion cycles (the network thickness is 0.418 mm). The dotted lines (first loading and unloading cycles) and solid lines (12th cycle) represent the theoretical prediction. (b) The plot of absolute residual strain versus the number of cycles. *Source*: Slobodian et al. [78]. Reproduced with permission of AIP Publishing

each cycle reduces by each cycle and tend to converge to an asymptotic value. It is also inferred that stress–strain hysteresis loop shape of MWCNT network films is also stabilized as the residual strain values converge to the asymptotic value, which can be explained by stabilization of CNT rearrangement. According to those results, Slobodian and Saha proposed that noncovalent entangled CNT network materials which are preprocessed sufficiently by compressive loading cycles can be employed for sensing compressive stress [78].

One of the solutions to strengthen noncovalent entangled CNT networks is to employ binder agent through the network that will improve the mechanical interaction between individual CNT units or bundles of CNTs by avoiding intertube sliding and increasing load transfer capability. In this aspect, Worsley et al. utilized carbon nanoparticles as binder agents to generate cross-linked networks with enhanced mechanical strength while preserving high electrical conductivity [79]. In their CNT foam structures, carbonaceous cross-links are formed through application of organic sol–gel chemistry on the aqueous media consisting of well-dispersed SWCNTs. Organic binders that are yielded by the introduction of sol–gel precursors in the network assembly are converted to carbon by drying and pyrolyzing the assembly. Closer examination of the CNT foams manufactured by Worsley et al. using SEM and TEM showed that sol–gel polymer practically nucleated on the CNT surfaces, which resulted in coating of the CNTs by a thin layer of carbon. Mechanical characterization of CNT foams was performed by nanoindentation tests subjected to both continuous and load–unload indentations. Figure 2.8 depicts the variation of elastic moduli of several nanoporous materials including CNT foams with respect to density. Comparison among other nanoporous materials points out significant mechanical performance of CNT foams for a given density value in terms of stiffness. For instance, for the density value of 100 mg/cm^3, elastic modulus of CNT

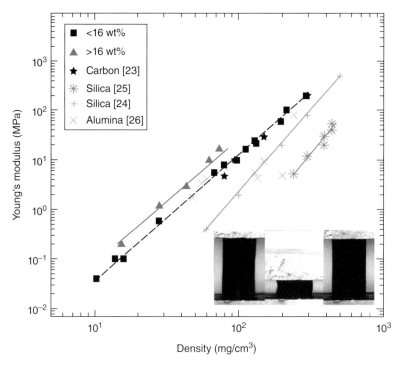

Figure 2.8 Dependence of Young's modulus on the density for monolithic CNT foams compared to carbon, silica, and alumina aerogels. The inset shows the sequence of uniaxial compression of a monolith (30 mg/cm³ and 55 wt% CNT content), illustrating the super-elastic behavior with complete strain recovery after compression to strains as large as 76%. *Source*: Worsley et al. [79]. Reproduced with permission of AIP Publishing

foams with high CNT content is approximately 12 and 3 times those of conventional silica and carbon aerogels, respectively. Furthermore, partial load and unload indentations also showed super-plastic behavior of CNT foams with complete recovery of 76% straining as demonstrated in Figure 2.8 [79].

Another technique developed for the reinforcing CNT networks by Kaskela et al. has utilized carbon vacuum arc to coat CNT networks with a 5- to 25-nm-thick carbon layer [80]. They demonstrated, for the first time, the carbon coating process on stand-alone pristine SWCNT networks through a physical vapor deposition without a supporting substrate. Deposition process of carbon on the CNT networks is realized by performing pulsed vacuum arc discharging method while monitoring it through *in situ* resistance measurements. Similar to the study presented by Worsley et al., nanoindentation tests on both coated and noncoated CNT networks were established to explore the mechanical improvement due to coating. According to the load–deflection curves for loading and unloading stages that are obtained through nanoindentation tests as depicted in Figure 2.9, coated CNT networks exhibit higher stiffness against indentation as well as larger elastic recovery with higher resilience. Furthermore, wear resistance of the coated and noncoated CNT networks was compared by scanning probe microscopy (SPM) and SEM images, and superiority of coated networks against to noncoated networks was demonstrated after multiple wear scans [80].

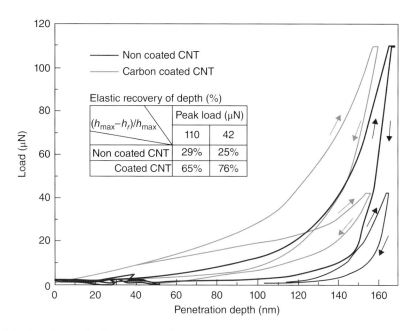

Figure 2.9 Loading–unloading curves of the carbon-coated (red) and noncoated (blue) SWCNT networks on silicon. The elastic recovery of indentation depth is calculated from the maximum penetration depth h_{max} and residual depth h_r. The coated SWCNT networks not only demonstrate higher resistance to the indenter penetration but also show larger elastic recovery. *Source*: Kaskela et al. [80]. Reproduced with permission of Elsevier

There are also studies investigating mechanical characteristics of the monolithic CNT network structures consisting of purified CNTs. Among those studies, Kim et al. reported a fabrication method for ultra-low density (i.e., 7.3 mg/mL) CNT foams, or specially named as CNT aerogels by their developers, with exceptionally high SSA (i.e., 1291 m²/g) and investigated their electrical and mechanical properties as summarized in Figure 2.10(a)–(e) [81]. In order to perform mechanical characterization, cylindrical macro specimens of CNT aerogels of 5 mm diameter and 3 mm length within density range of 7.3–18.5 mg/mL were tested under displacement-controlled uniaxial compression and tensile loadings. Illustrative example for both compressive and tensile stress–strain graphs are given in Figure 2.10(c) for the aerogel specimens with density of 9.9 mg/mL. It is observed that aerogel specimens can be compressed by larger than 90% straining. Apart from that, through all density spectrum studied, Poisson's ratio of the specimens under compression is measured very close from 0 up to 60% straining, which indicates densification of the aerogels without expanding normal to the loading direction owing to the highly porous architecture of the aerogels with fibrous cell walls. In addition to that, compressive stress–strain curves of aerogels at all density ranges demonstrated three distinct deformation regimes that are also observed in open-cell foams: (1) a linear elastic zone $\varepsilon < 9\%$, (2) a plateau zone $9\% \leq \varepsilon \leq 60\%$, and (3) densification zone $\varepsilon \geq 60\%$ in which the stress increases abruptly due to densification (Figure 2.10(c)). In the context of cyclic loading, it is observed that CNT aerogels deformed plastically after two to three compressive cycles at approximately 9% straining as depicted

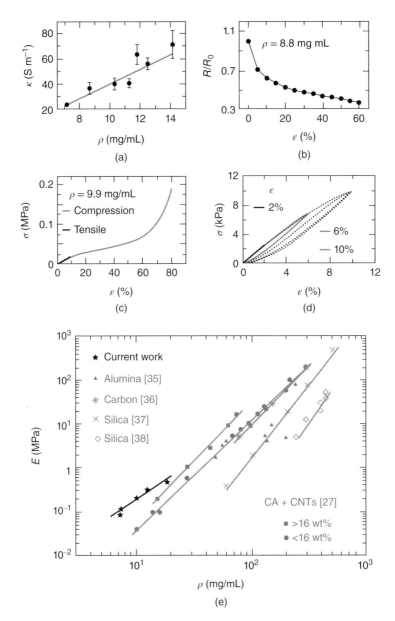

Figure 2.10 Electrical and mechanical properties of the aerogels. (a) Aerogels were electrically conducting which suggests that aerogel fabrication did not damage nanotubes. (b) R/R_0 decreased dramatically ($\approx 50\%$) with an increase in ε to $\approx 15\%$. The decrease was much smaller ($\approx 15\%$) for an increase in ε from $\approx 15\%$ to 60%. (c) Stress (σ) versus strain (ε) of aerogels under compression showed three distinct regions: Hookian, plateau, and densification. The Young's moduli (E) for tensile and compressive loading were similar. (d) Even in the Hookian regime, the aerogels plastically deformed when compressed for more than two to three cycles. The plastic deformation became more pronounced near $\varepsilon = 9\%$. (e) E of our aerogels at low density were larger than E of other reported aerogels. E of these aerogels had a power law dependence of ≈ 2 on ρ, which further confirmed the open-cell structure of the aerogels. *Source*: Kim et al. [81]. Reproduced with permission of John Wiley

in Figure 2.10(d). Another finding from their mechanical experiments is the similarity of Young's modulus values determined from tensile and compression tests of aerogel specimens with same density, which suggests the isotropic character of CNT aerogels. For instance, the compressive Young's modulus of specimens with density of 9.9 mg/mL is calculated as approximately 0.22 MPa in the elastic regime (i.e., $\varepsilon < 9\%$), while the corresponding value for tension is found to be nearly 0.21 MPa. As also shown in Figure 2.10(e), the elastic modulus of their CNT aerogels is larger than other aerogels including alumina, silica, and carbon aerogels in the low density regime. Furthermore, it is also founded that the Young's modulus of CNT aerogels possesses a power law (i.e., $\sim$2) relationship to their density as in the case of open-cell foams [81].

As a different approach among the other experimental studies, Rahatekar et al. researched the dependence of mechanical behavior of CNT networks on the length scale of individual CNTs constructing the network [82]. For this purpose, CNT/epoxy composite samples were prepared by employing short (i.e., 4 μm) and long (i.e., 60 μm) MWCNTs in distinct models. Rheological properties of the network models were determined by measuring the linear viscoelastic shear modulus and steady shear viscosity. Based on their detailed examination, they concluded that in the CNT networks composed of long nanotubes, the main load-bearing structural units are the individual nanotubes. On the other hand, in the CNT networks with the short CNTs, resistance to mechanical loading is contributed by the CNT aggregation. More importantly, another observation reported is that at reasonably higher mass fractions of MWCNTs, CNT networks with both short and long CNTs have approximately the same shear modulus although microstructures and deformation mechanisms are not matched. This observation may help to synthesize CNT networks consisting of shorter CNTs with equivalent mechanical properties instead of using ultra-long CNTs, which are associated with possible health effects [83].

2.2.2 Covalently Bonded CNT Network Materials

Fu et al. introduced a new technique to synthesize covalently bonded CNT network based on the growth of CNTs on a template [37]. In this technique, individual CNTs are grown on nickel foam by the CVD method, and following the growth process, nickel core is removed by etching. Because Ni foam has a 3D cross-linked cellular morphology, the CNTs grown on the template result in interconnected CNT networks. Synthesis process that is also discussed in Section 2.1.2 is outlined in Figure 2.11. Manufactured CNT network specimens are mechanically characterized using atomic force microscopy (AFM)-based bending tests. As shown in Figure 2.11(a)–(c), bending tests are established for both single CNTs and CNT networks suspended on the test fixture by applying a loading at the middle section of the CNTs using the AFM tip. In Figure 2.11(d), force–displacement curves for both single CNT and CNT network are given. Relying on the bending tests, Young's modulus of single CNTs is calculated by utilizing the continuum mechanics bending formula $F = (192EI\delta)/l^3$, where F, δ, l, E, and I are applied force, deflection of CNT, length of suspended CNT, equivalent Young's modulus, and second area moment of CNT cross section, respectively. Calculated values of the Young's modulus of CNTs reported are in good agreement with the theoretical calculations in the literature. A more important result from their study regarding the CNT network behavior is that CNT networks are much stiffer than the single suspended CNTs as depicted in Figure 2.11(d), which is referred to as the effective load-transfer capability in cross-linked CNT networks. As

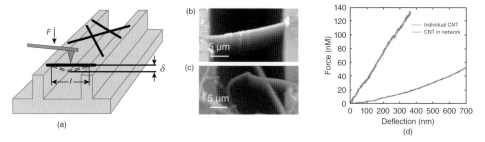

Figure 2.11 Mechanical tests on the individually suspended CNT and the interconnected CNT network. (a) Schematic of the test structure and principle. F is the force loaded at the midspan of the single CNT, δ is the deflection of the CNT and l is the span of the CNT. (b) and (c) AFM image of the individually suspended CNT and the suspended CNT network, respectively. (d) Approximately linear force–deflection curves of the individual CNT and the CNT in the network obtained from the mechanical tests. *Source*: Fu et al. [37]. Reproduced with permission of Wiley-VCH

a quantitative example, for the deflection of 200 nm, a loading of 8 nN is required to apply on a single CNT, while it is 78 nN for the CNT in a network.

In another experimental study, Satti et al. manufactured covalently cross-linked 2D CNT networks by using polymer molecules (i.e., poly(allylamine)) through a chemical approach with the purpose of developing strong composites [84]. By using their proposed technique, 2D CNT-based buckypaper structures consisting of covalent cross-linking of CNTs are synthesized in macroscopic dimensions with thickness range of 20–40 μm. Meantime, 2D CNT networks are also manufactured without covalent cross-links for comparison. Stress–strain curves obtained through tensile experiments of cross-linked and noncross-linked buckypaper composites have shown significant increase in tensile strength of cross-linked models compared to noncovalent cross-linked models. The average Young's modulus of cross-linked buckypaper films (i.e., 8.7 ± 1.8 GPa) is determined to be approximately 9 times higher than the value for noncross-linked films. In addition, the average tensile strength of cross-linked films (i.e., 36.1 ± 5.7 MPa) is found to be more than 4 times higher than noncross-linked specimens with nearly the same polymer content. Furthermore, it is also reported that the ultimate strain values have been enhanced by cross-linked composites compared to both noncross-linked films and pristine networks. Based on these experimental results that suggest mechanical enhancement through inclusion of cross-links into the CNT films, it is concluded that much stronger adhesion between CNTs results in intensive load–displacement transfer between individual CNTs, which in turn mitigates the slippage between CNTs and increases the network affinity [84].

The study established by Hashim et al. exploited a new method for bulk synthesis of covalently bonded CNT networks via boron-induced nanojunctions and examined their dynamic mechanical properties as depicted in Figure 2.12(a)–(d) [39]. In this context, by introducing boron atoms that act as atomic welding agents, 3D architectured randomly oriented and entangled CNTs were shown to be synthesized without yielding considerable amorphous carbon. As a consequence of boron doping, not only covalent junctions between CNTs are formed but also straight form of CNTs is altered and elbow-like morphologies are observed as illustrated in Figure 2.12(b). Dynamic mechanical analysis (DMA) was established at different compressive strain levels (i.e., 9%, 10%, 60%) on the bulk specimens of CNT networks with different densities. Dynamic mechanical tests performed at 60% strain

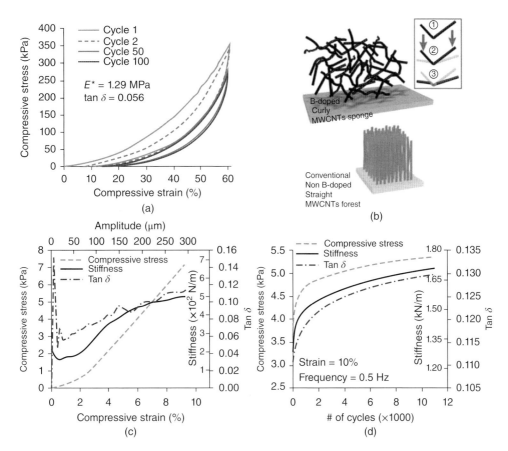

Figure 2.12 (a) Stress–strain curves up to 60% strain at 0.5 Hz and room temperature for cycles 1(blue), 2 (red), 50 (green), and 100 (purple) on a sample with $\rho = 5\ 27\ \text{mg/cm}^3$; (b) a computer graphics model shows the entangled random network as compared to conventional CNT arrays. The "elbow" defects aid in recovery via spring-back loading mode, shown in steps 1–3, to overcome the vdW "sticking" force of contacting tubes upon compression, which give rise energy dissipation and high tan δ values; (c) DMA on sponge with density $\approx 25\ \text{mg/cm}^3$ using multistrain mode at 1 Hz for 250 cycles; (d) DMA on sponge with density $\approx 20\ \text{mg/cm}^3$ over 11,000 cycles. *Source*: Hashim et al. [39]. Reproduced with permission of Nature Publishing Group

level yielded complex modulus (E^*) and ratio of loss modulus E'' to storage modulus E' (i.e., tan $\delta = E''/E'$) as 1.26 MPa and 0.058, respectively. The high value of tan δ that is often called damping implies the capability of high energy absorption and energy dissipation under cyclic loads. Compressive stress–strain curves obtained from the cyclic compression of specimens with 27 mg/mL density at 60% strain are shown in Figure 2.12(a). Moreover, 100 cycles of compression dynamic loadings at 60% strain resulted in only 18% permanent deformation. Furthermore, DMA realized on CNT foams with density of 25 mg/mL in multi-axial strain mode for 250 cycles at 9% strain yielded damping (i.e., tan δ) value of approximately 0.11 as shown in Figure 2.12(c). As a further study, they performed 11,000 cycles of compression at 10% strain and recorded a gradual improvement in damping (i.e., tan δ) from 0.11 to 0.13

as well as an increase in compressive stress and stiffness (see Figure 2.12(d)), which may be indicative of CNT alignment along the loading direction.

2.3   Theoretical Approaches Toward CNT Network Modeling

CNT network materials including 2D CNT films or 3D CNT sponges, aerogels, or foams have attracted tremendous interest owing to their ultra-lightweight, high flexibility, and controllable mechanical properties together with unique thermal and electrical properties. From the point of engineering, multifunctional characteristics of CNT networks hold great promise in a large variety of applications as discussed in Section 2.1.3. However their effective utilization in such applications requires in-depth understanding of their mechanical behavior and structure–property relationship. CNT network materials, as other fibrous network materials such as biological cytoskeleton networks, spider webs, or cellulose in paper, represent complicated morphologies due to their multi-scale hierarchical structures. In comparison with polymer network materials, which are also constructed from microfibers and nanofibers with dynamic interfaces representing flexible character, the mechanical behavior of CNT networks can be identified as semiflexible because of their relatively larger persistence length which is on the order of 10–100 µm [85].

The majority of theoretical studies was focused on the investigation of individual CNTs and their nanocomposites [86–89]. Due to the lack of a proper method for the generation of random CNT networks and computational limitations, there are limited number of studies that use numerical models and computational methods for investigating CNT networks. Moreover, existing studies [87] that employ numerical network models mostly use ordered networks, which do not include geometrical irregularities and other possible imperfections, such as bond rearrangements, at junctions. Furthermore, in these models, only one type of junctions is modeled. A recently published study [90] investigates the mechanical behavior of short SWCNT aggregates composed of randomly dispersed nonintersected CNTs by molecular mechanics.

2.3.1   Ordered CNT Networks

Regarding mechanical behavior of CNT films, Lu et al. performed molecular dynamics (MD) simulations as the first time to evaluate the mechanical properties of CNT films by examining the effects of entanglements between CNT units [91]. Force–strain behavior of CNT films subjected to tensile loading was obtained by carrying out coarse-grained molecular dynamics (CGMD) simulations, and it is reported that network entanglements significantly decrease the tensile strength of the network comparing to the CNT films with straight CNTs. Several reasons have been shown to explain this behavior. One reason is that CNT films composed of straight CNTs (ST-CNTs) are mostly directed along the loading direction while entangled CNTs (ET-CNTs) are randomly oriented, which lowers the load-bearing capacity. Secondly, in ST-CNT films, as CNTs are mostly aligned in one direction, they can be more closely packed than those in the ET-CNT film, which in turn increase the density and larger intertube area to elevate load transfer between CNTs. In their study, Lu et al. also identified local deformation modes of ET-CNT films based on MD simulation such as stretching, straightening, bundling, unwinding, and sliding.

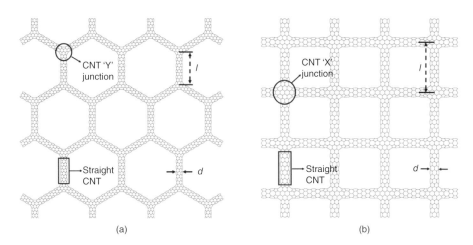

Figure 2.13 The configuration of the (a) super-graphene (SG)@(6, 0), (b) super-square (SS)@(6, 0) CNT networks and their building blocks (6, 0) SWCNTs, CNT Y-junctions, and X-junctions. *Source*: Li et al. [92]. Reproduced with permission of Elsevier

In another theoretical study focusing on the mechanical properties of CNT networks, perfectly organized conceptual CNT network architectures, which are composed of multiple layers of crossbar and hexagonal SWCNT nets (see Figure 2.13(a) and (b)), are investigated by molecular mechanics (MM) and MD simulations [86]. Coluci et al. performed static calculations based on molecular mechanics to calculate elastic modulus and bulk modulus of the multilayered network structures, which are shown to be smaller than those of graphite, with different geometrical parameters. Their results indicate very high flexibility of the networks under small deformations while the degree of flexibility decreases for larger deformations. It is also demonstrated that as the network characteristics including nanotube type and nanotube arrangement are modified, the mechanical properties of the networks can be tuned. In addition to static calculations, impact absorbing capacity of the proposed CNT networks subjected to collision of an energetic particle is presented through MD simulations.

As the computational resources become more powerful and cost-efficient, atomistic simulation techniques (e.g., MD simulation) are becoming a very useful tool not only for the modeling and analyzing of existing nanomaterials but also for the conceptual design and validation of novel nanostructures. In this regard, 3D CNT networks, namely "super"-carbon nanotubes based on CNTs were proposed by Coluci et al. [93] as inspired by the self-assembly ability of branching CNTs. Super-CNTs are formed by rolling of the "super"-sheets, which are constructed by SWCNTs connected by Y-like junctions, into seamless cylinder in analogous with the conceptual formation of CNTs from graphene sheets. In the following years, several works concentrated on the mechanical properties of super-CNTs [94, 95] or super-graphene sheets [92, 94, 96].

For example, Wang et al. studied mechanical properties of super-honeycomb structures, which are also called super-graphene (SG) sheets and computationally constructed by repeating and shifting perfect Y junctions periodically in a plane by using finite element method (FEM) with equivalent continuum thin-shell model [96]. Widely used values for the equivalent continuum shell parameter values of 0.066 nm, 5.5 TPa, and 0.19 for the respective thickness, elastic modulus, and the Poisson's ratio [97] were taken as homogeneous shell parameters. According to their results, it has been demonstrated that mechanical characteristics such as tensile modulus and Poisson's ratio are directly related to the number of junctions along the transverse direction to the loading direction. Apart from this, Coluci and Pugno worked on the stretching, twisting, and fracture characteristics of ordered CNT networks in the form of super-CNTs by employing classical MD simulations based on reactive empirical bond-order potential [95]. In this context, fracture process for the super-CNTs with different chiralities has been traced under the torsional and tensile loadings to observe the effect of chirality on the mechanical behavior of super-CNTs. It has been demonstrated that flexibility of super-CNTs in both tensile and torsional deformations is higher than the flexibility of individual SWCNTs constituting super-CNTs [95].

In addition to the super-CNTs and super-graphene sheets in which the individual CNTs are connected via Y-type connections, a theoretical design of another type of 2D CNT network called the super-square (SS) CNT is presented by Li et al. [87]. In their study, they investigated the deformation mechanisms of SS CNT networks by realizing a comparison with super-honeycomb (SH) CNT networks. For the purpose of examining the effect of chirality, networks are constructed by two types of CNT units with chiralities of $(6, 0)$ and $(4, 4)$. Instead of performing MD simulations that are under time and size limitations, molecular structural mechanics (MSM) method in which the chemical bond is simplified as a structural continuum beam element with a circular cross section was used to explore the mechanical behavior of SS CNT networks. Continuum equivalent material characteristics including tensile, flexural, and torsional stiffness values were determined by the energy equivalence principle [98]. Modeling the SS CNT network as continuum equivalent space frame structure, uniaxial tensile experiments were done within one of the commercial finite element (FE) software. Molecular mechanics based FE simulations have shown that deformation mechanism in SS CNT networks is dominated by stretching (see Figure 2.14), whereas in SH CNT networks bending mode plays an important role in tensile loading. Furthermore, tensile modulus of SS CNT networks was found to be higher than bending-dominated SH networks.

2.3.2 Randomly Organized CNT Networks

One of the studies focusing on the modeling of random-natured CNT networks was presented by Zaeri et al. [99]. In their study, they employed MSM and FEM to investigate the mechanical behavior of CNT network materials. In accordance with that, 2D random network of CNTs was modeled by random networks of curves representing the individual CNTs in the network. Figure 2.15 illustrates the FE model of 2D random network with boundary conditions of tensile loading at the top and displacement constraints at the bottom. Comparisons between the results obtained through MSM method together with FEM and those of other theoretical and experimental studies have shown the reliability of the technique in addition to its simplicity and adaptability for the modeling of nanostructures [99].

Macroscopic mechanical behavior of CNT networks strongly depends on the mechanics of individual building blocks (i.e., CNTs), their percolation through the network, and the interactions between the CNTs such as entanglement and covalent bonding [100–105]. As the deformation develops under applied loads, the local structure of the network is also reorganized, which in turn modifies the mechanical properties of the network interactively. In a recent study taking into account the stochastic nature of CNT networks, structural evolution

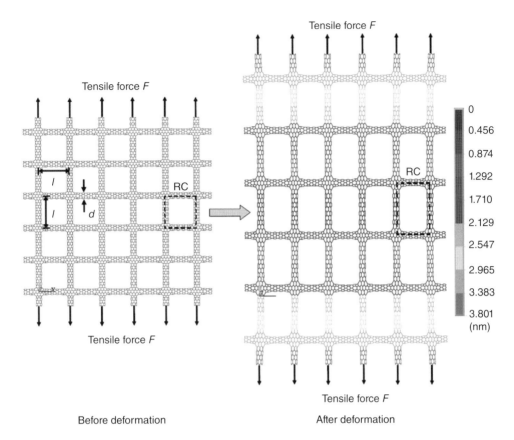

Figure 2.14 The SS@(6, 0) under 50% uniaxial tensile strain before and after the deformation. The dash box in the SS@(6, 0) is a representative cell (RC), which could be used to trace the deformation behavior of SS@(6, 0). The different colors represent the contour of displacement in the deformed SS@(6, 0) structure. *Source*: Li et al. [87]. Reproduced with permission of Elsevier

and corresponded effects on the mechanical performance of noncovalent 2D CNT network materials have been investigated through CGMD simulations [106]. Continuum equivalent mesoscopic models of 2D CNT networks were obtained by bead–spring approach in which discrete beads interact with each other by bond and bond-angle springs. Overall structure of their coarse-grained model for a 2D CNT network is depicted in Figure 2.16.

Thermally equilibrated models were subjected to uniaxial loading, and microstructural deformation mechanisms affecting the overall macroscopic behavior were examined [106]. According to that, in the initial phase of the tensile loading, an affine deformation character was recognized, which was explained by local stretching of van der Waals binding sites. As the tensile loading progresses, the Poisson effect causes transversely compression of the network structures. At this stage, it is noticed that some of CNT units start bundling to bear higher tensile loads under the effect of transverse contraction, which causes increase in contact surface area between CNTs. Xie et al. also reported that due to bundling process, deformation character turned into nonaffine deformation and significant evolutions were observed on the network topology. Figure 2.17(a)–(d) presents the microstructural evolution

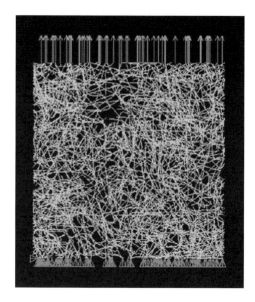

Figure 2.15 Lateral tensile loading of random curvilinear buckypaper (37% porosity). *Source*: Zaeri et al. [99]. Reproduced with permission of Elsevier

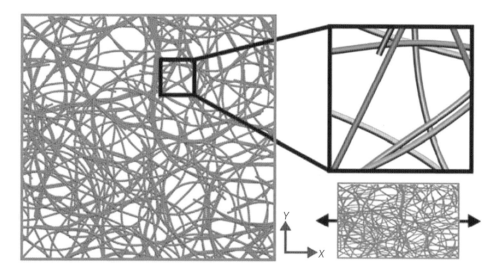

Figure 2.16 A coarse-grained model for carbon nanotube networks. The uniaxial load is applied to the *x*-direction of a super-cell with two-dimensional in-plane periodic boundary conditions. The inset shows local structure of crossings. In the molecular dynamics simulation, intrafiber elasticity is described by a bond stretching and a bond angle bending term, reproducing the Young's modulus and bending rigidity of a carbon nanotube. The interfiber binding is captured by a Lennard–Jones type pair interaction between coarse-grained beads. *Source*: Xie et al. [106]. Reproduced with permission of the Royal Society of Chemistry

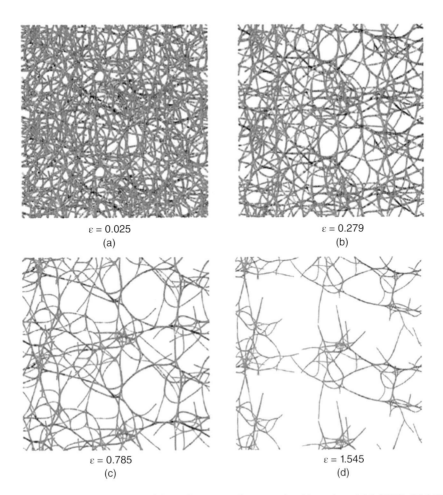

ε = 0.025
(a)

ε = 0.279
(b)

ε = 0.785
(c)

ε = 1.545
(d)

Figure 2.17 Structural evolution of the carbon nanotube network with strain at (a) 0.025%, (b) 0.279%, (c) 0.785%, and (d) 1.545%. The color maps stress amplitude along the tensile direction. The blue color represents zero or compressive stress, while other colors indicate the tensile stress state and the red color corresponds with the maximum stress value. The network structure is duplicated in the y-direction for a better illustration. *Source*: Xie et al. [106]. Reproduced with permission of the Royal Society of Chemistry

of the 2D CNT network model subjected to tensile loading together with stress color plots indicating the tensile and compression stress state. After a detailed discussion about the microstructural evolution and its effects on the mechanical behavior, they concluded that the mechanical properties of CNT networks could be improved by avoiding intertube sliding and enhancing the network deformation affinity. Another way of improving mechanical behavior was proposed to introduce chemical bonding between CNTs as covalent cross-links by a small ratio (i.e., 10%) of total contacts. Uniaxial test simulations resulted in more uniform stress distribution, demonstrating enhanced affinity and nonlocal failure of the material [106].

In a more recent study, another coarse-grained model, in which discrete beads and angle springs were utilized to account for bond stretching and bending, respectively, have been adopted by Wang et al. to formalize the mechanical performance of cross-linked semiflexible fibrous CNT networks [107]. Their coarse-grained model was also upgraded by adding a reactive cross-linking mechanism that considers the breakage and reformation of cross-links employing the stiffness and strength parameters fitted from first-principle calculations. Thus, the force within a cross-link can be modified depending on the distance between interfiber distances. Comparison between the networks with covalent and noncovalent cross-links has demonstrated that covalently cross-linking of CNT units resulted in networks with nearly 58 times higher strength. While van der Waals interactions that are relatively weaker cannot prevent sliding of CNTs on other CNTs or detaching of CNTs from each other, covalent cross-links provide much stronger mechanical response due to resistivity supplied reinforcement against sliding and detaching of CNTs. According to the simulation results, it is also pointed out that the failure of the network is initiated by local fracture zones where the cross-links are catastrophically broken.

In addition to introducing chemical cross-links between CNTs to enhance mechanical behavior by increasing affinity and load-transfer capability of the network, another idea that suggests noncovalent cross-links by presenting binder materials is also investigated [108]. This approach is actually inspired by the microstructure and behavior of biological materials such as collagenous bone fibrils or cellulosic wood in which relatively stiff fibrous units are cross-linked by soft binder matrix to maintain both high strength and toughness. With this proposal, Bratzel et al. investigated the effects of polymeric binders added into CNT-based nanofibers through atomistic-based multiscale simulations. Inclusion of disordered polymer phase was supposed to improve the adhesion between neighbor CNTs. By employing a coarse-grained model, they performed two types of numerical tests, that is, pull-out tests that perform pulling of a single CNT from the bundle and CNT bundle tensile tests, by systematically altering the cross-link density and cross-link lengths. Based on the results of computational experiments, better enhancements are observed with shorter cross-links and a cross-link density of nearly 20% by weight is reported as the best. Furthermore, it is also quantitatively shown that certain combinations of cross-link density and length parameter would produce optimum mechanical response in terms of toughness and strength [108].

Importance of interaction between CNT units has also been underlined in a different theoretical study presented by Wang et al. [14] that investigates refining the mechanical properties of CNT networks by the inclusion of physical mobile and discrete binders into the network. According to this study, a coarse-grained model consisting beads that interact through bond and angle springs is used in their numerical simulations. Stretching and bending of CNTs were defined by continuum equivalent spring constants, and their contributions into total energy could also be traced to identify the effect of different deformation mechanisms on the overall behavior. In their coarse-grained model, torsional deformation mode of CNTs was discarded due to the 2D nature of CNT network films being investigated. Spatial distribution of the binders among the fibers at different stages of mesoscale simulation including equilibration and loading cycles is given in Figure 2.18(a)–(d).

In addition to bending and stretching modes, van der Waals interactions between beads located on distinct CNTs were represented by Lennard–Jones potential with a cutoff distance of 3 nm. Parameters employed in CGMD simulations such as equilibrium interbead distance, tensile and bending stiffness parameters, Lennard–Jones potential parameters are fitted to full atomistic MD simulations [104, 109, 110]. Binders added into the network were also modeled as beads with a parameter depending on the interfacial energy between CNTs.

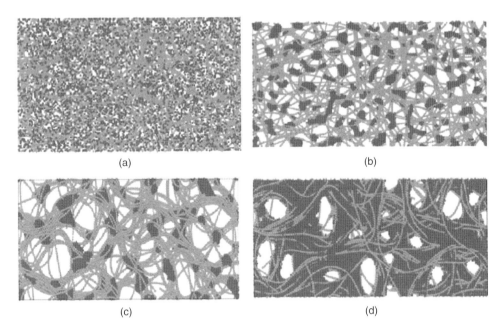

(a)   (b)

(c)   (d)

Figure 2.18   The spatial distribution of binders (blue) in a CNT network (red), with a weight fraction $f_w = 33.3\%$ (a)–(c) and 65.5% (d) at different stages of the CGMD simulation: (a) initial setup, (b) after structural relaxation, (c) and (d) after five loading cycles. *Source*: Wang et al. [14]. Reproduced with permission of Elsevier

Through CGMD simulations, it has been demonstrated that the inclusion of binders has remarkable effects on the enhancement of mechanical behavior of the CNT networks in terms of their stiffness and strength. Furthermore, energy absorption capacity and viscoelastic characteristics of the networks are shown to be improved by binder particles. Specifically, binding energy and weight fraction of the binders are stated to be decisive parameters for those enhancements. It is also determined that above critical weight fractions, degree of mechanical improvement is decreased. With regard to microstructural evolution of the network with mobile binders, accumulation of the binders around CNT junctions is noticed, which plays an important role to enhance deformation affinity due to more homogeneous load transfer. In summary, studies on the enhancement of CNT networks by the utilization of binders suggest that both mechanical and dynamical properties can be controlled effectively. Therefore, networks could be optimized for specific applications by selecting binders with specific physical and chemical properties.

2.4   Molecular Dynamics Study of Heat-Welded CNT Network Materials

In this section, a recently published theoretical study by the authors on the mechanical properties of 3D CNT networks with covalently bonded nanojunctions has been presented [111]. Atomistic-based CNT network models are tested under tensile loading through MD simulations to capture the detailed deformation mechanisms that affect the mechanical response. For this purpose, we first discuss an automated stochastic algorithm developed to generate atomistic models of random-natured CNT networks with certain controls on several topological parameters including cross-link density, distance between cross-links, as well as length and chirality of CNTs [112]. Covalent cross-links between individual CNTs are formed through heat welding method that has already been employed in the literature to generate covalent junctions [113–115]. The resulting atomistic models are employed in MD simulations of uniaxial tensile loading experiments to determine mechanical properties such as the Young's modulus, yield strength, and ultimate strain values. In addition to these, CNT network specimens with different cross-link densities are utilized to examine the effects of cross-link density on the mechanical performance.

2.4.1   A Stochastic Algorithm for Modeling Heat-Welded Random CNT Network

In both 2D and 3D CNT network structures excluding the networks with aligned architecture, individual CNTs are randomly oriented and chemically or physically cross-linked with each other representing a disordered network arrangement. For an atomistic-based modeling, the random characteristics of the networks such as random orientations and interrelations of individual CNTs can be considered using a stochastic algorithm developed for the modeling of CNT networks. Using this algorithm, it has been demonstrated that atomistic models of 2D/3D CNT network structures with several types of covalently linked CNT junctions (i.e., X, Y, or T type) can be generated.

In the context of this algorithm, two main cyclic loops are constructed to generate atomistic CNT network models with cross-link nodes, where individual CNTs are positioned sufficiently close to be heat welded in a post-processing step. In the main loop, individual candidate CNTs selected from a virtual library consisting of different types of CNTs are inserted and positioned in the design space through random rotation and translation operations. In the next sub-loop, compatibility of the candidate CNTs is checked through a set of user-defined constraints. For instance, a distance control is applied to check if the candidate CNT is within the cross-link neighborhood of a specific target CNT or to check if it is not closer to any other CNT than a predefined parameter. If all design constraints are satisfied by the candidate CNT, temporary location and orientation of the candidate CNT in the design space is permanently approved. On the other hand, if any one of the constraints is not satisfied, another candidate CNT is picked up to be examined for the constraints until a suitable CNT configuration is accomplished. The main loop is kept active until the point where the number of CNTs is acceptable for the required size of the specimen.

As shown in Figure 2.19(a) and (b), individual CNTs in the network are represented by 3D line segments passing through the central points of the cylindrical geometry of CNTs to maintain computational efficiency. Therefore, for simplicity, each 3D line segment is represented by the coordinates of end points, which are the central points of the circles at the end of CNTs. In this regard, all calculations and operations including distance and angle calculations, rotation, and translation operations are performed by utilizing only the end-point coordinates of the line segments instead of the atomic coordinates. Cross-link nodes on the CNTs are determined via random points on the line segments that are generated via 3D parametric line equations

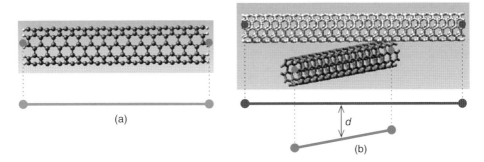

Figure 2.19 (a) Line segment representation of CNT, (b) minimum distance between CNTs by using line segment representation. *Source*: Kirca et al. [112]. Reproduced with permission of Elsevier

constructed by the tip coordinates and line segment parameter t varying in range [0,1]. The line segment parameter t is employed for further segmentation of line segments to position the candidate CNTs on the target CNT more evenly.

A classical molecular dynamics code that is publicly available, large-scale atomic/molecular massively parallel simulator (LAMMPS), is employed for all the numerical simulations including heat welding and tensile loading simulations [116]. For the modeling of the interaction between carbon atoms throughout the CNT network, the adaptive intermolecular reactive empirical bond order (AI-REBO) potential proposed by Stuart et al. [117] is used with a 2.0 Å cutoff distance. AI-REBO potential is capable of representing covalent bond forming and breaking mechanisms; therefore, investigations on the heat welding and mechanical deformation of hydrocarbon structures can be performed efficiently [86, 88, 94, 95, 112, 118, 119]. Furthermore, the AI-REBO potential also incorporates the van der Waals long-range atomic interactions and torsional terms for σ-bond in addition to the potential terms of the Brenner second-generation reactive empirical bond order (REBO) [120], which in turn yields better accuracy.

In order to generate covalent nanojunctions, welding spot fusion regions are defined by using spherical volumes centered at the cross-link node. Temperature of the atoms residing inside those spherical fusion regions is progressively increased to a reference temperature by using the Nosé–Hoover thermostating scheme. On the other hand, the range of reference temperature levels that needs to be maintained for a proper welding process is a controversial subject. For instance, in the literature some studies show that the reference temperature levels should be selected between 2000 K and 3000 K [113, 121–126]. Even though thermostating at high temperatures for welding purpose enhances bonding characteristics and provides a more physical way for the covalent bond formation, computational expense of this welding method is extremely high, especially for the atomistic models with large number of atoms. As a different strategy with significant computational savings, in the overlapped region between intersecting CNTs some spherical volumes are defined to remove atoms and thus dandling bonds that promote welding at relatively much lower temperatures (i.e., 600 K) are created. Figures 2.20 and 2.21 show some examples of junctions obtained through this heat welding procedure within a 3D CNT network. Because the stiffness of covalently bonded junctions plays a primary role of transferring the load within the network, the morphology of the junctions directly influence the mechanical properties of the CNT network structures. In this regard, some studies are reported

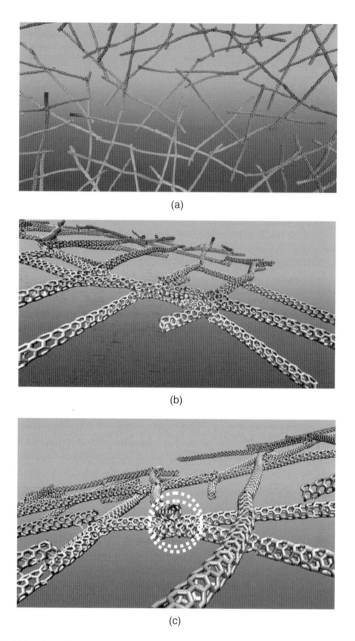

Figure 2.20 (a) Top view of a welded 2D network, (b) and (c) junctions formed after annealing. *Source*: Kirca et al. [112]. Reproduced with permission of Elsevier

in the literature based on the investigation of the effects of different types of junctions on the mechanical characteristics of CNT network structures [94, 125]. In the study presented in this section, CNT network models contain several types of CNT junctions including X, Y, or T that exists as can be seen from Figure 2.21.

Figure 2.21 Example of junctions throughout 3D CNT network. *Source*: Kirca et al. [112]. Reproduced with permission of Elsevier

As depicted in Figure 2.20(a), by the effect of thermal vibrations of CNTs and stable covalent bond formations at the cross-links, initially straight and nonbonded individual CNTs gain curvatures around the covalently bonded junctions, which are relatively stiffer portion of CNTs acting as hinge points for the bending deformation. Furthermore, Figures 2.20(b) and (c) and 2.21 illustrate some examples of covalent junctions in 2D and 3D CNT network samples, respectively, which are established by the application of the heat welding on the cross-links initially in contact. In these figures, different types of junctions including X, Y, and T can be created by the heating process applied by MD simulations. After heat welding, the atomistic models of covalently bonded CNT network models can be employed for the numerical experiments by MD simulations for any purpose.

For the purpose of investigating mechanical properties of covalently bonded CNT networks, several network models are generated including individual CNTs with a chirality of (5, 0) (armchair) and a length of 20 nm to be used in tensile experiments. The reason behind using same type of CNTs having approximately a diameter of 0.4 nm with (5, 0) chirality in the numerical atomistic models is just for the sake of simplicity and minimizing computational cost. As the CNT diameter increases, not only the number of atoms increases significantly but also the maximum welding temperature for the covalent CNT junctions increases as well, resulting in much longer welding process. In order to investigate the effects of average cross-link density, which is referred to as the expected number of CNTs cross-linked with another CNT item, four different types of CNT network models having cross-link densities per CNT of 5, 6, 7, and 8 are generated. Therefore, the only topological parameter that is varied in the generation of the specimens is the average cross-link density that is expected to have impacts on the mechanical behavior of network structures. In Figure 2.22(a) and (b),

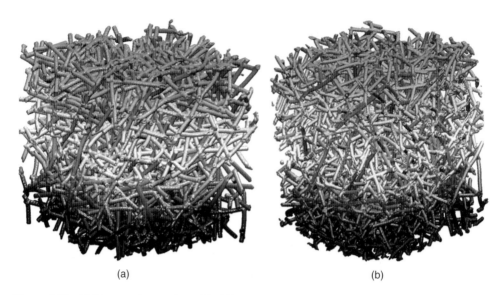

(a)   (b)

Figure 2.22 CNT network samples with different cross-link densities (a) 5 cross-links and (b) 7 cross-links per CNT. *Source*: Celebi et al. [111]. Reproduced with permission of Elsevier

a perspective view of two different types of 3D CNT network models having 230,196 (5 cross-links) atoms and 439,142 (7 cross-links) atoms is represented.

2.4.2   Tensile Behavior of Heat-Welded CNT Networks

Following the generation of atomistic models for the CNT networks, numerical experiments under tensile loading are carried out via classical MD simulations. Tensile simulations are conducted by applying tensile displacements to the group of atoms located at one end of the atomic structure, while keeping a group of atoms at the other side fixed during the loading process. The group of atoms on which tensile displacements are applied are selected from the upper region of the structure over 1/10 of the total specimen length. Prior to the application of tensile loading, in order to find the equilibrium configurations of the atomistic structures, energy minimization process is established to minimize the overall energy of the system. Minimization process can be considered as elimination of spurious residual stresses resulted from artificial modeling process to reach relaxed and more stable equilibrium condition. After establishing the energetically minimized configurations, each CNT network specimen is also thermally equilibrated by using the Langevin thermostat at 300 K for a total of 10 ps with a time step size of 0.5 fs. Following the thermalization period, tensile loading is applied with specific strain increments along the loading direction.

Because strain rate is a very important parameter directly affecting the mechanical response of materials, a rational strain rate selection must be conducted to obtain more accurate and stable results. In this regard, a reasonable and time-efficient strain rate value is determined by Mylvaganam et al. [127] for MD simulations of CNTs as $1.3 \times 10^9 \, \text{s}^{-1}$. Moreover, a feasible and appropriate strain rate also depends on the temperature, length, and chirality of the CNTs. For example, in order to avoid hardening observed at high strain rates, Liu et al. [125] employ a strain rate value (i.e., 10^8 s^{-1}), which is lower than a specified critical value of 10^{10} s^{-1} at 600 K for a nanojunction composed of zigzag (5,0) CNTs. According to the previous studies, for the sake of maximizing the computational efficiency and preventing hardening issue, a strain rate value of $1 \times 10^9 \text{ s}^{-1}$ is selected for the tensile simulations of CNT networks. Based on the selected strain rate, tensile displacement increments of 0.5 Å are applied to the end region of the network. The system is equilibrated at 300 K for 1 ps in each loading step. Each tensile displacement loading is applied within a period of 1000 time steps where the time step is kept fixed at 0.001 ps [126]. For each strain increment, tensile stresses are calculated based on the atomic forces averaged over the last 200 time steps of overall 1000 time steps with the purpose of decreasing the thermal noise due to thermal fluctuations. In other words, at each strain increment, atomic forces are averaged for each atom in the last 0.2 ps of the 1 ps load duration. By employing the computed time-averaged atomic forces on the cross section of the specimen, stresses based on continuum approach are determined simply by calculating the total force per cross-sectional area.

Tensile deformation process of CNT network structures can be traced at the atomic scale by taking snapshots (i.e., atomic coordinates) at different simulation time steps, which allows for exploring the local deformation modes that dictate the overall deformation mechanisms. In this respect, several snapshots of CNT network specimens with cross-link density of 7 are presented in Figure 2.23 to illustrate the overall tensile deformation process up to the rupture of the specimen. One important observation abstracted from the snapshots taken successively from the tensile simulations is that tensile strain increments induce bending deformation of CNT segments (i.e., part of individual CNTs between two consecutive cross-link points) around cross-link points, which in turn align the CNT segments along the loading direction. Thus, it can be substantiated that the initial deformation mode is dominated by aligning of CNT segments along the loading direction by bending deformation. This initial bending dominated phase is accompanied by local stretching mode that is due to the CNT segments already aligned along the loading direction. There are also cases in which the bending rigidity of CNT segments around the corresponding cross-link point may be sufficiently high to avoid bending and initiates stretching deformation without previous bending. Therefore, the overall deformation mechanism of CNT network structures is initially composed of the combination of two local deformation modes, which are primary bending and secondary stretching modes.

In the primary bending mode, cross-links act as hinge points due to relatively higher stiffness. Therefore, CNT segments tend to bend until the alignment of CNT segments is completed. After all nonaligned CNT segments are aligned along the loading direction, full stretching deformation mode commences and takes place until the specimen fails by successive ruptures of individual CNTs.

Because CNT segments are randomly distributed throughout the network, most of the CNT segments are not completely aligned along the loading direction. Thus, although both alignment and stretching of CNT segments occur simultaneously during the tensile deformation process, the dominant deformation mode is the alignment of CNTs. Within this context, the snapshots up to 275 ps, as shown in Figure 2.23, present the alignment-dominant deformation mode. As the tensile deformation progresses, the number of segments with local bending deformation decreases, which also mitigates the dominancy of alignment mode, and the network structure evolves into full stretching mode in which the CNT segments are deformed only in the loading direction with no bending. As an example, snapshots taken at time steps of 325 and 425 ps illustrate the instances of fully stretched deformation mode.

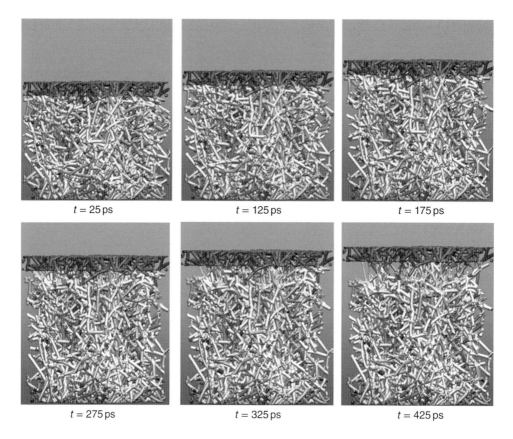

| $t = 25\,\mathrm{ps}$ | $t = 125\,\mathrm{ps}$ | $t = 175\,\mathrm{ps}$ |
| $t = 275\,\mathrm{ps}$ | $t = 325\,\mathrm{ps}$ | $t = 425\,\mathrm{ps}$ |

Figure 2.23 Several instances of CNT network specimen with cross-link density of 7 at different time points in tensile-loading simulation. *Source*: Celebi et al. [111]. Reproduced with permission of Elsevier

It should be stressed that the bending deformation mode of CNT segments is governed by the bending stiffness of CNTs as well as stiffness of cross-links acting as hinges. Stiffness of the cross-links, on the other hand, depends on the bond configurations within the junctions. Therefore, the parameters such as arrangement, orientation, and the number of covalent bonds per junction determine the character of the junction [126]. Local deformation mode is determined by the relative values of junction stiffness and bending rigidity of CNT segments. For instance, if the stiffness of a cross-link is higher than the bending stiffness of the relevant CNT segments, tendency of CNT segment to bend around the cross-link increases as well. Therefore, the struggling between bending stiffness of CNT segments and rigidity of cross-link region dictate the type of deformation mode.

After a certain level of alignment mode, CNT segments are locked for further bending deformation, and bending mode is transformed into stretching mode. In the full-stretching mode, all CNT segments are stretched along the loading direction and no CNT segments remain active in the bending mode. Full-stretching mode in which all CNT segments are strained in the same direction is ended by the successive failure of CNT segments in the close proximity of load application region.

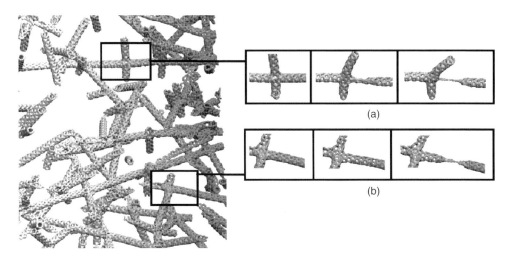

Figure 2.24 Local failure types at different junction locations: (a) tensile failure at junctional region, (b) tensile failure at nonjunctional, straight CNT section. *Source*: Celebi et al. [111]. Reproduced with permission of Elsevier

For the further examination of local failure modes, snapshots taken from the network specimens are zoomed around several junctions. Detailed examinations of the snapshots indicate that some of the local failures are initiated from the junctional regions, while some of them nucleate from the straight portions of CNT segments. In this regard, two different local failure types are illustrated by the snapshots provided in Figure 2.24(a) and (b). Although no direct relationship is observed between local failure mechanisms and overall failure of the sample, detailed inspection of the snapshots captured for all specimens demonstrates that the length of CNT segments between two consecutive cross-links affects the local failure mode. For example, it is noticed that as the length of CNT segments increases, the tendency of local failure mode at straight CNT portion increases as well. Because length of CNT segments are directly related to the number of cross-links per CNT, or namely cross-link density, local failure modes depend indirectly on the cross-link density. Accordingly, as the cross-link density decreases, the number of local straight failures increases due to longer CNT segments.

As a result of considerable number of simulations, the effects of cross-link density on the mechanical behavior of CNT networks are evaluated by determining several mechanical properties including Young's modulus, yield strength, and ultimate tensile strain of the specimens with different cross-link densities. As a consequence of numerical tensile experiments, the primary output is the load versus displacement curve, which is compiled and transformed into stress–strain curves. In this regard, stress–strain curves are obtained by performing classical continuum definitions of stress and strain (i.e., $\sigma = F/A$, $\varepsilon = \Delta L/L$). Figure 2.25 shows the stress–strain curves of all CNT network specimens with different cross-link densities derived directly from load–displacement relationship. By the evaluation of stress–strain curves depicted in Figure 2.25, it can be inferred that the increase in cross-link density leads to stiffness enhancement of the networks. The specimen with the largest Young's modulus is with the highest cross-link density. Another insight gained from stress–strain curves is that by

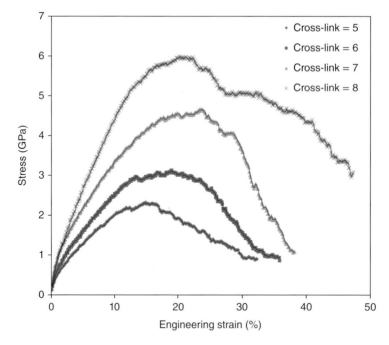

Figure 2.25 Stress–strain curves of CNT network specimens with different cross-link densities. *Source*: Celebi et al. [111]. Reproduced with permission of Elsevier

Table 2.1 Summary of quantitative results from MD results

Cross-link density	Density (g/cm^3)	Modulus (GPa)	Yield strength (MPa)	Ultimate strain (%)
5	0.45	14.55	2006	32
6	0.53	18.96	2935	36
7	0.61	24.35	4248	38
8	0.83	37.00	5200	47

Source: Celebi et al. [111]. Reproduced with permission of Elsevier

increasing the cross-link density, yield, and ultimate strength values of the specimens are also improved. Quantitative results calculated for all specimens with different cross-link densities are given in Table 2.1.

   One of the reasons that explain the enhancement of mechanical properties is the increased network percolation by increasing the cross-link density. As the number of cross-links per CNT (i.e., cross-link density) increases, the degree of interconnectivity of individual CNTs is also upgraded resulting in more homogeneous load distribution throughout the network. Due to more uniform distribution of overall load on the CNT segments in the network, load-bearing capacity of the networks can be improved by increasing the number of segments that sustain the total applied load.

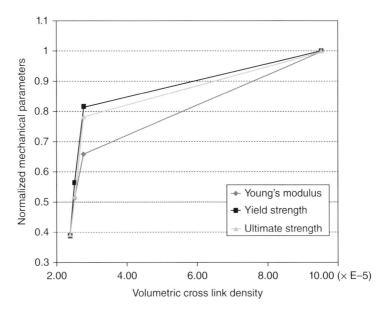

Figure 2.26 Variation of normalized tensile mechanical properties with volumetric cross-link density. *Source*: Celebi et al. [111]. Reproduced with permission of Elsevier

From a different point of view, increase in the number of cross-links per individual CNT also yields increase in the mass density of the specimens. However, rate of mass density increase is noticed to be less than climb rate of stiffness, which in turn indicates the super-linear change of stiffness-to-weight ratio. The effect of volumetric cross-link density, which is defined as the ratio of the total number of cross-links to the bulk volume of the test specimens, on the normalized tensile mechanical properties is demonstrated in Figure 2.26. According to this, all aforementioned mechanical properties such as tensile Young's modulus, yield, and ultimate strengths are enhanced by elevating the volumetric cross-link density. It is also interpreted from Figure 2.26 that at low-density regime, mechanical properties are linearly proportional to the change in volumetric cross-link density. Furthermore, Figure 2.26 clearly depicts that volumetric density change is much more effective on the mechanical behavior in the low-density regime ($<4 \times 10^{-5}$) comparing to higher density regime. Therefore, as the volumetric density increases, the amount of enhancement on the mechanical characteristics gets lesser, which points out that mechanical properties converge to the limiting values of the corresponding bulk solid as the cross-link density approaches to infinity.

References

[1] Kaiser, A.B., Skakalova, V. and Roth, S. (2008) Modelling conduction in carbon nanotube networks with different thickness, chemical treatment and irradiation. *Physica E*, **40**, 2311–2318.
[2] Snow, E.S., Novak, J.P., Campbell, P.M. and Park, D. (2003) Random networks of carbon nanotubes as an electronic material. *Applied Physics Letters*, **82**, 2145–2147.

[3] Hu, L., Hecht, D.S. and Gruner, G. (2004) Percolation in transparent and conducting carbon nanotube networks. *Nano Letters*, **4**, 2513–2517.

[4] Cao, Q. and Rogers, J.A. (2009) Ultrathin films of single-walled carbon nanotubes for electronics and sensors: a review of fundamental and applied aspects. *Advanced Materials*, **21**, 29–53.

[5] Topinka, M.A., Rowell, M.W., Goldhaber-Gordon, D. et al. (2009) Charge transport in interpenetrating networks of semiconducting and metallic carbon nanotubes. *Nano Letters*, **9**, 1866–1871.

[6] Li, J., Lu, Y., Ye, Q. et al. (2003) Carbon nanotube sensors for gas and organic vapor detection. *Nano Letters*, **3**, 929–933.

[7] Stampfer, C., Helbling, T., Obergfell, D. et al. (2006) Fabrication of single-walled carbon-nanotube-based pressure sensors. *Nano Letters*, **6**, 233–237.

[8] Bondavalli, P., Legagneux, P. and Pribat, D. (2009) Carbon nanotubes based transistors as gas sensors: state of the art and critical review. *Sensors and Actuators B*, **140**, 304–318.

[9] Wang, Y., Zhou, Z., Yang, Z. et al. (2009) Gas sensors based on deposited single-walled carbon nanotube networks for DMMP detection. *Nanotechnology*, **20**, 345502.

[10] Wu, Z., Chen, Z., Du, X. et al. (2004) Transparent, conductive carbon nanotube films. *Science*, **305**, 1273–1276.

[11] Pasquier, A.D., Unalan, H.E., Kanwal, A. et al. (2005) Conducting and transparent single-wall carbon nanotube electrodes for polymer–fullerene solar cells. *Applied Physics Letters*, **87**, 203511.

[12] Barnes, T.M., van de Lagemaat, J., Levi, D. et al. (2007) Optical characterization of highly conductive single-wall carbon-nanotube transparent electrodes. *Physical Review B*, **75**, 235410.

[13] Aliev, A.E., Oh, J., Kozlov, M.E. et al. (2009) Giant-stroke, superelastic carbon nanotube aerogel muscles. *Science*, **323**, 1575–1578.

[14] Wang, C., Wang, L. and Xu, Z. (2013) Enhanced mechanical properties of carbon nanotube networks by mobile and discrete binders. *Carbon*, **64**, 237–244.

[15] Bryning, M.B., Milkie, D.E., Islam, M.F. et al. (2007) Carbon nanotube aerogels. *Advanced Materials*, **19**, 661–664.

[16] Huang, S., Cai, X. and Liu, J. (2003) Growth of millimeter-long and horizontally aligned single-walled carbon nanotubes on flat substrates. *Journal of the American Chemical Society*, **125** (19), 5636–5637.

[17] Zhao, M.Q., Tian, G.L., Zhang, Q. et al. (2012) Preferential growth of short aligned, metallic-rich singlewalled carbon nanotubes from perpendicular layered double hydroxide film. *Nanoscale*, **4** (7), 2470–2477.

[18] Che, Y., Wang, C., Liu, J. et al. (2012) Selective synthesis and device applications of semiconducting singlewalled carbon nanotubes using isopropyl alcohol as feedstock. *ACS Nano*, **6** (8), 7454–7462.

[19] Han, S., Liu, X. and Zhou, C. (2005) Template-free directional growth of single-walled carbon nanotubes on a- and r-plane sapphire. *Journal of the American Chemical Society*, **127** (15), 5294–5295.

[20] Huang, L., Wind, S.J. and O'Brien, S.P. (2003) Controlled growth of singlewalled carbon nanotubes from an ordered mesoporous silica template. *Nano Letters*, **3** (3), 299–303.

[21] Han, Z.J., Mehdipour, H., Li, X. et al. (2012) SWCNT networks on nanoporous silica catalyst support: morphological and connectivity control for nanoelectronic, gas-sensing, and biosensing devices. *ACS Nano*, **6** (7), 5809–5819.

[22] Wongwiriyapan, W., Honda, S., Konishi, H. et al. (2005) Direct growth of single-walled carbon nanotube networks on alumina substrate: a novel route to ultrasensitive gas sensor fabrication. *Japanese Journal of Applied Physics Part 1*, **44** (11), 8227–8230.

[23] Kaur, S., Ajayan, P.M. and Kane, R.S. (2006) Design and characterization of three-dimensional carbon nanotube foams. *Journal of Physical Chemistry B*, **110** (42), 21377–21380.

[24] Huang, S. (2003) Growing carbon nanotubes on patterned submicron-size SiO2 spheres. *Carbon*, **41** (12), 2347–2352.

[25] Jo, J.W., Jung, J.W., Lee, J.U. and Jo, W.H. (2010) Fabrication of highly conductive and transparent thin films from single-walled carbon nanotubes using a new non-ionic surfactant via spin coating. *ACS Nano*, **4** (9), 5382–5388.

[26] Liu, Q., Fujigaya, T., Cheng, H.-M. and Nakashima, N. (2010) Free-standing highly conductive transparent ultrathin single-walled carbon nanotube films. *Journal of the American Chemical Society*, **132** (46), 16581–16586.

[27] Shi, Z., Chen, X., Wang, X. et al. (2011) Fabrication of superstrong ultrathin free-standing single-walled carbon nanotube films via a wet process. *Advanced Functional Materials*, **21** (22), 4358–4363.

[28] Skákalová, V., Kaiser, A.B., Woo, Y.S. and Roth, S. (2006) Electronic transport in carbon nanotubes: from individual nanotubes to thin and thick networks. *Physical Review B*, **74** (8), 085403.

[29] Jack, D.A., Yeh, C.S., Liang, Z. et al. (2010) Electrical conductivity modeling and experimental study of densely packed SWCNT networks. *Nanotechnology*, **21** (19), 195703.

[30] Nasibulin, A.G., Kaskela, A., Mustonen, K. et al. (2011) Multifunctional free-standing single-walled carbon nanotube films. *ACS Nano*, **5** (4), 3214–3221.

[31] Zhang, M., Fang, S., Zakhidov, A.A. et al. (2005) Strong, transparent, multifunctional, carbon nanotube sheets. *Science*, **309** (5738), 1215–1219.

[32] Yao, N. and Lordi, V. (1998) Young's modulus of single-walled carbon nanotubes. *Journal of Applied Physics*, **84**, 1939–1944.

[33] Filleter, T., Bernal, R., Li, S. and Espinosa, H.D. (2011) Ultrahigh strength and stiffness in cross-linked hierarchical carbon nanotube bundles. *Advanced Materials*, **23** (25), 2855–2860.

[34] Fonseca, A.F., Borders, T., Baughman, R.H. and Cho, K.J. (2010) Load transfer between cross-linked walls of a carbon nanotube. *Physical Review B*, **81** (4), 045429.

[35] Kis, A., Csanyi, G., Salvetat, J.P. et al. (2004) Reinforcement of single-walled carbon nanotube bundles by intertube bridging. *Nature Materials*, **3** (3), 153–157.

[36] Peng, B., Locascio, M., Zapol, P. et al. (2008) Measurements of near-ultimate strength for multiwalled carbon nanotubes and irradiation-induced crosslinking improvements. *Nature Nanotechnology*, **3** (10), 626–631.

[37] Fu, Y., Carlberg, B., Lindahl, N. et al. (2012) Templated growth of covalently bonded three-dimensional carbon nanotube networks originated from graphene. *Advanced Materials*, **24** (12), 1576–1581.

[38] Shan, C., Zhao, W., Lu, X.L. et al. (2013) Three-dimensional nitrogen-doped multiwall carbon nanotube sponges with tunable properties. *Nano Letters*, **13**, 5514–5520.

[39] Hashim, D.P., Narayanan, N.T., Romo-Herrera, J.M. et al. (2012) Covalently bonded three-dimensional carbon nanotube solids via boron induced nanojunctions. *Scientific Reports*, **2**, 363.

[40] Sumpter, B.G., Huang, J., Meunier, V. et al. (2009) A theoretical and experimental study on manipulating the structure and properties of carbon nanotubes using substitutional dopants. *International Journal of Quantum Chemistry*, **109**, 97–118.

[41] Dunlap, B.I. (1992) Connecting carbon tubules. *Physical Review B*, **46**, 1933–1936.

[42] Kahaly, M.U. (2009) Defect states in carbon nanotubes and related band structure engineering: a first-principles study. *Journal of Applied Physics*, **105**, 024312.

[43] Sharifi, T., Nitze, F., Barzegar, H.R. et al. (2012) Nitrogen doped multi walled carbon nanotubes produced by CVD-correlating XPS and Raman spectroscopy for the study of nitrogen inclusion. *Carbon*, **50**, 3535–3541.

[44] Romo-Herrera, J.M., Cullen, D.A., Cruz-Silva, E. et al. (2009) The role of sulfur in the synthesis of novel carbon morphologies: from covalent Y-junctions to sea-urchin-like structures. *Advanced Functional Materials*, **19**, 1193–1199.

[45] Romo-Herrera, J.M., Sumpter, B., Cullen, D.A. et al. (2008) An atomistic branching mechanism for carbon nanotubes: sulfur as the triggering agent. *Angewandte Chemie International Edition*, **47**, 2948–2953.

[46] Valles, C., Perez-Mendoza, M., Castell, P. et al. (2006) Towards helical and Y-shaped carbon nanotubes: the role of sulfur in CVD processes. *Nanotechnology*, **17**, 4292–4299.

[47] Gui, X., Wei, J., Wang, K. et al. (2010) Carbon nanotube sponges. *Advanced Materials*, **22**, 617–621.

[48] Novak, J.P., Snow, E.S., Houser, E.J. et al. (2003) Nerve agent detection using networks of single-walled carbon nanotubes. *Applied Physics Letters*, **83** (19), 4026–4028.

[49] Snow, E.S., Campbell, P.M., Ancona, M.G. and Novak, J.P. (2005) High-Mobility Carbon-Nanotube Thin-Film Transistors on a Polymeric Substrate. *Applied Physics Letters*, **86** (3), 033105.

[50] Snow, E.S., Novak, J.P., Lay, M.D. et al. (2004) Carbon nanotube networks: nanomaterial for macroelectronic applications. *Journal of Vacuum Science & Technology B*, **22** (4), 1990–1994.

[51] Zhou, Y.X., Gaur, A., Hur, S.H. et al. (2004) P-channel, N-channel thin film transistors and P-N diodes based on single wall carbon nanotube networks. *Nano Letters*, **4** (10), 2031–2035.

[52] Cao, Q., Kim, H.S., Pimparkar, N. et al. (2008) Medium-scale carbon nanotube thin-film integrated circuits on flexible plastic substrates. *Nature (London)*, **454** (7203), 495–U4.

[53] Kocabas, C., Hur, S.H., Gaur, A. et al. (2005) Guided growth of large-scale, horizontally aligned arrays of single-walled carbon nanotubes and their use in thin-film transistors. *Small*, **1** (11), 1110–1116.

[54] Kocabas, C., Meitl, M.A., Gaur, A. et al. (2004) Aligned arrays of single-walled carbon nanotubes generated from random networks by orientationally selective laser ablation. *Nano Letters*, **4** (12), 2421–2426.

[55] Kocabas, C., Shim, M. and Rogers, J.A. (2006) Spatially selective guided growth of high-coverage arrays and random networks of single-walled carbon nanotubes and their integration into electronic devices. *Journal of the American Chemical Society*, **128** (14), 4540–4541.

[56] Reuss, R.H., Chalamala, B.R., Moussessian, A. et al. (2005) Macroelectronics: perspectives on technology and applications. *Proceedings of the IEEE*, **93** (7), 1239–1256.

[57] Snow, E.S., Perkins, F.K., Houser, E.J. et al. (2005) Chemical detection with a single-walled carbon nanotube capacitor. *Science*, **307** (5717), 1942–1945.

[58] Arico, A.S., Bruce, P., Scrosati, B. et al. (2005) Nanostructured materials for advanced energy conversion and storage devices. *Nature Materials*, **4** (5), 366–377.

[59] Baughman, R.H., Zakhidov, A.A. and de Heer, W.A. (2002) Carbon nanotubes – the route towards applications. *Science*, **297**, 787–792.

[60] Hu, J., Odom, T.W. and Lieber, C.M. (1999) Chemistry and physics in one-dimension: synthesis and properties of nanowires and nanotubes. *Accounts of Chemical Research*, **32**, 435–445.

[61] Leroy, C.M., Carn, F., Trinquecoste, M. et al. (2007) Multiwalled-carbon-nanotube-based carbon foams. *Carbon*, **45**, 2317–2320.

[62] Lau, P.H., Takei, K., Wang, C. et al. (2013) Fully printed, high performance carbon nanotube thin-film transistors on flexible substrates. *Nano Letters*, **13** (8), 3864–3869.

[63] Zhong, J., Yang, Z.Y., Mukherjee, R. et al. (2013) Carbon nanotube sponges as conductive networks for supercapacitor devices. *Nano Energy*, **2** (5), 1025–1030.

[64] Kumar, S., Cola, B.A., Jackson, R. and Graham, S. (2011) A review of carbon nanotube ensembles as flexible electronics and advanced packaging materials. *Journal of Electronic Packaging*, **133**, 1–12.

[65] Saran, N., Parikh, K., Suh, D.S. et al. (2004) Fabrication and characterization of thin films of single-walled carbon nanotube bundles on flexible plastic substrates. *Journal of the American Chemical Society*, **126** (14), 4462–4463.

[66] Gui, X.C., Li, H.B., Wang, K.L. et al. (2011) Recyclable carbon nanotube sponges for oil absorption. *Acta Materialia*, **59** (12), 4798–4804.

[67] Sayago, I., Santos, H., Horrillo, M.C. et al. (2008) Carbon nanotube networks as gas sensors for NO2 detection. *Talanta*, **77**, 758–764.

[68] Nabeta, M. and Sano, M. (2005) Nanotube foam prepared by gelatin gel as a template. *Langmuir*, **21**, 1706–1708.

[69] Zhang, H., Liu, Y., Kuwata, M. et al. (2015) Improved fracture toughness and integrated damage sensing capability by spray coated CNTs on carbon fibre prepreg. *Composites Part a: Applied Science and Manufacturing*, **70**, 102–110.

[70] Tai, N.H., Yeh, M.K. and Liu, J.H. (2004) Enhancement of the mechanical properties of carbon nanotube/phenolic composites using a carbon nanotube network as the reinforcement. *Carbon*, **42**, 2735–2777.

[71] Zeng, Z.P., Gui, X.C., Lin, Z.Q. et al. (2013) Carbon nanotube sponge–array tandem composites with extended energy absorption range. *Advanced Materials*, **25**, 1185–1191.

[72] Wang, M., Wang, J., Chen, Q. and Peng, L.M. (2005) Fabrication and electrical and mechanical properties of carbon nanotube interconnections. *Advanced Functional Materials*, **15**, 1825–1831.

[73] Frankland, S.J.V., Harik, V.M., Odegard, G.M. et al. (2003) The stress–strain behavior of polymer–nanotube composites from molecular dynamics simulation. *Composites Science and Technology*, **63**, 1655–1661.

[74] Thostenson, E.T. and Chou, T.W. (2006) Carbon nanotube networks: sensing of distributed strain and damage for life prediction and self healing. *Advanced Materials*, **18**, 2837–2841.

[75] Li, Y., Qiu, X., Yin, Y. et al. (2009) The specific heat of carbon nanotube networks and their potential applications. *Journal of Physics D: Applied Physics*, **42**, 155405.

[76] Dimaki, M., Bøggild, P. and Svendsen, W. (2007) Temperature response of carbon nanotube networks. *Journal of Physics: Conference Series*, **61**, 247–251.

[77] Slobodian, P., Riha, P., Lengalova, A. and Saha, P. (2011) Compressive stress-electrical conductivity characteristics of multiwall carbon nanotube networks. *Journal of Materials Science*, **46**, 3186–3190.

[78] Slobodian, P. and Saha, P. (2011) Stress-strain hysteresis of a carbon nanotube network as polymer nanocomposite filler under cyclic deformation. *AIP Conference Proceedings*, **1375**, 224–231.

[79] Worsley, M.A., Kucheyev, S.O., Satcher, J.H. et al. (2009) Mechanically robust and electrically conductive carbon nanotube foams. *Applied Physics Letters*, **94** (7), 073115.

[80] Kaskela, A., Koskinen, J., Jiang, H. et al. (2013) Improvement of the mechanical properties of single-walled carbon nanotube networks by carbon plasma coatings. *Carbon*, **53**, 50–61.

[81] Kim, K.H., Oh, Y. and Islam, M.F. (2012) Mechanical and thermal management characteristics of ultrahigh surface area single-walled carbon nanotube aerogels. *Advanced Functional Materials*, **23**, 377–383.

[82] Rahatekar, S.S., Koziol, K.K., Kline, S.R. et al. (2009) Length-dependent mechanics of carbon nanotube networks. *Advanced Materials*, **21** (8), 874–878.

[83] Fry, D., Langhorst, B., Kim, H. et al. (2005) Anisotropy of sheared carbon-nanotube suspensions. *Physical Review Letters*, **95**, 038304.

[84] Satti, A., Perret, A., McCarthy, J.E. and Guńko, Y.K. (2010) Covalent crosslinking of single-walled carbon nanotubes with poly(allylamine) to produce mechanically robust composites. *Journal of Materials Chemistry*, **20** (37), 7941–7943.

[85] Fakhri, N., Tsyboulski, D.A., Cognet, L. et al. (2009) Diameter-dependent bending dynamics of single-walled carbon nanotubes in liquids. *Proceedings of the National Academy of Sciences of the United States of America*, **106**, 14219.

[86] Coluci, V.R., Dantas, S.O., Jorio, A. and Galvao, D.S. (2007) Mechanical properties of carbon nanotube networks by molecular mechanics and impact molecular dynamics calculations. *Physical Review B*, **75**, 075417.

[87] Li, Y., Qiu, X., Yang, F. et al. (2009) Stretching-dominated deformation mechanism in a super square carbon nanotube network. *Carbon*, **47**, 812–819.

[88] Hall, L.J., Coluci, V.R., Galvao, D.S. et al. (2008) Sign change of Poisson's ratio for carbon nanotube sheets. *Science*, **320**, 504–507.

[89] Jin, Y. and Yuan, F.G. (2003) Simulation of elastic properties of single-walled carbon nanotubes. *Composites Science and Technology*, **63**, 1507–1515.

[90] Lan, M. and Waisman, H. (2012) Mechanics of SWCNT aggregates studied by incremental constrained minimization. *Journal of Nanomechanics and Micromechanics*, **2**, 15–22.

[91] Lu, W.B., Liu, X., Li, Q.W. and Byun, J.H. (2013) Mechanical behavior and structural evolution of carbon nanotube films and fibers under tension: a coarse-grained molecular dynamics study. *Journal of Applied Mechanics*, **80**, 051015, 1–9.

[92] Li, Y., Qiu, X., Yin, Y. et al. (2010) The elastic buckling of super-graphene and super-square carbon nanotube networks. *Physics Letters A*, **374** (15–16), 1773–1778.

[93] Coluci, V.R., Galvao, D.S. and Jorio, A. (2006) Geometric and electronic structure of carbon nanotube networks: 'super'-carbon nanotubes. *Nanotechnology*, **17** (3), 617–621.

[94] Liu, X., Yang, Q.S., He, X.Q. and Mai, Y.W. (2011) Molecular mechanics modeling of deformation and failure of super carbon nanotube networks. *Nanotechnology*, **22**, 475701–475711.

[95] Coluci, V.R. and Pugno, N. (2010) Molecular dynamics simulations of stretching, twisting and fracture of super carbon nanotubes with different chiralities: towards smart porous and flexible scaffolds. *Journal of Computational and Theoretical Nanoscience*, **7**, 1294–1298.

[96] Wang, M., Qui, X. and Zhang, X. (2007) Mechanical properties of super honeycomb structures based on carbon nanotubes. *Nanotechnology*, **18**, 075711.

[97] Yakobson, B.I., Brabec, C.J. and Bernholc, J. (1996) Nanomechanics of carbon tubes: instabilities beyond linear response. *Physical Review Letters*, **76**, 2511–2514.

[98] Li, C.Y. and Chou, T.W. (2003) A structural mechanics approach for the analysis of carbon nanotubes. *International Journal of Solids and Structures*, **40**, 2487–2499.

[99] Zaeri, M.M., Ziaei-Rad, S., Vahedi, A. and Karimzadeh, F. (2010) Mechanical modelling of carbon nanomaterials from nanotubes to buckypaper. *Carbon*, **48** (13), 3916–3930.

[100] Picu, R.C. (2011) Mechanics of random fiber networks—a review. *Soft Matter*, **7**, 6768–6785.

[101] Deng, F., Ito, M., Noguchi, T. et al. (2011) Elucidation of the reinforcing mechanism in carbon nanotube/rubber nanocomposites. *ACS Nano*, **5**, 3858–3866.

[102] Kasza, K.E., Rowat, A.C., Liu, J. et al. (2007) The cell as a material. *Current Opinion in Cell Biology*, **19**, 101–107.

[103] Volkov, A.N. and Zhigilei, L.V. (2010) Structural stability of carbon nanotube films: the role of bending buckling. *ACS Nano*, **4**, 6187–6195.

[104] Cranford, S.W. and Buehler, M.J. (2010) In silico assembly and nanomechanical characterization of carbon nanotube buckypaper. *Nanotechnology*, **21**, 265706–265712.

[105] Deng, L., Trepat, X., Butler, J.P. et al. (2006) Fast and slow dynamics of the cytoskeleton. *Nature Materials*, **5**, 636–640.

[106] Xie, B., Liu, Y.L., Ding, Y.T. et al. (2011) Mechanics of carbon nanotube networks: microstructural evolution and optimal design. *Soft Matter*, **7**, 10039–10047.

[107] Wang, C., Gao, E., Wang, L. and Xu, Z. (2014) Mechanics of network materials with responsive crosslinks. *Comptes Rendus Mecanique*, **342**, 264–272.

[108] Bratzel, G.H., Cranford, S., Espinosa, H. and Buehler, M.J. (2010) Bioinspired noncovalently crosslinked fuzzy carbon nanotube bundles with superior toughness and strength. *Journal of Materials Chemistry*, **20**, 10465–10474.

[109] Buehler, M.J. (2011) Mesoscale modeling of mechanics of carbon nanotubes: self-assembly, self-folding, and fracture. *Journal of Materials Research*, **21** (11), 2855–2869.

[110] Cranford, S., Yao, H., Ortiz, C. and Buehler, M. (2010) A single degree of freedom 'lollipop' model for carbon nanotube bundle formation. *Journal of the Mechanics and Physics of Solids*, **58** (3), 409–427.

[111] Celebi, A.T., Kirca, M., Baykasoglu, C. et al. (2014) Tensile behavior of heat welded CNT network structures. *Computational Materials Science*, **88**, 14–21.

[112] Kirca, M., Yang, X. and To, A.C. (2013) A stochastic algorithm for modeling heat welded random carbon nanotube network. *Computer Methods in Applied Mechanics and Engineering*, **259**, 1–9.

[113] Meng, F.Y., Shi, S.Q., Xu, D.S. and Yang, R. (2004) Multiterminal junctions formed by heating ultrathin single-walled carbon nanotubes. *Physical Review B*, **70**, 125418.

[114] Piper, N.M., Fu, F., Tao, J. et al. (2011) Vibration promotes heat welding of single walled nanotubes. *Chemical Physics Letters*, **502**, 231–234.

[115] Yang, X., Han, Z., Li, Y. et al. (2012) Heat welding of non-orthogonal X-junction of single-walled carbon nanotubes. *Physica E*, **46**, 30–32.

[116] Plimpton, S. (1995) Fast parallel algorithms for short-range molecular dynamics. *Journal of Computational Physics*, **117**, 1–19.

[117] Stuart, S.J., Tutein, A.B. and Harrison, J.A. (2000) A reactive potential for hydrocarbons with intermolecular interactions. *The Journal of Chemical Physics*, **112**, 6472–6486.

[118] Faria, B., Silvestre, N. and Canongia Lopes, J.N. (2013) Tension–twisting dependent kinematics of chiral CNTs. *Composites Science and Technology*, **74**, 211–220.

[119] Faria, B., Silvestre, N. and Canongia Lopes, J.N. (2011) Interaction diagrams for carbon nanotubes under combined shortening–twisting. *Composites Science and Technology*, **71**, 1811–1818.

[120] Brenner, D.W., Shenderova, O.A., Harrison, J.A. et al. (2002) A second-generation reactive empirical bond order (REBO) potential energy expression for hydrocarbons. *Journal of Physics: Condensed Matter*, **14**, 783–802.

[121] Meng, F.Y., Shi, S.Q., Xu, D.S. and Chan, C.T. (2006) Rendering of huge point-sampled geometry based on LOD control and out-of-core techniques. *Modelling and Simulation in Materials Science and Engineering*, **14**, 1–8.

[122] Meng, F.Y., Shi, S.Q., Xu, D.S. and Yang, R. (2006) Size effect of X-shaped carbon nanotube junctions. *Carbon*, **44**, 1263–1266.

[123] Meng, F.Y., Shi, S.Q., Xu, D.S. and Chan, C.T. (2006) Surface reconstructions and stability of X-shaped carbon nanotube junction. *The Journal of Chemical Physics*, **124**, 024711.

[124] Meng, F.Y., Liu, W.C. and Shi, S.Q. (2009) Tensile deformation behavior of carbon nanotube junctions. *NSTI-Nanotechnology*, **3**, 450–453.

[125] Liu, W.C., Meng, F.Y. and Shi, S.Q. (2010) A theoretical investigation of the mechanical stability of single-walled carbon nanotube 3-D junctions. *Carbon*, **48**, 1626–1635.

[126] Stormer, B.A., Piper, N.M., Yang, X.M. et al. (2012) Mechanical properties of SWCNT X-junctions through molecular dynamics simulation. *International Journal of Smart and Nano Materials*, **3**, 33–46.

[127] Mylvaganam, K. and Zhang, L.C. (2004) Important issues in a molecular dynamics simulation for characterising the mechanical properties of carbon nanotubes. *Carbon*, **42**, 2025–2032.

3

Mechanics of Helical Carbon Nanomaterials

Hiroyuki Shima[1] and Yoshiyuki Suda[2]

[1]*Department of Environmental Sciences, University of Yamanashi, Kofu, Yamanashi, Japan*
[2]*Department of Electrical and Electronic Information Engineering, Toyohashi University of Technology, Toyohashi, Aichi, Japan*

3.1 Introduction

3.1.1 Historical Background

"Helical carbon nanomaterial" is the generic term for long, thin, and helically twisted nanocarbons. They are seemingly nanoscopic analogues of "telephone cords" attached to a traditional phone receiver. Similar to a phone cord, the long spiral geometry of these nanomaterials enables them to be stretched a lot in their axial direction and be flexible under bending. For instance, the thinnest-type helical carbon nanotube (abbreviated as "HN-Tube"; see Section 3.1.2) was predicted to undergo elastic elongation at strains up to 60%, as suggested by numerical simulations [1]. More surprisingly, it was experimentally found that an amorphous-type helical carbon nanofiber (abbreviated as "HN-Fiber"), whose size is significantly larger than that of the thinnest-type helical nanotube, can be elastically extended up to about three times its original coil length [2]. These facts demonstrate that helical carbon nanomaterials have excellent mechanical strength and flexibility primarily due to both their helical morphology and toughness resulting from the strong chemical bonding between constituent carbon atoms. Figure 3.1(a)–(b) gives an atomistic model of a helical carbon nanomaterial [3].

Helical carbon nanomaterials have a long history, which dates back to their serendipitous finding by Davis et al. in the 1950s [4]. During the deposition of carbon in the brickwork of a blast furnace, Davis et al. observed by chance that the carbon was getting deposited as minute vermicular growths that could penetrate the considerable thicknesses of the brickwork. Two years later, Hofer et al. reported the growth of carbon filaments with fiber diameters of 10–200 nm [5]. They used the catalytic deposition technique based on the following

Advanced Computational Nanomechanics, First Edition. Edited by Nuno Silvestre.
© 2016 John Wiley & Sons, Ltd. Published 2016 by John Wiley & Sons, Ltd.

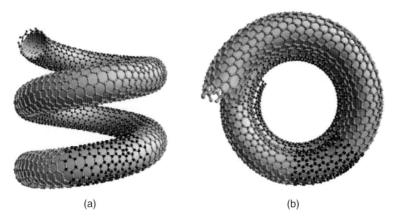

(a)   (b)

Figure 3.1   An atomistic model of a thinnest-type helical carbon nanomaterial. The side-view (a) and top-view (b) of a portion of a long spiral structure are highlighted. *Source*: Milošević et al. [3]. Reproduced with permission of John Wiley and Sons

chemical reaction:

$$2CO \rightarrow C + CO_2 \tag{3.1}$$

where Ni, Co, and Fe were used as catalysts. Afterward, in 1973, Boehm succeeded in developing twisted filaments that frequently formed double or even triple helices [6] by feeding metal carbonyl into a CO injection stream in the chemical reaction given by (3.1). These twisted filaments are precursors of helical carbon nanomaterials that have been under stimulated investigation in the realm of advanced material science.

In the early stages, the above-mentioned carbon fibers were regarded as a curious by-product; efforts were thus focused on their prevention rather than on their synthesis [7, 8]. Eventually, in the 1990s, the discovery of carbon nanotubes (CNTs) sparked a renewed interest in carbon fibers and tubes, especially those with a spring-like morphology [9, 10]. Nowadays, a wide variety of helical carbon nanomaterials have been fabricated, mainly using chemical vapor deposition (CVD) [11]. Figure 3.2 shows a diagram of CVD for fabricating helical carbon nanomaterials. The preference of CVD for the synthesis of helical carbon nanomaterials is in contrast with the case of CNTs, as laser ablation is considered the most reliable method to prepare high-quality specimens of the latter.

The spiral geometry of helical carbon nanomaterials enables them to be used for the development of novel, nanosized mechanical devices ranging from resonating elements and nanosprings to reinforced fibers in high-strain composites [12]. High-sensitive tactile sensors are also expected to be fabricated using helical carbon nanomaterials, which may provide high resolutions of up to femtograms [13]. However, quantitative determination of their mechanical properties as well as performance in actual applications remains largely unexamined. In particular, the experimental difficulty in the synthesis of the thinnest-type helical nanotubes (i.e., those based on a monoatomic graphitic layer) has counteracted the discovery of their physical properties, although an intriguing interplay between their mechanical and other (electric, magnetic, optical, or chemical) properties holds promise. Against this backdrop, advanced computational techniques will play an important role in exploring the physics of helical carbon nanomaterials, especially for those holding the promise for use in the next-generation nanodevices that have been unfeasible by the current fabrication techniques.

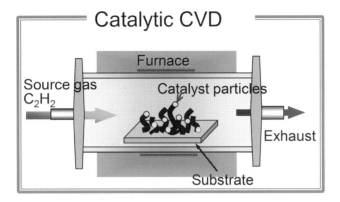

Figure 3.2 Diagram of catalytic chemical vapor deposition (CVD) for fabricating helical carbon nanomaterials

3.1.2   Classification: Helical "Tube" or "Fiber"?

Before discussing the main issue, it is practically important to establish a clear one-to-one correspondence between the terminology and the specific class of helical carbon nanomaterials. This is because during the last two decades, diverse descriptive terminology has been used for expressing the family of helical carbon nanomaterials. However, there has been no broadly consistent definition of their terminology, which causes slight controversies in the relevant field. For instance, one may call a sufficiently thin and graphite-layer-based helical nanocarbons as "coiled carbon nanotubes," while someone else may call an identical material as "carbon nanosprings," or some other name. Such contradictions can be resolved by providing a proper classification of helical carbon nanomaterials, as earlier suggested [14, 15].

In this chapter, we choose the following two terms for categorizing various entities contained in the family of helical carbon nanomaterials. The first one is

"Helical carbon nanotube," abbreviated as "HN-Tube"

This class of nanomaterials is obtained by coiling the tubular axis of a single-walled or few-walled CNT, whose tubular axes are originally straight (Figure 3.3(a)). The cause of helicity in HN-Tubes is attributed to the periodic distribution of topological defects (five- or seven-membered carbon rings) embedded in the twisted graphene cylinder. It has been thought that the insertion of defects releases the mechanical strain induced by twisting the tubular axis, and this may stabilize the regular coil formation of HN-Tubes (see Section 3.2.1 for details).

The second class of helical carbon nanomaterials, which we will discuss in this chapter, is

"Helical carbon nanofiber," abbreviated as "HN-Fiber"

By using "Fiber" instead of "Tube," we mean that HN-Fibers may not consist of graphitic hollow tubes (Figure 3.3(b)). They are typically amorphous fibers with no concentric tubular structures, having larger spatial dimensions compared to HN-Tubes. Table 3.1 summarizes the dimensional comparison between HN-Tubes and HN-Fibers. The definitions of geometric parameters such as tube/fiber diameter, coil diameter, and coil pitch are schematically illustrated in Figure 3.4.

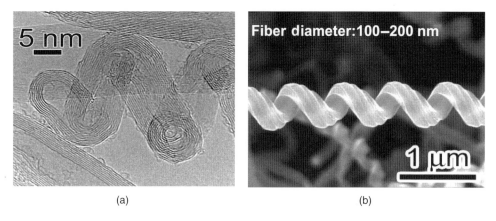

(a)   (b)

Figure 3.3 Microscopic images of helical carbon nanomaterials. (a) A helical carbon nano-"tube" (HN-Tube) composed of seven-layered graphitic hollow structure. (b) A helical carbon nano-"fiber" (HN-Fiber) having amorphous solid-core structure

Table 3.1 Classification of helical carbon nanomaterials adopted in this chapter

	Helical carbon nanotube (HN-Tube)	Helical carbon nanofiber (HN-Fiber)
Tube/fiber diameter	1–20 nm	100–200 nm
Coil diameter	2–50 nm	200–500 nm
Coil pitch	1–50 nm	100–1000 nm
Structure	Entirely graphitic; single-walled or multiwalled	Amorphous or partly graphitic
Cause for helicity	Inclusion of pentagon/heptagon pairs into graphene sheets	Unbalanced extrusion of carbon from a catalyst surface

It should be emphasized that the classification shown in Table 3.1 is not at all a unique one. Still, a variety of terms are currently adopted depending on the authors' preference, as well as on the geometry and internal structure of the material being considered. Only a few examples of other descriptive terminology reported in the literature are given in Table 3.2. Hence, when referring to these materials, the reader should be careful with regard to definitions in the terminology used. Another caution arises from that fact that in addition to these nanoscale materials, micrometer-scale helical carbon materials also exist. They are commonly called "carbon microcoils" or "coiled carbon microfibers" [33, 34]; they have a fiber diameter of ~1 μm and a coil diameter of 2–10 μm. The growth mechanism and physical properties of carbon microcoils (CMCs) have been very actively explored in the last decade; nevertheless, owing to the significant deviation of their size from the nanometer scale, we put a disclaimer that the subject of CMCs will be omitted in the subsequent discussion.

3.1.3 Fabrication and Characterization

Experimental efforts in synthesizing multiwalled HN-Tubes of high quality have been triggered by the first transmission electron microscopy (TEM) observation in 1994 [19, 35].

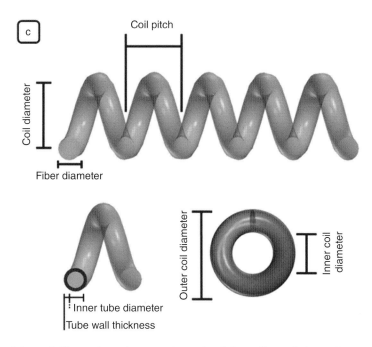

Figure 3.4   Schematic illustrations of parameters used to define coil morphology. *Source*: Shaikjee and Coville [16]. Reproduced with permission of Elsevier

Table 3.2   Various terminologies used in previous literature for expressing helical carbon nanomaterials

Helical carbon nano-tube (HN-Tube)	Helically coiled cage form of graphitic carbon [17]
	Helix-shaped graphite nanotube [18]
	Coil-shaped carbon nanotubule [19]
	Helically coiled cage of graphitic carbon [20]
	Helically coiled carbon nanotube [3, 21–23]
	Helical carbon nanotube [24–26]
	Nanotube coil [27]
	Carbon nanocoil [1, 28]
	Coiled carbon nanotube [29]
	Carbon nanosprings [30]
Helical carbon nano-fiber (HN-Fiber)	Coiled carbon nanotube [14]
	Helically coiled carbon nanofiber [31]
	Helically coiled carbon nanowire [32]

Catalytic decomposition of acetylene (C_2H_2) was employed to fabricate 10-walled HN-Tubes having 30 nm in coil pitch and 18 nm in fiber diameter. The structural analysis based on electron diffraction method indicated that these HN-Tubes are multiwalled, hollow, and polygonized such that they consist of short straight segments. Since the first synthesis, many researchers have tried to make up these materials [36]. Chemical vapor deposition has been primarily chosen as production method [37–41], while laser evaporation [42] and opposed

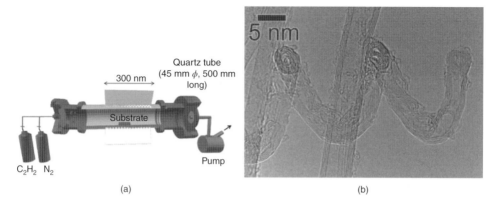

(a)   (b)

Figure 3.5 (a) Experimental setup for HN-Tube synthesis performed by Lim et al. (b) High-resolution TEM image of a three-walled HN-Tube synthesized with C_2H_2/N_2 ratio of 0.01. *Source*: Lim et al. [45]. Reproduced with permission of IOP Publishing

flow flame combustion method [43] have also been reported to date. Recently, the diameter control of multiwalled HN-Tubes via tuning the particle size of the nanoscale catalysts was suggested [44].

In 2013, Lim et al. succeeded to take a microscopic photograph of three-walled HN-Tubes as presented in Figure 3.5(a) and (b) [45]. The synthesis was carried out in a tubular electric furnace into which the mixture of nitrogen (N_2) and acetylene (C_2H_2) gases was injected at high temperature. Lim et al. found that the HN-Tube growth was optimized by heating the furnace at 700 °C and lowering the C_2H_2/N_2 ratio to 0.01; under these conditions, the obtained HN-Tubes exhibit a high graphitization degree and significant reduction in the tube diameter, resulting in nanotubular structures with three graphene layers at the tip.

Among the family of helical carbon nanomaterial, especially difficult to synthesize are those of single-walled HN-Tube with both tubular diameter and pitch length down to 1 nm. So far, there has been only one report of such ultrathin HN-Tube. Biró et al. [46] insisted the presence of single-walled HN-Tubes in their scanning tunneling microscopy (STM) image, and proposed the corresponding atomistic model. The structure is formed by connecting alternately the two kinds of CNT segments with (6,6) and (10,0) chiralities, both of which are approximately 0.8 nm in tube diameter. The coil diameter and the coil pitch are estimated to be 2.3 and 1.2 nm, respectively. These values are close to the lower limits of the geometric parameters listed in Table 3.1.

3.2 Theory of HN-Tubes

3.2.1 Microscopic Model

In this section, we review the theoretical and computational studies on HN-Tube mechanics, mainly the atomistic modeling and numerical simulations of single-walled HN-Tubes. As previously mentioned, single-walled HN-Tubes have been rarely synthesized in experiments, and thus, satisfactory measurements of their mechanical response to external load have not been carried out. In contrast, numerical simulations of their equilibrium structures and dynamic behaviors have been actively conducted using state-of-the-art computational techniques. This is partly because the structures of single-walled HN-Tubes are relatively simpler than those of HN-Fibers, which are large-scale nanomaterials endowed with amorphous structures.

A possible way to construct atomistic models of single-walled HN-Tubes is based on defect insertion into linear CNT segments. This idea was initiated in the early 1990s, just 2 years after the boost of CNT research by Dunlap [47] and Ihara et al. [17] independently. They each proposed several structural models for single-walled HN-Tubes and discussed the relationships between the geometric parameters and the energetics. The models assume that the regular coil formation results from the periodic distribution of topological defects (five- and seven-membered carbon rings) in the predominantly hexagonal carbon network, as implied by the defect-induced curvature theory [48, 49]. In this sense, nonhexagonal carbon rings in HN-Tubes are essential supplies in contrast to the cases of CNTs.

The key concept proposed by Dunlap [47] is a knee structure, which is formed by the insertion of pentagonal and heptagonal defects at the junction of two CNT segments. In Figure 3.6(a), insertion of a pentagonal defect into a perfect graphitic cylinder causes locally a conical surface of a hexagonal network with an apical five-membered ring. Conversely, a heptagonal defect insertion causes a saddle-shaped surface with a seven-membered ring at the center. The combination of pentagonal and heptagonal defect insertions, therefore, stabilizes the knee structure (Figure 3.6(b)). Afterward, Fonseca et al. [51] generalized Dunlap's concept and showed the availability of a long spiral structure by connecting the knee segments in such a way that consecutive knees are joined out of plane.

Independent of the seminal work mentioned earlier, Ihara et al. [17] showed that structures that included pentagons and heptagons gave a variety of HN-Tubes that were thermodynamically and energetically stable. Figure 3.7(a)–(c) illustrates a few examples of the atomistic model proposed by Ihara et al. Structural stability of the proposed atomistic model has been examined by numerical simulations; the cohesive energies turned out to be ca. 7.4 eV/atom, which is slightly higher than that of a fullerene molecule (7.29 eV/atom) [17, 52]. It is

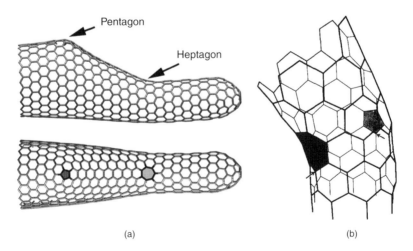

(a)   (b)

Figure 3.6 (a) Schematics of defect-induced surface curvature in a graphitic cylinder. *Source*: Terrones [49]. Reproduced with permission of American Chemical Society (b) A model of knee structure. The pentagon and heptagon rings responsible for bending the tube are visualized by the shaded areas. *Source*: Lambin et al. [50]. Reproduced with permission of Elsevier

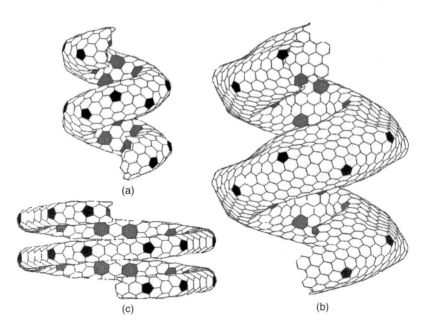

Figure 3.7 Microscopic models of single-walled HN-Tubes predicted by Ihara et al. in 1993. (a) Helix C_{360}, (b) helix C_{1080}, and (c) helix C_{540}. Fivefold and sevenfold rings shaded appear in the outer and inner ridge lines, respectively, amid a background of the sixfold rings. *Source*: Ihara et al. [17]. Reproduced with permission of American Physical Society

interesting to note that curvature and topology of HN-Tubes lead to an enhancement of molecular hydrogen absorption on their external surfaces [53]. The enhanced storage originates from very high surface area and a large number of topological defects, implying the utility of HN-Tubes as energy storage nanomaterials.

Recently, another atomistic model of single-walled HN-Tubes was suggested by Liu et al. [1]. The model is based on the combination of a pair of pentagons and a pair of heptagons. It starts from a piece of straight CNT depicted in Figure 3.8(a). In two opposite sides of the CNT segment, one pair of pentagons and one pair of heptagons are individually introduced at blue and red positions, respectively, marked in Figure 3.8(a). Upon relaxation, the CNT segment is bent around the nonhexagonal rings in order to release the strain energy (see Figure 3.8(b)). After preparing a number of segments, we connect one by one with a certain rotating angle to make a combined structure spiral (see Figure 3.8(c)) that serves as a building block. Repetitive operation of this connecting procedure results in a seamless spiral geometry as demonstrated in Figure 3.8(d). The index (n, n) in the plot means that the HN-Tube is constructed from many pieces of the straight (n, n) nanotube. By changing the tube length at the two ends of the building block segment, or by varying the nanotube diameter, we can control coil diameter, coil pitch, and tubular diameter of the HN-Tube obtained.

It should be remarked that the construction procedure based on the inclusion of pentagon/heptagon pairs can only be used to explain single-walled or at best double-walled HN-Tubes. A study by Setton and Setton [54] suggested that for many-walled HN-Tubes, pentagon and heptagon pairs would have to be arranged along the helical path, or alternatively other kinds of defects would need to be considered. More critically, the idea cannot be

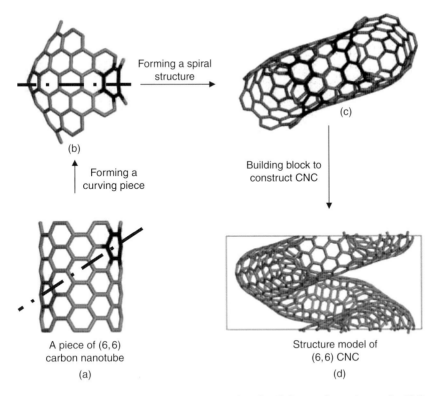

Figure 3.8 Liu's procedure of constructing a HN-Tube of a (6,6) type from pieces of a (6,6) carbon nanotube (a)–(d). *Source*: Liu et al. [1]. Reproduced with permission of Springer

used to fully explain helicity in HN-Fibers, since their spatial dimensions exceeds by far over those of HN-Tubes. In addition, internal structure of HN-Fibers is a solid core with no seamless graphitic layers and range from the amorphous to highly crystalline depending on the position. Hence, it is natural to conclude that the helical nature of HN-Fibers is caused by the unequal extrusion of carbon from a catalyst surface [18].

3.2.2   Elastic Elongation

Elastic deformation of HN-Tubes under axial loading has been examined by Liu et al. using tight-binding calculations [1]. Atomistic model demonstrated in Figure 3.8 was used to evaluate the spring constants, Young's moduli, and elastic limits of a series of single-walled HN-Tubes composed of (n, n) CNT segments with $5 \leq n \leq 8$. The computed spring constants ranged between 15 N/m and 45 N/m, showing a monotomic increase with n. The Young's moduli were found in-between 3 GPa and 6 GPa, without clear n-dependent trend. The obtained values of Young's moduli are lower by two orders of magnitudes than the original CNT, indicating that HN-Tubes are much softer than CNTs because of their spiral shapes.

The simulations by Liu et al. also revealed that HN-Tubes exhibit elastic deformations, accepting strains up to about 60% in elongation and 20-35% in compression. Above such elastic limits, the HN-Tubes will undergo plastic deformation. A noteworthy finding was that the average C–C bond length is very robust under external strains of both directions. Figure 3.9(a) plots the changes in the average C–C bond lengths of HN-Tubes during elongation and compression. With elongation strain up to 50%, the increase in average bond length is only less than 1%. Similarly, the average bond length is very slightly reduced under compression. This phenomenon can be understood by the substantial strength of sp^2 chemical bonds. In order to avoid significant changes of C–C bond lengths, the relative orientations of neighboring C–C bonds (i.e., bond angles) alter during compression or elongation. Figure 3.9(b) shows the population of bond angle distribution for a (5,5)-type HN-Tube. Clearly the peak width increases during elongation or compression, justifying the scenario mentioned earlier.

3.2.3   Giant Stretchability

Better understandings on the elongation process of HN-Tubes are obtained by monitoring their instant structural developments under tension [55]. Figure 3.10 shows a series of snapshots of a HN-Tube during axial deformation in tension. A set of tension tests were performed in the numerical simulations to rupture at a low temperature (1 K) to evaluate the overall deformation behavior of HN-Tubes. Atomistic models were constructed consisting of (5,3) single-walled CNT segments.

In the equilibrium state of $\varepsilon = 0$, the coil pitch of the HN-Tube remains uniform and constant along the helical axis. The parallel dash lines depicted in Figure 3.10(a) and (b) indicate the intertube spacing d. The value of d in the equilibrium state equals to 3.35 Å, while after stretching, d becomes nonuniform along the axis, showing slight variation due to detachment of a part of adjacent intertubes as explained later.

As the applied strain continues to increase, stepwise detachment of adjacent intertubes occurs one after another, without a dramatic increase in the stress at the inside of the helix (Figure 3.10(c)-(e)). This stepwise detachment continues up to 31% in stain; beyond it, the HN-Tube becomes subjected to homogeneous deformation in the axial direction, which induces significant stress concentration situated inside. After 82% strain, mechanical instability emerges in a gradual manner, causing two bucking patterns (inset of Figure 3.10) in which applied stress is highly concentrated on the inner surface.

For larger strain application, an increase in stress on the inner edge of the HN-Tube is clearly observed (Figure 3.10(f) and (g)). In addition, the top-view motifs lose axial symmetric shape when the strains increase beyond 82%; this symmetry breaking indicates the formation of buckles as highlighted in the inset of Figure 3.10. The inset shows two buckle modes: twisting (left) and collapse (right).

With further increasing the strain, it is observed in Figure 3.10(h) and (i) that the release in the load is due to local fracture initiated at defects that are marked by the dashed ovals in Figure 3.10(i). The local fracture is immediately arrested because of the loss in concentrated stresses near the fracture points. Further increases in strain are accompanied by a succession of such localized fractures. Finally, the highly elongated and fractured HN-Tubes become straight and taut (Figure 3.10(j)). Continued fracture events of the carbon ring structures eventually lead to the stretch of a single monatomic chain (Figure 3.10(k)). Due to the large amount of damage to the structure of the HN-Tubes at this level, the broken segments do not recoil when unloaded.

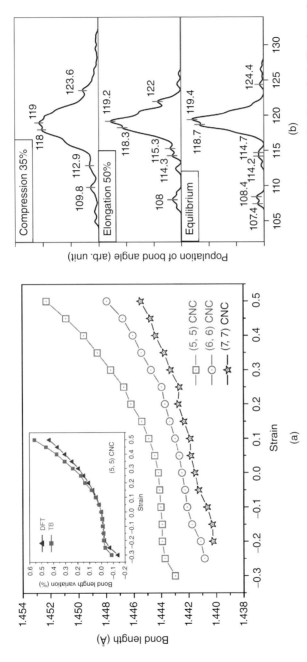

Figure 3.9 (a) Variation of average C–C bond length in HN-Tubes under elongation (positive strain) and compression (negative). The inset shows the percentages of bond length variation with regard to the equilibrium state for a (5,5)-type HN-Tube obtained by density-functional theory (DFT)-based and tight-binding (TB)-based calculations. (b) Bond angle distribution of (5,5)-type HN-Tube under large elongation (50%) and compressive strain (35%) compared to the equilibrium state. *Source*: Liu et al. [1]. Reproduced with permission of Springer

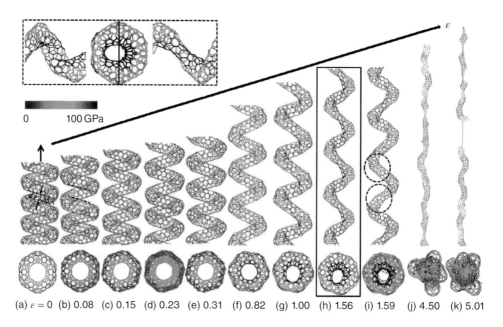

(a) $\varepsilon = 0$ (b) 0.08 (c) 0.15 (d) 0.23 (e) 0.31 (f) 0.82 (g) 1.00 (h) 1.56 (i) 1.59 (j) 4.50 (k) 5.01

Figure 3.10 Structural evolution of a HN-Tube during tension loading. The degree of stress concentration is represented by color tone. Inset at the top-left shows two buckle modes: twisting (left) and collapse (right), which lead to the asymmetry appeared in the top-view motifs at a strain of 156% (see the bottom panel (h)). At larger strains, a series of break-arrest events leads to the tube straightening, reaching 500% strain to complete rupture. *Source*: Wu et al. [55]. Reproduced with permission of American Chemical Society

Figure 3.11(a) shows the resulting global stretching force–elongation curves for a (5,3) HN-Tube. In the first stage (ca. $0 \le \varepsilon \le 0.8$), a steep increase in stretching force occurs. From each initial linear stretch–extension curve, the stiffness of the (5,3) HN-Tubes with four turns was calculated to be 12.57 nN/nm. For clarity, an enlarged view of the initial stretch–elongation curves is plotted in Figure 3.11(b). In the second stage of deformation ($0.8 \le \varepsilon \le 4.0$), a significant straightening of HN-Tubes occurs, which leads to the characteristic saw-tooth pattern in the tension force–elongation curves. The number of saw-tooth steps is greater than the number of helical turns, indicating the occurrence of multiple structural transformations of HN-Tubes during elongation. The third stage ($\varepsilon \ge 4.0$) is associated with a rapid increase in load, followed by a complete drop of the load. Overall, the whole tensile force–elongation curves show strong nonlinearity, indicating complex deformation of HN-Tubes due to a combination of bending, torsion, and tension stresses in the HN-Tubes.

3.2.4 Thermal Transport

We have seen that helical morphology of HN-Tubes results from periodic insertion of topological defects into a linear graphitic cylinder, which causes the loss of hexagonal lattice symmetry in the original graphitic sheet. This symmetry breaking is expected to affect the thermal conductivity of the system, because both phonon transport and phonon-phonon scattering in crystalline systems are governed by the spatial symmetry of the underlying lattice structure.

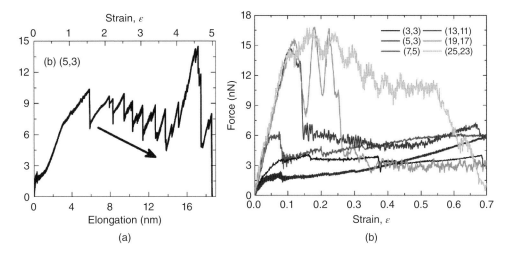

Figure 3.11 (a) Tensile force versus elongation (strain) curves of the (5,3)-typed HN-Tube. The peaks of stepwise loads display declining trend. The last two peaks correspond to the final rupture. (b) Enlarged view of the curves at the small strain region for various-typed HN-Tubes. *Source*: Wu et al. [55]. Reproduced with permission of American Chemical Society

A feasible impact of HN-Tubes' geometric parameters on their thermal conductivity was theoretically deduced by Popović et al. [56]. They considered a wide temperature range and heat conductor lengths from 10 nm to 10 mm, in which the topological coordinate method [3] was used to model HN-Tubes. They found that thermal conductivity in single-walled HN-Tubes is slightly lower than that in single-walled CNTs, while phonon-scattering mechanism is largely different between the two nanomaterials.

Figure 3.12 presents the length dependence of the room temperature thermal conductivity of a HN-Tube. Since HN-Tubes do not have translational periodicity and contain thousands of atoms within a single coil pitch, line group symmetry-based techniques [57] were applied in computations. The result for a straight (7,1) CNT having similar tubular diameters to the HN-Tubes is also displayed for comparison. Dashed lines show the results obtained in the long wavelength approximation. Thermal conductivity saturates to finite value: $\kappa = 3300$ W/K m (matching the conductivity of the purified diamond [58]) for the HN-Tube and $\kappa = 4300$ W/K m for the CNT. Despite the apparent similarity in the κ-curves between CNTs and HN-Tubes, the microscopic mechanisms that determine the thermal transport through the materials are largely different.

The difference in the thermal transport mechanism is outlined below. Generally in ideal crystalline structures with no boundaries, the thermal transport is regulated only by the phonon–phonon scattering due to the lattice anharmonicity. The scattering can be described by considering three phonon Umklapp processes [59], in which (i) a phonon having an eigenfrequency $\omega(\boldsymbol{k})$ leaves the state with the helical quantum number $\boldsymbol{k}$ as being absorbed by another phonon from the heat flux, and (ii) a phonon comes to the state $\boldsymbol{k}$, due to the decay of a phonon from the state $\boldsymbol{k}''$. For these two types of Umklapp processes, energy conservation laws are as follows:

$$\omega(\boldsymbol{k}) \pm \omega(\boldsymbol{k}')\text{-}\omega(\boldsymbol{k}'') = 0$$

where the upper (lower) sign corresponds to the first (second) process.

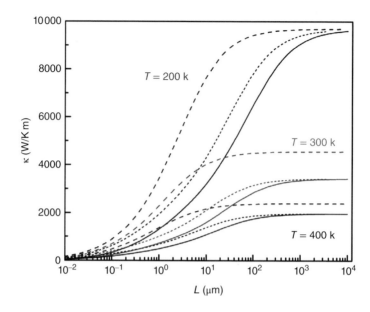

Figure 3.12   Length dependence of thermal conductivity of a single-walled HN-Tube (solid line) and a single-walled CNT (dashed line) at different temperatures. Dotted lines indicate thermal conductivity of the HN-Tube along the helical axis. *Source*: Popović et al. [56]. Reproduced with permission of Elsevier

In the case of HN-Tubes, three phonons which pertain to any combination of the branches can take part in an Umklapp process, provided that the energy is conserved and the condition imposed on the quasi-momenta is fulfilled:

$$k \pm k' = k'' + K$$

where K is a reciprocal lattice vector. On the other hand, in the case of the straight CNTs, the scattering phase space is severely restricted [60] due to the chirality-dependent quasi-angular momenta selection rules, there is no such reduction of the scattering phase space. Consequently, Umklapp scattering which involves three acoustic phonons is not allowed [60]. This situation is in contrast with the case of HN-Tubes, where all the phonon-scattering channels are open. Such a difference in the phonon-scattering mechanism between HN-Tubes and CNTs does not cause drastic change in the length dependence of the thermal conductivity, as found in Figure 3.12. However, in a perspective of nanoelectromechanical system (NEMS) applications, better understandings on the thermal transport mechanism in HN-Tubes should be critically important to secure optimal performance of nanomechanical devices based on HN-Tubes.

3.3   Experiment of HN-Fibers

3.3.1   Axial Elongation

This section gives a brief review on the mechanical response of HN-Fibers under loading. Differently from the ultrathin HN-Tubes, HN-Fibers commonly have a large dimension both in fiber diameter and coil diameter. Hence, manipulation and measurements in an apparatus are feasible owing to the advanced experimental techniques.

The seminal work on a direct tensile-loading test, which allows to determine the spring constants of HN-Fibers, was carried out by Chen et al. in 2003 [61]. In their experimental set-up, a HN-Fiber with fiber diameter of ca. 130 nm is clamped between two opposed atomic force microscope (AFM) cantilevers, in which the left cantilever is the stiffer and the right one is the more compliant. After reaching the maximum 33% relative elongation, the load was slowly released, and the HN-Fiber relaxed to its original length and geometry. Remarkably, no plastic deformation was identified after the HN-Fiber was unloaded, which indicates the high elasticity of the HN-Fibers.

The spring constant k of a HN-Fiber is defined by the total applied load divided by the total elongation as same as macroscopic coils. At low-strain levels, the HN-Fiber behaves as an elastic spring with a spring constant $k = 0.12$ N/m, with a characteristic upturn in k at higher strain. It is noteworthy that the spring constant obtained earlier is much smaller than those of other nanomaterials having coiled geometry: the superlattice ZnO nanohelices (=4 N/m) [62], metal-coated Si nanocoils (=8.75 N/m) [63], and an array of tightly packed SiO nanocoils (=590 N/m) [64] are by far stiffer than HN-Fibers.

Quite recently, Yonemura et al. performed successfully real-time deformation measurements of HN-Fibers under axial tension [65]. In order to develop better nanomanipulation techniques, it is crucially important to secure real-time measurements of HN-Fibers deformation beyond the linear elastic regime. Yonemura's work clarified the HN-Fibers' responses to prolonged stretching, which include initial elastic elongation to the large-scale deformation in the plastic regime and subsequent tensile fracture followed by post-fracture contraction and the release of the applied strain.

Figure 3.13(a) and (b) shows the experimental system for measuring the spring constant [65]. The spring table had two leaf springs with a specific spring constant of 12 ± 0.4 N/m. The spring-table substrate moved with the W probe because the leaf springs deformed when a tensile load was applied to the HN-Fiber. During the measurement, the W probe was tilted at an angle of $\theta = 38°$ relative to the substrate. When a force was applied to the HN-Fiber along the W-probe axis, the force detected by the leaf springs was the cosine component of the actual force on the HN-Fiber. The detected force was therefore divided by $\cos(\theta)$ to estimate the force applied to the HN-Fiber, and a k value was obtained. The displacements of both HN-Fiber and substrate were evaluated using scanning ion microscopy (SIM) equipped within the focused ion beam (FIB) apparatus, which enabled the values to be determined with an accuracy of 90%.

Figure 3.14 shows the real-time data of a HN-Fiber tensile test performed using a spring table in the FIB chamber. The horizontal axis shows the elapsed time, and the vertical axes represent the axial strain of the HN-Fiber (left) and the applied force (right). A series of three SIM images, shown in Figure 3.14, offers the visualization of the geometric evolution of the HN-Fiber under a tensile load. These images were captured in the free state ($t = 0$ s), the maximum elongation point ($t = 910$ s), and a post-fracture state ($t = 960$ s). A constant tensile-extension speed was carefully maintained in all tensile tests.

In the free state, the HN-Fiber had a coil length of 2200 nm and 14 turns. As soon as the force was applied, the strain on the HN-Fiber grew almost linearly with time until fracture, indicating brittleness (such as cast iron or glass) of the HN-Fibers fabricated for these experiments. The fracture point was estimated to be 106% strain, at which point the applied force reached 14 µN. After fracturing, the HN-Fiber rapidly contracted to release the applied strain, and the substrate

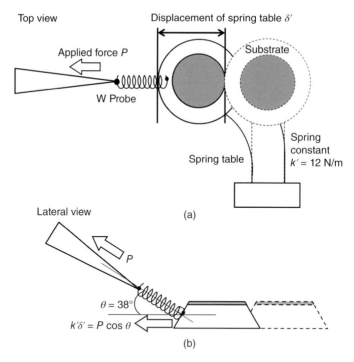

Figure 3.13 Diagram of experimental apparatus used in the spring-constant measurements done by Yonemura et al. in 2015. Top-view (a) and lateral-view (b) are shown. *Source*: Yonemura et al. [65]. Reproduced with permission of Elsevier

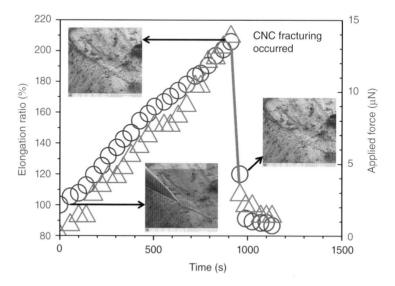

Figure 3.14 Real-time observation of HN-Fiber tensile tests performed in the FIB chamber. The three SIM images offer the visualization of variations in the coil geometry over time while under a steadily increasing tensile load. *Source*: Yonemura et al. [65]. Reproduced with permission of Elsevier

returned to its original position. At the final (fully relaxed) state ($t = 1130\,s$), the HN-Fiber had a coil length of 1910 nm and 9 turns, both of which are smaller than their original values because fragments of the HN-Fiber were ejected during fracture. The spring constant of the HN-Fibers was measured by observing the movements of the HN-Fiber and substrate using micrographs.

Yonemura et al. also evaluated the stress-strain curve of nine HN-Fibers with different fiber/coil diameters, determining their elastic limit to be ca. 15% in axial strain. The spring constant k of nine HN-Fibers in the elastic region was measured, showing 0.9 N/m at the minimum and 4.8 N/m at the maximum. The average value for the nine HN-Fibers was 1.8 N/m, which is in fair agreement with previously reported results [61]. Though the spring constant depends on the number of turns n, the dependence can be removed by evaluating the normalized spring constant nk; for the nine HN-Fibers, an average of nk was found to be 16.0 N/m.

3.3.2   Axial Compression

Similar to extension behaviors, axial compression of HN-Fibers is an important subject of research in view of both academic interest and practical device designing. On this issue, Poggi et al. [66] demonstrated that a 1100 nm length of coil could undergo compression/buckling/decompression repeatedly with a limiting compression of 400 nm.

Figure 3.15 presents the cantilever deflection signal observed when a nanospring-tipped cantilever was brought into and out of contact with the gold substrate. Scanner movement was large enough to encompass both the compression of the nanosprings and mechanical contact between the silicon cantilever tip and the substrate. The distance between the point of contact

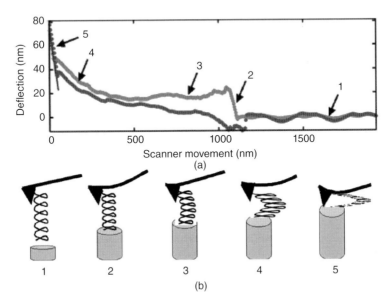

Figure 3.15   (a) Cantilever deflection versus scanner movement. (b) Illustration of the postulated response of the HN-Fiber during axially compressing loading. *Source*: Poggi et al. [66]. Reproduced with permission of American Chemical Society

of the substrate to the nanospring and to the silicon tip is ca. 1100 nm (Figure 3.15(a)), which compares well with the length of nanospring extending above the silicon tip. Figure 3.15(b) presents the magnitude of cantilever deflection versus scanner movement. The force applied to the nanospring was computed by multiplying the cantilever deflection by the cantilever-beam spring constant. The postulated response of the nanotube during vertical loading is illustrated in Figure 3.15. Prior to the contact of the nanospring with the substrate, no deflection is observed (item 1). During the steeply rising portion (item 2), the cantilever is deflecting rapidly with scanner movement commensurate with axial compressive loading of the nanosprings until buckling occurs. Additional movement of the scanner results in an increased buckling with little change in cantilever deflection (item 3). When the nanospring reaches its buckling limit, the slope of the beam deflection versus scanner movement increases (item 4) until the substrate contacts the silicon tip onto which the nanospring is appended (item 5). Repeated compression/buckling/decompression of the nanosprings was very reproducible.

While researchers considered the mechanical response of individual coils or springs (nano and micro), Daraio et al. [67] examined the response characteristics of a foam-like forest of coiled CNTs. An ensemble of vertically aligned nanocoils arranged in bundles, which is called a carbon nanocoils forest, may serve as a shock-wave mitigation materials and an energy-absorbing device. To this end, mechanical response of a forest against high strain rate deformation has been studied using a simple and conventional experimental approach [67].

Figure 3.16(a)–(c) shows that the setup for measuring the mechanical response consists of impacting ball and the array of nanocoils. The thickness and density of the forest are 100 μm and 100 μm^{-2}, respectively, in which the nanocoils are arranged in bundles due to van der Waals force attraction between them. The single coiled nanotubes in the forest had a narrow diameter distribution around 20 nm with a coiling pitch of 500 nm. Against the forest, high strain rate impacts were generated by dropping a stainless steel bead (2 mm diameter, mass of 0.03 g) from various heights at room temperature. The calculated velocities of the impacts varied between 0.2 and 2.0 m/s.

The typical contact force–time response after a small amplitude impact (0.2 m/s striker velocity) on the surface of the film is shown in Figure 3.16(d), curve 1. Repeated experiments demonstrate the identical behavior of contact force. For comparison, the contact response on the bare substrate (without the forest) under identical impact conditions was also measured and presented in Figure 3.16(d), curve 2. The presence of the forest dramatically changes the slope of the contact force and attenuates the amplitude of the pulse. A perfectly elastic response is also noticeable from the symmetry of curve 1. Such a resilient system could find applications in microelectromechanical/nanoelectromechanical systems and actuators as well as in the coating for protection purposes.

It is worthy to note that the deformation of this forest under vertical impacts (see Figure 3.17) exhibits a strongly nonlinear contact interaction law [68]. This fact means that the nanocoils respond to dynamic loading as perfect elastic nonlinear springs that fully recover their original lengths under the impact conditions. The nonlinear behavior is fully described by considering the entanglement of adjacent nanocoils in the superior part of the forest surface [68]. This entanglement among neighbors is due to the bending of the coil tips produced by the ball impact.

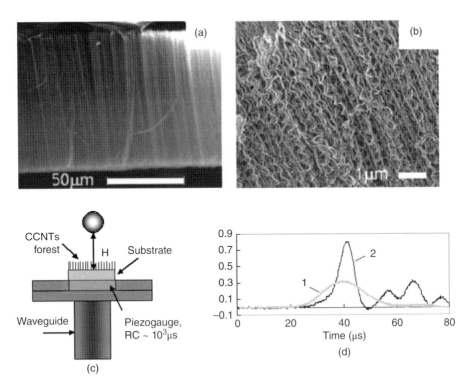

Figure 3.16 (a) and (b) Low- and high-magnification SEM pictures showing as-grown forest of HN-Fibers. (c) Experimental setup used for testing the forest of HN-Fibers. (d) Experimentally determined force versus time response obtained for the forest (curve 1) and the bare quartz substrate (curve 2) when impacted with a 2-mm-diameter steel bead (0.02 g) dropped from a height of 2 mm. *Source*: Daraio et al. [67]. Reproduced with permission of AIP Publishing LLC

3.3.3   *Resonant Vibration*

If you pinch the edge of a long thin macroscopic spring, and then swing it in the transverse direction with an appropriate swing speed, you will find the spring resonate and show transverse vibration with a large amplitude. A similar transverse resonance has been observed in nanoscopic counterparts too. It was demonstrated by Saini et al. [69] in 2014 for a few HN-Fibers with different coil geometries.

Similar to macroscopic springs, the flexural spring constants of HN-Fibers are expected to be lower than their axial spring constants due to high aspect ratio (i.e., the ratio of length-to-coil diameter of the HN-Fiber). Hence at room temperature, the probability that ambient thermal energy actuates transverse resonance modes is much higher than that for axial resonance modes. Figure 3.18(a) shows the resonant transverse modes of a HN-Fiber, whose shape is characterized by 137 nm in fiber diameter, 290 nm in coil diameter, and 876 nm in coil pitch. A direct visual examination under the SEM revealed that the HN-Fiber

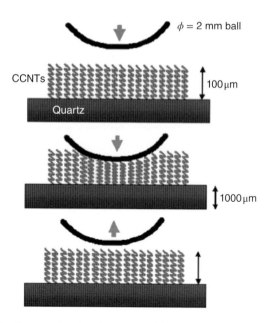

Figure 3.17   Schematic diagram showing the stages of interaction and full recovery of the HN-Fiber array during and after the impact. *Source*: Daraio et al. [67]. Reproduced with permission of AIP Publishing LLC

resonated at frequency $f_1 = 30$ kHz at its first mode (Figure 3.18(b)) and at $f_2 = 190$ kHz at its second mode (Figure 3.18(c)).

The story does not end. When applying a higher driving force to the singly clamped HN-Fiber, the coupling between bending and shearing moduli becomes significant, which results in a nonplanar circular resonant mode instead of purely transverse/longitudinal modes. Figure 3.19(a) gives an experimental evidence of such circular resonance. Interestingly, as shown in Figure 3.19, we found that (i) the HN-Fibers resonated in a nonplanar circular mode when the driving voltage was increased >8 V_{ac} (or the gap distance is reduced <5 mm) and (ii) simultaneous with the observation of the circular mode, the HDR amplitude signal exhibited a bifurcation of an otherwise single transverse resonance peak (Figure 3.19(b)). Through visual comparison, we confirmed that the lower (higher) frequency peak corresponded only to the planar transverse (nonplanar circular) resonance. As we varied the driving frequencies (from 85 to 95 kHz), we observed the onset of the nonplanar modes between 90 and 94 kHz and a clear rotation of the plane of resonance (i.e., circular mode as seen in Figure 3.19(a)). During the reverse frequency sweep, we observed a typical hysteresis (as indicated by the hatched area in Figure 3.19(b) associated with mechanical nonlinearities (i.e., nonplanar circular mode). The hysteresis was more prominent for the higher frequency peak or circular mode (gray) relative to the lower frequency peak or transverse mode (green). Although nonplanar modes were reported previously in some straight nanocantilevers [70] (SiC nanowires, multiwalled CNTs and Si nanocantilevers [71]), the necessary driving force was sufficiently large to destroy the samples impeding detailed studies of their nonlinear mechanical behavior. The helically coiled geometry plays a pivotal role in enabling the actuation of nonplanar modes at significantly lower driving force.

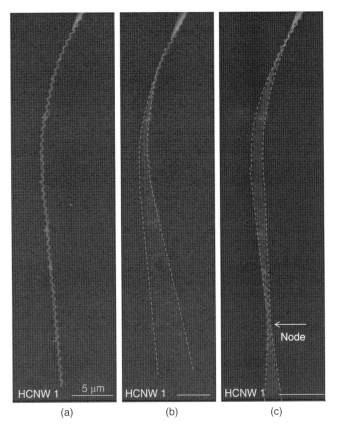

Figure 3.18 SEM images of a singly clamped HN-Fibers. (a) Off resonance, (b) first and (c) second transverse modes of a HN-Fiber at driving frequency $f_1 = 30$ kHz and $f_2 = 190$ kHz. The dotted lines serve as a guide to the eye. *Source*: Saini et al. [69]. Reproduced with permission of Nature Publishing Group

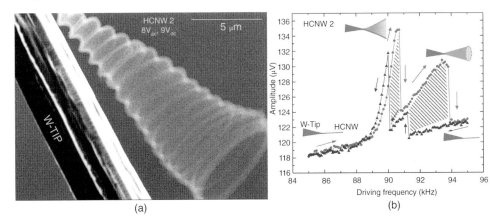

Figure 3.19 (a) SEM image of a HN-Fiber resonating in circular mode at a driving signal 93 kHz. (b) Harmonic detection of resonance (HDR) signal shows a bifurcation in the resonance signal. The peak at 90 kHz corresponds to an in-plane transverse resonance that occurs before the onset of the circular mode (peak at 93.5 kHz). *Source*: Saini et al. [69]. Reproduced with permission of Nature Publishing Group

3.3.4   Fracture Measurement

Earlier efforts on the measurement of HN-Fiber mechanics have been focused mainly on their elastic/plastic deformation and transverse vibration. There should be another important mechanical behavior called "fracture" under extreme loading. For macroscopic objects such as pipes and beams, in fact, fracturing mechanism is one of the main subject in the structural mechanics engineering, in which initial crack propagation, the degree of stress concentration under loading, and postfractured geometry are to be considered [72]. In contrast, mechanical failure processes of HN-Fibers has remained to be addressed; thus far, only one experimental attempt on torsion fracture of HN-Fibers was carried out by Yonemura et al. in 2012 [73].

The part figures in Figure 3.20(A) show tensile test micrographs obtained by Yonemura et al. By carefully maintaining constant tensile speeds for all tensile tests, they observed the elongation behavior of HN-Fibers in the FIB instrument using SIM. The part figure (c) of Figure 3.20(A) shows that the coil pitch of the HN-Fiber returns to its original length after fracturing, thus confirming that the HN-Fiber is a spring. Maximum stretch ratio on the verge of fracture ranged from 130% to 230% for eight HN-Fibers, with 150% in the average.

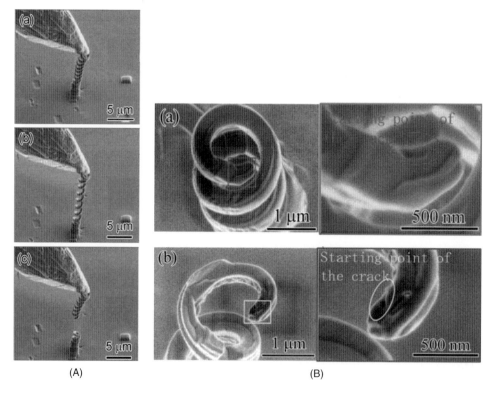

(A)   (B)

Figure 3.20   (A) Scanning ion microscopy images of HN-Fibers on a Si substrate in an FIB instrument, HN-Fiber with (a) relaxed length, (b) maximum stretching, and (c) after fracturing. The HN-Fiber fractures after stretching to 150% of its relaxed length. (B) (a) and (b) SEM micrographs of the fractured surfaces of two different HN-Fibers. *Source*: Yonemura et al. [73]. Reproduced with permission of AIP Publishing LLC

Fractured surfaces of the externally elongated HN-Fiber are demonstrated in the part figures in Figure 3.20(B). The stretch by ca. 150%, before fracturing. The ovals depicted in the rightmost two images in Figure 3.20(B) highlight hollow areas on the fractured sections. We consider these hollow areas to be the points where the fracture originates, corresponding to the maximum stress point in the cross section of the fiber. Industrial steel coil springs also exhibit the latter fracture mechanism [74], that is, the fracture mechanism is the same for macroscopic coil springs (used for automobiles) and HN-Fibers.

Using the fracture mechanism of industrial coil springs, we show that stress mainly generates on the inner edge of the coil wire cross section. Figure 3.20 shows SEM images of the fractured surfaces of two different HN-Fiber specimens. The red or blue circles in Figure 3.20 show hollow areas on the fractured sections. We consider these hollow areas to be the points where the fracture originates, corresponding to the maximum stress point on the coil surface area. Industrial steel coil springs also exhibit the latter fracture mechanism [74], that is, the fracture mechanism is the same for macroscopic coil springs and HN-Fibers.

3.4   Perspective and Possible Applications

3.4.1   Reinforcement Fiber for Composites

A possible application of helical carbon nanomaterials is effective carbon fillers in reinforcement composites. The effectiveness of the nanomaterials as a reinforcing material was implied by Yoshimura et al. [75], who examined the impact of CMCs embedding on the mechanical properties of epoxy resin composites.

Two kinds of epoxy resin were used as a polymer matrix, into which CMCs having 0.1-1 μm in fiber diameter and 1-10 μm in coil diameter were embedded with various volume fraction. The Young's modulus as well as the tensile strength of the epoxy resins could be improved by the addition of just 2% CMCs. When compared to macroscopic carbon fiber-reinforced resins, the carbon microcoil/epoxy resin showed better reinforcement capabilities. This improvement is attributed to the large specific surface area of the spring-shaped CMCs, implying further enhanced reinforcement by embedding nanoscale analogues. Yoshimura et al. also suggested that the CMCs tended to extend with the polymer matrix and break only when an excessive load was applied. In contrast, carbon fibers can be pulled out of the matrix due to the lack of interfacial adhesion.

3.4.2   Morphology Control in Synthesis

The use of helical carbons in technological applications will be dependent on our ability to control the coil morphology of these materials. It includes control of the coil diameter, coil pitch, and fiber/tube diameters. Since the growth of carbon nanomaterials can be controlled by varying temperature, gas environment, and the type of catalyst, the alteration of any of these variables is expected to result in a significant change in the type and amount of helical carbon nanomaterials formed. To achieve this control, an understanding of the growth mechanism and the role played by the various parameters is needed. To date, control over the synthesis of a specific type of helical carbon nanomaterial has been met with only limited success.

Toward the mass production of carbon nanocoils, it is crucial to understand the relation between the geometric structure and synthesis methods. In fact, a wide variety of synthesis

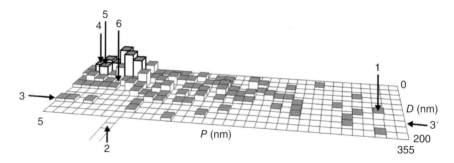

Figure 3.21 Three-dimensional plot of a number of coils versus coil diameter (D) and pitch (P) for coiled nanotubes grown by chemical vapor deposition. The two primary stability islands are highlighted in bold lines. *Source*: Szabó et al. [76]. Reproduced with permission of Elsevier

techniques for carbon nanocoils has been reported thus far as mentioned in Section 3.1.3. Hence, possibility of controlling the coiled geometry should be examined in order to accomplish the mass production. The need has motivated a statistical analysis of the characteristic parameters, coil diameters and pitch, of coils produced under various experimental conditions [76].

The plot in Figure 3.21 shows the statistical distribution of helical carbon nanomaterials grown by CVD. A three-dimensional plot was used to analyze the number of coils exhibiting a certain pair of coil diameter and pitch value. It should be borne in mind that the data come from three different laboratories and from a number of at least 15 experimental cycles. Hence, there can be no correlation in coiled geometry between samples generated by different experiments. Nevertheless, the plot clearly shows the presence of specific regions in which the number of coils produced is significantly higher than in the neighboring regions. The most preferred configuration is delineated by the crossing of the region of 50-70 nm pitch and 20-60 nm diameter. In addition, a slightly less pronounced region is found between pitch 30-50 nm and the same diameters.

To be noted is the fact that a fraction of 24.5% of the total number of coils is found in the two stability regions; this has to be compared with 2% that would correspond to the same area if a uniform distribution of coils over the entire area of the plot would be found. The analysis of the coil diameter distribution of Figure 3.21 shows that half of the coils are found in the range of 50-70 nm, while again, half of the coils have a pitch in the range of 30-80 nm. Since the analysis is done for an ensemble of nanocoils produced by different experiments, the possibility of accidental coincidences should be ruled out. Therefore, these data can be considered as an indirect proof that the way in which helical carbon nanomaterials are coiled has an intrinsic, structural origin, and it is not decided by particular conditions under which coiling occurs in actual experiments.

References

[1] Liu, L.Z., Gao, H.L., Zhao, J.J. and Lu, J.P. (2010) Superelasticity of carbon nanocoils from atomistic quantum simulations. *Nanoscale Research Letters*, **5**, 478–483.
[2] Motojima, S., Kawaguchi, M., Nozaki, K. and Iwanaga, H. (1990) Growth of regularly coiled filaments by Ni catalyzed pyrolysis of acetylene, and their morphology and extension characteristics. *Applied Physics Letters*, **56**, 321–323.

[3] Milošević, I., Popović, Z.P. and Damnjanović, M. (2012) Structure and stability of coiled carbon nanotubes. *Physica Status Solidi B*, **249**, 2442–2445.

[4] Davis, W.R., Slawson, R.J. and Rigby, G.R. (1953) An unusual form of carbon. *Nature*, **171**, 756.

[5] Hofer, L.J.E., Sterling, E. and McCartney, J.T. (1955) Structure of the carbon deposited from carbon monoxide on iron, cobalt and nickel. *The Journal of Physical Chemistry*, **59**, 1153–1155.

[6] Boehm, H.P. (1973) Carbon from carbon monoxide disproportionation on nickel and iron catalysts: morphological studies and possible growth mechanism. *Carbon*, **11**, 583–590.

[7] Baker, R.T.K., Gadsby, G.R., Thomas, R.B. and Waite, R.J. (1975) The production and properties of filamentous carbon. *Carbon*, **13**, 211–214.

[8] Qin, Y., Zhang, Z. and Cui, Z. (2004) Helical carbon nanofibers with a symmetric growth mode. *Carbon*, **42**, 1917–1922.

[9] Zhang, M. and Li, J. (2009) Carbon nanotube in different shapes. *Materials Today*, **12**, 12–18.

[10] Qi, X., Qin, C., Zhong, W. et al. (2010) Large-scale synthesis of carbon nanomaterials by catalytic chemical vapour deposition: a review of the effects of synthesis parameters and magnetic properties. *Materials*, **3**, 4142–4174.

[11] Li, D., Pan, L., Wu, Y. and Peng, W. (2012) The effect of changes in synthesis temperature and acetylene supply on the morphology of carbon nanocoils. *Carbon*, **50**, 2571–2580.

[12] Motojima, S., Chen, X., Yang, S. and Hasegawa, M. (2004) Properties and potential applications of carbon microcoils/nanocoils. *Diamond and Related Materials*, **13**, 1989–1992.

[13] Volodin, A., Buntinx, D., Ahlskog, M. et al. (2004) Coiled carbon nanotubes as self-sensing mechanical resonators. *Nano Letters*, **4**, 1775–1779.

[14] Xie, J., Mukhopadyay, K., Yadev, J. and Varadan, V.K. (2003) Coiled carbon nanotubes: their structural and electrical properties. Proc. SPIE 5055, Smart Structures and Materials 2003: Smart Electronics, MEMS, BioMEMS, and Nanotechnology, 223.

[15] Hanus, M.J. and Harris, A.T. (2010) Synthesis, characterisation and applications of coiled carbon nanotubes. *Journal of Nanoscience and Nanotechnology*, **10**, 2261–2283.

[16] Shaikjee, A. and Coville, N.J. (2012) The synthesis, properties and uses of carbon materials with helical morphology. *Journal of Advanced Research*, **3**, 195–223.

[17] Ihara, S., Itoh, S. and Kitakami, J. (1993) Helically coiled cage forms of graphitic carbon. *Physical Review B*, **48**, 5643–5647.

[18] Amelinckx, S., Zhang, X.B., Bernaerts, D. et al. (1994) A formation mechanism for catalytically grown helix-shaped graphite nanotubes. *Science*, **265**, 635–639.

[19] Zhang, X.B., Zhang, X.F., Bernaerts, D. et al. (1994) The texture of catalytically grown coil-shaped carbon nanotubules. *Europhysics Letters*, **27**, 141.

[20] Akagi, K., Tamura, R., Tsukada, M. et al. (1995) Electronic structure of helically coiled cage of graphitic carbon. *Physical Review Letters*, **74**, 2307–2310.

[21] Akagi, K., Tamura, R., Tsukada, M. et al. (1996) Electronic structure of helically coiled carbon nanotubes: relation between the phason lines and energy band features. *Physical Review B*, **53**, 2114–2120.

[22] Hou, H.Q., Jun, Z., Weller, F. and Greiner, A. (2003) Large-scale synthesis and characterization of helically coiled carbon nanotubes by use of Fe(CO)5 as floating catalyst precursor. *Chemistry of Materials*, **15**, 3170–3175.

[23] Fejes, D., Popović, Z.P., Raffai, M. et al. (2013) Synthesis, model and stability of helically coiled carbon nanotubes. *ECS Solid State Letters*, **2**, M21–M23.

[24] Gao, R.P., Wang, Z.L. and Fan, S.S. (2000) Kinetically controlled growth of helical and zigzag shapes of carbon nanotubes. *The Journal of Physical Chemistry B*, **104**, 1227–1234.

[25] Szabó, A., Fonseca, A., Nagy, J.B. et al. (2005) Synthesis, properties and applications of helical carbon nanotubes. *Fullerenes, Nanotubes, and Carbon Nanostructures*, **13**, 139–146.

[26] Cheng, J.P., Zhang, X.B., Tu, H. et al. (2006) Catalytic chemical vapor deposition synthesis of helical carbon nanotubes and triple helices carbon nanostructure. *Materials Chemistry and Physics*, **95**, 12–15.

[27] Bai, J.B. (2003) Growth of nanotube/nanofibre coils by CVD on an alumina substrate. *Materials Letters*, **57**, 2629–2633.

[28] Ju, S.-P., Lin, J.-S., Chen, H.-L. et al. (2014) A molecular dynamics study of the mechanical properties of a double-walled carbon nanocoil. *Computational Materials Science*, **82**, 92–99.

[29] Ghaderi, S.H. and Hajiesmaili, E. (2013) Nonlinear analysis of coiled carbon nanotubes using the molecular dynamics finite element method. *Materials Science and Engineering A*, **582**, 225–234.

[30] Feng, C., Liew, K.M., He, P. and Wu, A. (2014) Predicting mechanical properties of carbon nanosprings based on molecular mechanics simulation. *Composite Structures*, **114**, 41–50.

[31] Blank, V.D. and Kulnitskiy, B.A. (2004) Proposed formation mechanism for helically coiled carbon nanofibers. *Carbon*, **42**, 3009–3011.

[32] Wang, W., Yang, K.Q., Gaillard, J. et al. (2008) Rational synthesis of helically coiled carbon nanowires and nanotubes through the use of tin and indium catalysts. *Advanced Materials*, **20**, 179–182.

[33] Hwang, W.I., Kuzuya, T., Iwanaga, H. and Motojima, S. (2001) Oxidation characteristics of graphite micro-coils, and growth mechanism of the carbon coils. *Journal of Material Science*, **36**, 971–978.

[34] Motojima, S., Noda, Y., Hoshiya, S. and Hishikawa, Y. (2003) Electromagnetic wave absorption property of carbon microcoils in 12–110 GHz region. *Journal of Applied Physics*, **94**, 2325–2330.

[35] Ivanov, V., Nagy, J.B., Lambin, P. et al. (1994) The study of carbon nanotubules produced by catalytic method. *Chemical Physics Letters*, **223**, 329–335.

[36] Lau, K.T., Lu, M. and Hui, D. (2006) Coiled carbon nanotubes: synthesis and their potential applications in advanced composite structures. *Composites Part B: Engineering*, **37**, 437–448.

[37] Hayashida, T., Pan, L. and Nakayama, Y. (2002) Mechanical and electrical properties of carbon tubule nanocoils. *Physica B*, **323**, 352–353.

[38] Pradhan, D. and Sharon, M. (2002) Carbon nanotubes, nanofilaments and nanobeads by thermal chemical vapor deposition process, *Materials Science and Engineering B* **96**, 24-28.

[39] Takenaka, S., Ishida, M., Serizawa, M. et al. (2004) Formation of carbon nanofibers and carbon nanotubes through methane decomposition over supported cobalt catalysts. *The Journal of Physical Chemistry B*, **108**, 11464–11472.

[40] Xie, J., Mukhopadyay, K., Yadev, J. and Varadan, V.K. (2003) Catalytic chemical vapor deposition synthesis and electron microscopy observation of coiled carbon nanotubes. *Smart Materials and Structures*, **12**, 744–748.

[41] Pan, L., Hayashida, T., Zhang, M. and Nakayama, Y. (2001) Field emission properties of carbon tubule nanocoils. *Japanese Journal of Applied Physics*, **40**, L235–L237.

[42] Koós, A.A., Ehlich, R., Horváth, Z.E. et al. (2003) STM and AFM investigation of coiled carbon nanotubes produced by laser evaporation of fullerene. *Materials Science and Engineering: C*, **23**, 275–278.

[43] Saveliev, A.V., Merchan-Merchan, W. and Kennedy, L.A. (2003) Metal catalyzed synthesis of carbon nanostructures in an opposed flow methane oxygen flame. *Combustion and Flame*, **135**, 27–33.

[44] Hokushin, S., Pan, L. and Nakayama, Y. (2007) Diameter control of carbon nanocoils by the catalyst of organic metals. *Japanese Journal of Applied Physics*, **46**, 5383–5385.

[45] Lim, S.L., Suda, Y., Takimoto, K. et al. (2013) Optimization of chemical vapor deposition process for reducing the fiber diameter and number of graphene layers in multi walled carbon nanocoils. *Japanese Journal of Applied Physics*, **52**, 11NL04.

[46] Biró, L.P., Lazarescu, S.D., Thiry, P.A. et al. (2000) Scanning tunneling microscopy observation of tightly wound, single-wall coiled carbon nanotubes. *Europhysics Letters*, **50**, 494–500.

[47] Dunlap, B.I. (1992) Connecting carbon tubules. *Physical Review B*, **46**, 1933–1936.

[48] Cortijo, A. and Vozmediano, M.A.H. (2007) Effects of topological defects and local curvature on the electronic properties of planar graphene. *Nuclear Physics B*, **763**, 293–308.

[49] Terrones, M. (2010) Sharpening the chemical scissors to unzip carbon nanotubes: crystalline graphene nanoribbons. *ACS Nano*, **4**, 1775–1781.

[50] Lambin, P., Fonseca, A., Vigneron, J.P. and Lucas, A.A. (1995) Structural and electronic properties of bent carbon nanotubes. *Chemical Physics Letters*, **245**, 85–89.

[51] Fonseca, A., Hernadi, K., Nagy, J.B. et al. (1995) A model structure of perfectly graphitizable coiled carbon nanotubes. *Carbon*, **33**, 1759–1775.

[52] ZhongCan, O.Y., Su, Z.B. and Wang, C.L. (1997) Coil formation in multishell carbon nanotubes: Competition between curvature elasticity and interlayer adhesion. *Physical Review Letters*, **78**, 4055–4058.

[53] Gayathri, V., Devi, N.R. and Geetha, R. (2010) Hydrogen storage in coiled carbon nanotubes. *International Journal of Hydrogen Energy*, **35**, 1313–1320.

[54] Setton, R. and Setton, N. (1997) Carbon nanotubes: III. Toroidal structures and limits of a model for the construction of helical and s-shaped nanotubes. *Carbon*, **35**, 497–505.

[55] Wu, J., He, J., Odegard, G.M. et al. (2013) Giant stretchability and reversibility of tightly wound helical carbon nanotubes. *Journal of the American Chemical Society*, **135**, 13775–13785.

[56] Popović, Z.P., Damnjanović, M. and Milošević, I. (2014) Phonon transport in helically coiled carbon nanotubes. *Carbon*, **77**, 281–288.

[57] Damnjanović, M., Milošević, I., Dobardžić, E. et al. (2005) Symmetry based fundamentals of carbon nanotubes, in *Applied Physics of Nanotubes: Fundamentals of Theory, Optics and Transport Devices* (eds S.V. Rotkin and S. Subramoney), Springer-Verlag, Berlin.

[58] Berman, R. (1992) Thermal conductivity of isotopically enriched diamonds. *Physical Review B*, **45**, 5726–5728.

[59] Nika, D.L., Pokatilov, E.P., Askerov, A.S. and Balandin, A.A. (2009) Phonon thermal conduction in graphene: role of Umklapp and edge roughness scattering. *Physical Review B*, **79**, 155413.

[60] Lindsay, L., Broido, D.A. and Mingo, N. (2009) Lattice thermal conductivity of single-walled carbon nanotubes: beyond the relaxation time approximation and phonon-phonon scattering selection rules. *Physical Review B*, **80**, 125407.

[61] Chen, X., Zhang, S., Dikin, D.A. et al. (2003) Mechanics of a carbon nanocoil. *Nano Letters*, **3**, 1299–1304.

[62] Gao, P.X., Ding, Y., Mai, W.J. et al. (2005) Conversion of zinc oxide nanobelts into superlattice-structured nanohelices. *Science*, **309**, 1700–1704.

[63] Singh, J.P., Liu, D.L., Ye, D.X. et al. (2004) Metal-coated Si springs: Nanoelectromechanical actuators. *Applied Physics Letters*, **84**, 3657–3659.

[64] Kong, X.Y., Ding, Y., Yang, R. and Wang, Z.L. (2004) Single-crystal nanorings formed by epitaxial self-coiling of polar nanobelts. *Science*, **303**, 1348–1351.

[65] Yonemura, T., Suda, Y., Shima, H. et al. (2015) Real-time deformation of carbon nanocoils under axial loading. *Carbon*, **83**, 183–187.

[66] Poggi, M.A., Boyles, J.S., Bottomley, L.A. et al. (2004) Measuring the compression of a carbon nanospring. *Nano Letters*, **4**, 1009–1016.

[67] Daraio, C., Nesterenko, V.F., Jin, S. et al. (2006) Impact response by a foamlike forest of coiled carbon nanotubes. *Journal of Applied Physics*, **100**, 064309.

[68] Coluci, V.R., Fonseca, A.F., Galvão, D.S. and Daraio, C. (2008) Entanglement and the nonlinear elastic behavior of forests of coiled carbon nanotubes. *Physical Review Letters*, **100**, 086807.

[69] Saini, D., Behlow, H., Podila, R. et al. (2014) Mechanical resonances of helically coiled carbon nanowires. *Scientific Reports*, **4**, 5542.

[70] Perisanu, S., Barois, T., Ayari, A. et al. (2010) Beyond the linear and Duffing regimes in nanomechanics: circularly polarized mechanical resonances of nanocantilevers. *Physical Review B*, **81**, 165440.

[71] Villanueva, L.G., Karabalin, R.B., Matheny, M.H. et al. (2013) Nonlinearity in nanomechanical cantilevers. *Physical Review B*, **87**, 024304.

[72] Anderson, T.L. (2005) *Fracture Mechanics: Fundamentals and Applications*, 3rd edn, CRC Press.

[73] Yonemura, T., Suda, Y., Tanoue, H. et al. (2012) Torsion fracture of carbon nanocoils. *Journal of Applied Physics*, **112**, 084311.

[74] Wahl, A.M. (1963) *Mechanical Spring*, McGraw-Hill Book Company, New York.

[75] Yoshimura, K., Nakano, K., Miyake, T. et al. (2006) Effectiveness of carbon microcoils as a reinforcing material for a polymer matrix. *Carbon*, **44**, 2833–2838.

[76] Szabó, A., Fonseca, A., Nagy, J.B. et al. (2005) Structural origin of coiling in coiled carbon nanotubes. *Carbon*, **43**, 1628–1633.

4

Computational Nanomechanics Investigation Techniques

Ghasem Ghadyani[1] and Moones Rahmandoust[2,3,4]

[1]*Faculty of Mechanical Engineering, University Malaya, Kuala Lumpur, Malaysia*
[2]*Protein Research Center, Shahid Beheshti University, G.C., Velenjak, Tehran, Iran*
[3]*School of Engineering, Griffith University, Gold Coast Campus, Southport, Queensland, Australia*
[4]*Deputy Vice Chancellor Office of Research and Innovation, Universiti Teknologi Malaysia, Johor Bahru, Johor, Malaysia*

4.1 Introduction

Nanostructures are the construction facilities in nanoscience and nanotechnology. These structures function as major time-, size-, and energy-savors. They are too small for direct measurement and observation, and might be too outsized to be defined by quantum-mechanical computational approaches. Consequently, computer-based predictive numerical techniques and other simulation approaches are the best options to investigate their properties and behaviors, and hence, nanoscale procedures play an important role in their utilization, optimization, and analysis. Combination of theoretical approaches and major analysis of practical nanoscale components of nanostructures are other important benefits of computer-based numerical techniques. A computer-based numerical technique is composed of investigating nano-sized grain domains in nanostructured materials, and other nanostructural properties dominated to nanointerfaces. The significance of these computational methods, other than the clear perception they provide from nanotechnology and nanoscience, is based on the fact that they can provide principally direct data related to nanoscale model structures and consequently, lead to reduced energy, time periods, and sizes involved, which shall be very hard to achieve with techniques other than that [1].

Several empirical investigations have been developed to obtain the mechanical properties of nanostructures; however, extremely different results have been achieved due to differences in

Advanced Computational Nanomechanics, First Edition. Edited by Nuno Silvestre.
© 2016 John Wiley & Sons, Ltd. Published 2016 by John Wiley & Sons, Ltd.

the empirical methods used. Based on the measurement techniques employed, the experimental investigations can be divided into two major categories, namely, "direct" and "indirect" ones. The small dimensions of nanostructures and the placement of these structures in an appropriate testing configuration impose several difficulties on experimental studies. On the other hand, the application of a desired loading and the measurement of deformations at nanometer length scales are other major challenges faced by scientists in this field. The diversity in the experimental values reported can be attributed to the following main reasons:

− the lack of proper direct measurement techniques at nanometer scale;
− tremendous limitations on the specimen size;
− uncertainty in the data obtained from indirect measurements;
− inadequacy in the preparation techniques of test specimens and the lack of control over the alignment and distribution of nanostructures [1, 2].

The extremely scattered data obtained through experimental observations have therefore encouraged many researchers to pursue a variety of theoretical studies on the effective properties of nanotubes in order to both justify the observation of experimental data and provide the necessary information, which was not accessible by employing experimental methods. The theoretical and computational approaches can be divided into three main categories:

− crystal elasticity-based approaches or atomistic modeling [3–5];
− analytical continuum mechanics-based approaches [6];
− numerical continuum mechanics-based approaches, referred to as finite element methods (FEMs) [7].

Many mechanical, thermal, and electronic properties of the building blocks of a nanoscale material depend vitally on the size, shape, and the precise geometrical arrangement of the atoms within the block. Additionally, the construction of functional assemblies of nanostructures, that is, nanoscale devices, requires a deep understanding of the coupling and interaction of individual component nanostructures. In this chapter, we provide an account of the fundamental theories for the modeling techniques that are in use in computational nanoscience and computational nanotechnology, and the related fields of computational condensed matter physics, computational chemistry, and computational material sciences.

4.2 Fundamentals of the Nanomechanics

Quantum (molecular) mechanics and continuum (Newtonian) mechanics have been highly developed to describe material properties at small and large material scales, respectively. As we enter the world of nanotechnology, it becomes increasingly important to model phenomena at microscopic scales. Two alternative approaches, namely the "bottom-up" approach, based on quantum or molecular mechanics, and the "top-down" approach, based on continuum mechanics, are frequently used to model mechanical properties of nanostructured materials. Although much effort has been made to develop theories and approaches spanning multiple length scales [8, 9], the link between molecular and Newtonian descriptions of material properties is still not well established [10].

4.2.1 Molecular Mechanics

To study the mechanical properties of materials in atomic length scales, it is essential to accurately model atomic connections and interactions. Molecular mechanics and quantum mechanics are the two most commonly used methods to define these connections and interactions. Capture of the variation of the system energy, associated with changes in atomic coordination, is one of the key factors in both methods. Newton's law of motion is capable of predicting the interaction force exerted on each atom under the influence of its neighboring atoms. The motion of all atoms in the structure can be obtained using atomic positions. The interaction force, on the other hand, can be considered from the derivative of the system energy. Therefore, having the dynamic motion of atoms in the system, the mechanical behavior of the material can be obtained numerically at nanoscale using Newton's law. Determination of the energy based on calculations of the electronic structure of molecules is also possible by quantum mechanics. However, although accurate, even after simplifications such as employing semi-empirical methods, this process is usually very time consuming. Molecular mechanics, on the contrary, is a technique that is based on the Born–Oppenheimer approximation, and it is capable of speeding up the calculation time by ignoring the electronic structure of the system and by expressing the system energy only as a function of the atomic positions [8, 9].

4.2.2 Newtonian Mechanics

The motion of a system for point masses in three-dimensional space is studied by Newtonian mechanics. The six-dimensional group of Euclidean motions of space is the basic concept and theorem in Newtonian mechanics. The potential energy and mass points are specified in a mechanical system. In addition, the law of conservation of energy, stating that the total energy of an isolated system remains constant, solves problems such as the motion in a force fields by Newton's equations in mechanics [11]. Classical mechanics has a set of experimental facts, as follows:

1. Time is one dimensional and space is three dimensional with Euclidean representation.
2. All the rules of motion in all time and coordinate systems are similar.

Consider a volume V, composed of N atoms with energy E, which is confined to a nanoscale system. Usually N is of the order of 10^{23} and is very huge. The N atoms have 3N generalized *position* coordinates $(q_1, q_2, q_3, q_4, \ldots, q_{3N})$ and 3N generalized *momentum* coordinates $(p_1, p_2, p_3, p_4, \ldots, p_{3N})$, which are conjugate to the position coordinates. The arrangement of these 6N coordinates makes a particular microstate of the N-particle nanosystem. Three-dimensional space can describe the positions of these N atoms either by N points or by a single point in the 3N-dimensional space. Similarly, the momentum space can be represented by a single point in a 3N-dimensional space. Hence, one can combine the separate configuration and momentum spaces into a unified 6N-dimensional mathematical space, wherein a particular microstate is represented by just a single point with coordinates $(q_1,$ $q_1, q_2, q_3, q_4, \ldots, q_{3N}; p_1, p_2, p_3, p_4, \ldots, p_{3N})$. This 6N-dimensional space is named the phase space, or the Γ-space, whereas in case of a single point, it is named the representative point, the phase point, or the Γ-point. The Γ-point represents the dynamical state of the system, that is,

$$\Gamma = (q_1, q_2, q_3, q_4, \ldots, q_{3N}; p_1, p_2, p_3, p_4, \ldots, p_{3N}) \equiv (q_i, p_i) \qquad (4.1)$$

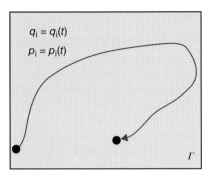

Figure 4.1   Motion through phase space

where $i = 1, \; 2, \ldots, 3N$. It is obvious that the first set of three coordinates, $i = 1, 2, 3$, refers to the first atom, and the second set, $i = 4, 5, 6$, to the second atom, and so on. Hence, first atom for instance can be shown in its own six-dimensional phase space, as $(\Gamma = (q_1, q_2, q_3; p_1, p_2, p_3))$. Now, if we consider just one atom, as a Γ-space, for any individual phase point named as the μ-point, the Γ-space can be generated by combining individual μ-points.

Straight movement through the phase space can be imagined in dynamics as a "simple" motion through the high-dimensional phase space [1, 10]. Figure 4.1 lets us facilitate the relationship between quantum mechanics and the simplified theoretical dynamics by illustrating the simple motion in phase space.

4.2.3   Lagrangian Equations of Motion

The equation of motion for a system of interacting material points, i.e., atoms and particles, can be most generally written in terms of a Lagrangian function L, as follows [11–13]:

$$L = K - U \tag{4.2}$$

$$L(\boldsymbol{q}, \dot{\boldsymbol{q}}) \equiv K(\boldsymbol{q}, \dot{\boldsymbol{q}}) - U(\boldsymbol{q}) \tag{4.3}$$

$$\frac{d}{dt}\left(\frac{\partial L}{\partial \dot{q}_\alpha}\right) - \frac{\partial L}{\partial q_\alpha} = 0 \tag{4.4}$$

Motion of the system is therefore obtained from the differential equations as described above. Here, vector quantities are expressed in bold script and q's are the generalized coordinates, that is, the variables that individually define the spatial positions of the atoms; and the superposed dot denotes their time derivative. As will be discussed in subsequent sections, Eq. (4.2) can be rewritten in terms of the generalized coordinates and moment, to be employed in statistical mechanics. The molecular dynamics (MD) simulations are most typically run in Cartesian coordinate systems, where Eq. (4.4) can be simplified to give

$$\frac{d}{dt}\frac{\partial L}{\partial \dot{r}_i} - \frac{\partial L}{\partial r_i} = 0; \; i = 1, 2, 3, \ldots, N \tag{4.5}$$

where $r_i = (x_i, \; y_i, \; z_i)$ is the radius vector of atom i, as shown in Figure 4.2, and N is the total number of atoms. The spatial volume employed by these N atoms is generally referred within MD domain [8, 14]. Due to the homogeneousness of time and space, as well as the isotropy

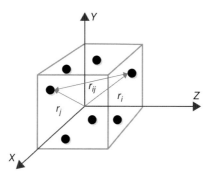

Figure 4.2 Coordinates in atomic system

of space in inertial coordinate systems, the equation of motion (4.5) must not depend on the initial time of observation, the origin of the coordinate system, and directions of its axes. These fundamental principles are equivalent to the requirements of the Lagrangian function that does not explicitly depend on time and direction of the radius and the velocity vectors r_i and $\dot{r}_i$, but instead, it only depends on the absolute value of the velocity vector r_i. In order to provide identical equations of motions in all inertial coordinate systems, the Lagrangian function must also fulfill the Galilean relativity principle, making the function adequate for all mentioned requirements [8, 11].

$$L = \sum_{i=1}^{N} \frac{m_i}{2} (\dot{x}_i^2 + \dot{y}_i^2 + \dot{z}_i^2) \equiv \sum_{i=1}^{N} \frac{m_i \dot{r}_i^2}{2} \tag{4.6}$$

where for a system of free, noninteracting particles, m_i is the mass of particle i.

Interaction between the particles can then be described by adding a particular function of atomic coordinates U to Eq. (4.3) that depends on the properties of the interaction. Such function is defined with a negative sign, in a way that the Lagrangian acquires the following form:

$$L = \sum_{i=1}^{N} \frac{m_i \dot{r}_i^2}{2} - U(r_1, r_2, \dots, r_N) \tag{4.7}$$

where the two terms represent the systems kinetic and potential energy, respectively (note the additive nature of the kinetic energy term). This gives the general structure of Lagrangian for a conservative system of interacting material points (atoms) in the Cartesian coordinate system. The Lagrangian equation, however, contains two important features, that is, the additive nature of the kinetic energy term and the absence of explicit time dependence. Based on the fact that the potential energy term only depends on spatial configuration of the particles, any change in this configuration results in an immediate effect on the motion of all particles within the simulated domain, where the inevitability of this assumption is due to the relativity principle. Indeed, if such a particle propagates with a finite speed, the equation would depend on the chosen inertial system, making the laws of motion or, in particular, the MD solutions dissimilar in various systems that would in turn contradict the relativity principle.

The equations of the motion can finally be written in the Newtonian system by substituting the obtained Lagrangian (4.7) in Eq. (4.5),

$$m_i \ddot{r}_i = -\frac{\partial U(r_1, r_2, \dots, r_N)}{\partial r_i} \equiv F_i, \ i = 1, 2, \dots, N \tag{4.8}$$

where the force F_i is commonly referred to as the internal force applied on atom i due to its environmental conditions. Equation (4.8) can further be solved for a given set of initial conditions to provide the trajectories of the atomic motion in the virtual system [14–16].

4.2.4 Hamilton Equations of a Γ-Space

Hamiltonian formulation is appropriate for application to statistical mechanics, quantum mechanics, and also classical perturbation theory. In this calculation, Newtonian and Lagrangian viewpoints take the q_i as the fundamental variable. Furthermore, Hamiltonian formulation seeks to work with first-order differential equations. In order to propagate Hamilton equations, let us consider Lagrangian treats q_i and $\dot{q}_i$ as distinct in following steps [14, 17].

$$L = K - U = \frac{1}{2}m\dot{q}_i^2 - U(q) \tag{4.9}$$

$$L = (q, \dot{q}, t) \tag{4.10}$$

where identify the generalized momentum as $p = \partial L / \partial \dot{q} = m\dot{q}$. We would like a formulation in which p is an independent variable, p_i is the derivative of the Lagrangian with respect to $\dot{q}_i$, and we are looking to replace p_i by Hamilton equations from Lagrangian equations,

$$L = (q, \dot{q}, t) \xrightarrow{\text{Transfer by Hamilton}} (q, p, t)$$

$$\frac{dp}{dt} = \dot{p} = \frac{\partial L}{\partial q} \tag{4.11}$$

$$dL = \frac{\partial L}{\partial q}dq + \frac{\partial L}{\partial \dot{q}}d\dot{q} = \dot{p}dq + pd\dot{q} \tag{4.12}$$

$$H = -(L - p\dot{q}) \tag{4.13}$$

$$dH = -(\dot{p}dq - \dot{q}dp) = -\dot{p}dq + \dot{q}dp \tag{4.14}$$

and the Hamilton's equations of motion will be

$$\dot{q} = +\frac{\partial H}{\partial p} \tag{4.15}$$

$$\dot{p} = -\frac{\partial H}{\partial q} \tag{4.16}$$

where the position and momentum coordinates of the atoms are time dependent.

$$H(q, p) = -\left[L(q, \dot{q}) - \sum p_i\dot{q}_i\right] = -K(q, \dot{q}) + U(q) + \sum \frac{\partial K}{\partial \dot{q}_j}$$

$$= -\sum a_i\dot{q}_i^2 + U(q) + \sum (2a_i\dot{q}_i)\dot{q}_j = \sum a_i\dot{q}_i^2 + U(q) = K + U = E \tag{4.17}$$

If the atoms of the system experience additional potentials, such as an external potential due to the walls of their enclosure, then those potentials have to be added to Eq. (4.17). For a conservative system for instance,

$$H(q, p) = K + U \tag{4.18}$$

For a constant motion, time derivative of the Hamiltonian function is equal to zero, that is,

$$\frac{\partial H}{\partial t} = -\dot{p}\frac{dq}{dt} + \dot{q}\frac{dp}{dt} = -\dot{p}\dot{q} + \dot{q}\dot{p} = 0 \tag{4.19}$$

When the initial state of the atoms and also the Hamiltonian function in the system are known, one can calculate the momentary positions and momentums of the atoms at all consecutive times by solving Eqs (4.15) and (4.16). Using this approach, the phase space trajectory of the atomic motion can be obtained, which is of particular importance in studying the dynamic evolution of bonds and atomic structure, as well as the thermodynamic states of the system. We note, however, that the Newtonian equation (4.8), followed from the Lagrangian function (4.2), can be more useful in studying specific details of the atomistic processes, especially in solids. The Newtonian equation is, on the other hand, commonly more appropriate for imposing external forces and constraints (e.g., periodic boundary conditions), as well as for postprocessing and visualization of the results. As time evolves, the Γ —point traces out a trajectory, named the phase space trajectory of the system. The evolution of the Γ —point in the phase space is governed by a pair of equations, known as Hamilton's equations of motion:

$$\dot{p}_i = \frac{\partial H(\{q_k, p_k\}, t)}{\partial p_i} \tag{4.20}$$

$$\dot{q}_i = -\frac{\partial H(\{q_k, p_k\}, t)}{\partial q_i} \tag{4.21}$$

$$H(\{p_i, q_i\}) = \sum_i K_i(p_i) + \sum_i \sum_{j>i} U_i(q_i, q_j) \tag{4.22}$$

If the atoms of the system experience additional potentials, such as an external potential due to their surrounding walls, then those potentials have to be added to (4.22). For a conventional system, where

$$H(\Gamma, t) = E \tag{4.23}$$

and a constant motion, that is,

$$\frac{\partial H}{\partial t} = 0 \tag{4.24}$$

$$H(p^N, r^N) = \sum_i \frac{p_i^2}{2m_i} + H_I(r^n) \tag{4.25}$$

where $r^N \equiv (r_1, r_2, r_3, \ldots, r_N)$ and $p^N \equiv (p_1, p_2, p_3, \ldots, p_N)$ stand for the coordinates and momenta of all the N atoms, respectively, and $H_I(r^n)$ is the total interaction Hamiltonian. From (4.25), Hamilton's equations are obtained as

$$\dot{r}_i = \frac{p_i}{2m_i} \tag{4.26}$$

$$\dot{p}_i = \frac{\partial H_I}{\partial r_i} = F_i \tag{4.27}$$

where F_i is the Newtonian force on atom i, satisfying Newton's equation of motion.

$$m_i\ddot{r}_i = F_i \tag{4.28}$$

To compute the atomic trajectories, one can solve either the set of 6N first-order differential equations (4.27) or the set of 3N second-order differential equations (4.28).

4.3   Molecular Dynamics Method

Molecular dynamics modeling of a system, composed of N isolated atoms and confined to a volume V, can be numerically studied via the classical MD simulation method, which considers time evolution. The number of atoms that can be involved in this method depends on the computational hardware power available. A nano-sized model system may be composed of several billions of interacting atoms or molecules that are simulated by MD simulation. The interactions, defined by applicable interatomic potentials, can be of either multibody or a pair-wise (i.e., two-body). These potentials characterize the physics of the nanosystems, which can be in any material phases, that is, gas, liquid, or solid. The purpose of a simulation is to achieve the properties of the nanostructure, based on time-averages of the instantaneous properties of atoms during the simulation period, as well as the changes in the state of nanostructure from one condition to another. Depending on the type of the system, given the initial coordinates and velocities of the atoms in the nanostructure, the subsequent motion of each individual atom is derived either by Newtonian dynamics or by Langevin accidental dynamics [18–21]. In a classical standard MD simulation, the modeled nanosystem is limited to a cell of volume V, with three basic assumptions:

1. The motion is described dynamically with a vector of instantaneous position and velocity describing the movement of the molecules or atoms, as a system of interacting material points.
2. The atomic interactions have strong dependence on the distances between individual atoms and their spatial orientation. This model is regularly referred to as the smooth sphere model, which is similar to the electron clouds of atoms.
3. The number of atoms in the system remains constant, and hence, there are no mass changes in the system.

4.3.1   Interatomic Potentials

Based on Eq. (4.5), a straightforward second-order ODE gives the general structure of the governing equations for MD simulations. However, when accurate representation of the atomic interactions within the simulated system is required, the potential function for Eq. (4.5) can be extremely complex. The complex nature of this interaction is due to complicated quantum effects taking place at the subatomic level. These quantum mechanics interactions are known responsible for major chemical properties, such as valence and bond energy properties, as well as the spatial arrangement (topology) of the interatomic bonds, their formation, and breakage. In order to achieve reliable results in molecular dynamic simulations, the classical interatomic potentials should accurately justify the quantum mechanical (QM) effects [14, 22].

The potential energy of an arbitrary geometry of a molecule is written as a superposition of various two-body, three-body, and four-body interactions. The potential energy is expressed as the sum of valence or bonded interactions with other nonbonded interactions, which has the following general structure [23]:

$$E = E_R + E_\theta + E_\phi + E_\omega + E_{\text{vdw}} + E_{\text{el}} \tag{4.29}$$

The valence interactions include bond stretching (E_R), angular distortions or bond angle bending (E_θ), dihedral angle torsion (E_ϕ), and inversion energy (E_ω), whereas the nonbonded interactions comprise the van der Waals (E_{vdw}) and electrostatic energy (E_{el}) terms, leading to a comprehensive form of energy as defined in Eq. (4.29).

The total force field that defines the bond stretching interaction as a harmonic oscillator is given by

$$E_R = \frac{1}{2}k_{ij}(r - r_{ij})^2 \tag{4.30}$$

or by the Morse potential function as

$$E_R = D_{ij}(e^{-\alpha(r-r_{ij})} - 1)^2 \tag{4.31}$$

where k_{ij} is the force constant in (kcal/mol)/Å^2, r_{ij} is the typical or natural bond length in angstroms (Å), and D_{ij} is the bond dissociation energy in kcal/mol. In addition,

$$\alpha = \left[\frac{k_{ij}}{2D_{ij}}\right]^{\frac{1}{2}} \tag{4.32}$$

The Morse function is a more accurate definition, since it implicitly includes harmonic terms near equilibrium (r_{ij}) and leads to a finite energy (D_{ij}) for breaking bonds. As for force field for calculations using the Morse stretch, the dissociation energy (D_{ij}) is set to $n = 70$ kcal/mol, where n is the bond order between particle centers i and j. The total force fields defining the bond stretch interaction for a harmonic oscillator are

$$E_\theta = \frac{1}{2}k_i^\theta\left(\theta_i - \theta_i^0\right)^2 \tag{4.33}$$

$$E_\phi = k_i^\theta[(1 + \cos(n_i\phi_i) - \delta_i)] \tag{4.34}$$

The natural bond length r_{ij}, on the other hand, is supposed to be the sum of the atom's specific single bond radii and a bond order correction, plus an electronegativity correction, as follows:

$$\mathbf{r}_{ij} = \mathbf{r}_i + \mathbf{r}_j + \mathbf{r}_{\text{BO}} + \mathbf{r}_{\text{EN}} \tag{4.35}$$

Given

$$U(\mathbf{r}_1, \mathbf{r}_2, \ldots, \mathbf{r}_n) = \sum_i V_i(\mathbf{r}_i) + \sum_{j>i} V_2(\mathbf{r}_i, \mathbf{r}_j) + \sum_{j>i,k>j} V_3(\mathbf{r}_i, \mathbf{r}_j, \mathbf{r}_k) + \ldots \tag{4.36}$$

where $\mathbf{r}_n$ is the radius vector of the nth particle and the function V_m is the body potential, a typical form of force fields can therefore be

$$U(\mathbf{r}_1, \mathbf{r}_2, \ldots, \mathbf{r}_n) = \sum_{\text{bonds},i} \frac{1}{2}k_i^b\left(r_i - r_i^0\right)^2$$

$$+ \sum_{\text{angles},i} \frac{1}{2}k_i^\theta\left(\theta_i - \theta_i^0\right)^2$$

$$+ \sum_{\text{torsions},i} k_i^\theta[(1 + \cos(n_i\phi_i) - \delta_i)]$$

$$+ \frac{1}{2}\sum_{\text{nonbond pairs},(i,j)}\sum\left\{4\varepsilon_{ij}\left[\left(\frac{\sigma}{r_{ij}}\right)^{12} \pm \left(\frac{\sigma}{r_{ij}}\right)^6\right] + \frac{q_iq_j}{\varepsilon r_{ij}}\right\} \tag{4.37}$$

The first term in Eq. (4.36) represents the energy due an external force field, such as electrostatic or gravitational energy in which the system is immersed; the second term shows the pair-wise or two-body interactions of the particles; and the third gives the three-body interaction component; and so on. In experimental study, the external field term is commonly ignored, while all the multibody effects are incorporated into V_2 or the two-body interaction term with the purpose of reducing computational calculations [24, 25].

4.3.1.1   Two-Body or Pair Potentials

At interatomic level, the electrostatic field generated by negatively charged electron clouds is neutralized by the surrounded positively charged atom nucleus. A probabilistic approach has been employed to estimate the probability densities at which the electrons can occupy particular spatial locations according to the QM description of electron motion. The term "electron cloud" is typically used regarding the spatial distributions of these probability densities.

The negatively charged electron clouds experience an atomic attraction, which increases as the distance between the nuclei decreases. After reaching some particular distance, however, which is referred to as the stability bond length, this attraction is equilibrated by the repulsive force due to the positively charged nuclei. Further reduction in the distance results in a fast growth of the consequential repulsive interactions. A variety of mathematical models are available to define the above physical phenomena. In 1924, Jones [26, 27] suggested the following potential function, known as two-body or pair-wise atomic interaction:

$$V(\mathbf{r}_i, \mathbf{r}_j) = V(\mathbf{r}) = 4\varepsilon \left[\left(\frac{\sigma}{r_{ij}} \right)^{12} - \left(\frac{\sigma}{r_{ij}} \right)^{6} \right] ; \; r_{ij} = |\mathbf{r}_i - \mathbf{r}_j| \qquad (4.38)$$

This model, which is normally referred to as the *Lennard–Jones* (L–J) potential, is represented in Figure 4.3. The Lennard–Jones potential is used in many simulations of great diversity of atomistic structures and procedures. Here, r_{ij} is the interatomic radius, σ is the collision diameter, that is, the distance at which $V(r) = 0$, and ε shows the bonding/dislocation minimum energy to happen for an atomic pair in balance. The first term of this potential

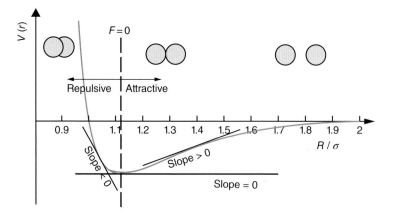

Figure 4.3   Lennard–Jones potential function

represents atomic repulsion, dominating at small separation distances, while the second term shows the attraction (bonding) potential between two atoms.

Since the square bracket magnitude is dimensionless, the choice of unit for V depends on the definition of ε. Usually, ε is in the range between 10^{-19} and 10^{-18} joules (J); thus, it is more appropriate to use a smaller energy unit, such as electron volt (eV) instead. $1\,\mathrm{eV} = 1.602 \times 10^{-19}\,\mathrm{J}$, which represents the work done by a single electron if accelerated within an electrostatic field of a unit voltage potential.

In order to eliminate one of the two coupled atoms from its equilibrium position ρ, to physical infinity, the energy ε, would be required. The collision diameter as $\rho = \sqrt[6]{2}\sigma$ is also known as the equilibrium bond length. In a usual atomistic system, the collision diameter σ is equal to several angstroms, where $1\,\text{Å} = 10^{-10}\,\mathrm{m}$. The corresponding force between the two atoms can then be expressed as a function of the interatomic distance by

$$F(r) = -\frac{\partial V(r)}{\partial r} = 24\frac{\varepsilon}{\sigma}\left[2\left(\frac{\sigma}{r}\right)^{13} - \left(\frac{\sigma}{r}\right)^{7}\right] \tag{4.39}$$

The potential and force function equations (4.38) and (4.39) versus the interatomic distance in Figure 4.4(a) are plotted using dimensionless quantities. Another common model for showing the two-body interactions, as shown in Figure 4.4(b), is the Morse potential function, introduced in the following equations:

$$V(r) = \varepsilon \left[e^{2\beta(\rho - r)} - 2e^{\beta(\rho - r)}\right] \tag{4.40}$$

$$F(r) = 2\varepsilon\beta \left[e^{2\beta(\rho - r)} - 2e^{\beta(\rho - r)}\right] \tag{4.41}$$

where β is an inverse length scaling factor, ρ and ε are the equilibrium bond length and displacement energy, respectively. In the Lennard–Jones model, the first term of this potential is repulsive and the second is attractive. In the atomic interaction modeling, the Morse potential function (4.40) has been modified for various types of materials and interfaces [25, 28]. In chemistry, physics, and engineering, the Lennard–Jones and Morse potentials are commonly used in MD simulations based on the two-body approximation.

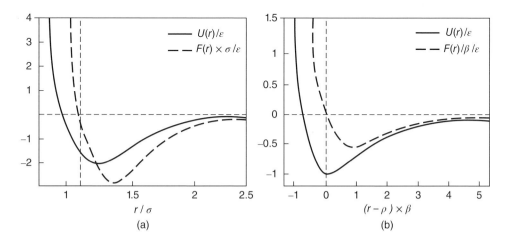

Figure 4.4 Two-body potentials and the interatomic forces: (a) Lennard–Jones, (b) Morse

4.3.1.2   Cut-Off Radius Approximation of Potential Functions

The approximation of the potential functions, such as in Eqs (4.40) and (4.41), is a significant concern in MD simulations. Note that for any N number of atoms, computing the internal forces for equations of motion in Eq. (4.8) for instance, using two-body interaction only, involves $(N^2 - N)/2$ terms; this corresponds to the case where the interaction of atom i with all other atoms $i \neq j$ is taken into account, making the computations quite expensive even for very small systems. Assuming that atom i interacts with just its nearest neighbors, the computation time can be reduced significantly. Therefore, a cut-off radius R is usually introduced and defined as the radius of maximum value of the truncated pair-wise potential, so that the equation can be written as the following [14]:

$$V^{tr}(r) = \begin{cases} V(r) \ r \leq R \\ 0 \ r > R \end{cases} \tag{4.42}$$

where each of the atoms interacts with only n atoms in its R-proximity. Thus, the evaluation of the internal two-body forces will be reduced to $nN/2$ terms only. To assure continuity (differentiability) of V, according to (4.8), a "skin" factor can be alternatively introduced for the truncated potential by means of a smooth step-like function f_c, which is referred to as the *cut-off function*,

$$V^{tr}(r) = f_c(r)V(r) \tag{4.43}$$

The cut-off function f_c assures a smooth and quick transition from 1 to 0 when the value of r approaches R, and is usually chosen to take the form of a simple analytical function at the interatomic distance r.

4.3.1.3   Multibody Interaction

Higher-order terms of potential functions (4.36) are usually used in simulations of complex molecular structures and solids, for example, for the calculation of chemical bond formation, their topology, and spatial arrangement, as well as the chemical valence of atoms are taken into account. However, practical application of the multibody interaction is particularly complex. As a result, all the multibody terms of orders higher than three are commonly ignored. Essentially, the three-body potential V_3 is intended to provide contributions to the potential energy due to the change of angle between radius vectors $r_{ij} = r_i - r_j$, in addition to the change of absolute values $|r_{ij}|$. This accounts for changes in molecular shapes and bonding geometries in atomic structures [8, 14, 29].

However, the general three-body potentials, such as V_3 in Eq. (4.36), have been criticized for the inability to describe energies of all possible bonding geometries [30], while a general form for a four- and five-body potential appears intractable, and would contain too many free parameters. As a result, a diversity of advanced two-body potentials has been proposed to efficiently explain the specifications of internal atomic environments by incorporating some particular multibody dependence inside function V_2, known as *bond-order functions*, rather than introducing the multibody potential functions $V_{m>2}$. These terms implicitly include the angular dependence of interatomic forces by introducing the so-called bond-order function, while the overall two-body formulation is preserved. Moreover, these potentials are usually treated as short-range potentials, accounting for interaction between each atom and its immediate neighbors only.

Some of the most common models of this type are the Tersoff potential [26, 27, 30, 31] for a class of covalent systems, such as carbon, silicon, and germanium, the Brenner [32–34] and reactive empirical bond order [34] potentials for carbon and hydrocarbon molecules, and the Finnis–Sinclair potential for metals [35]. Despite of the variation of existing potentials, all of them have the general specifications of the following expression:

$$V_2(\mathbf{r}_i, \mathbf{r}_j) \equiv V_{ij} = V_R(r_{ij}) - B_{ij}V_A(r_{ij}), \; r_{ij} = |\mathbf{r}_{ij}| \tag{4.44}$$

where the pair-wise attractive and repulsive interactions are considered in V_A and V_R, respectively, and the bond-order function B is intended to represent the multibody effects by accounting for the spatial arrangements of the bonds in an existing atom's neighborhood. For instance, the local environment potential for silicon by Tersoff [30] potential gives

$$V_{ij} = f_c(r_{ij})(Ae^{-\lambda_1 r_{ij}} - B_{ij}e^{-\lambda_2 r_{ij}}) \tag{4.45}$$

where

$$B_{ij} = (1 + \beta^n \zeta_{ij}^n)^{\frac{-1}{2n}} \tag{4.46}$$

$$\zeta_{ij} = \sum_{k \neq i,j} f_c(r_{ik})g(\theta_{ijk})e^{\lambda_2^3(r_{ij}-r_{ik})^3} \tag{4.47}$$

and

$$g(\theta) = 1 + \frac{c^2}{d^2} - \frac{c^2}{[d^2 + (h - \cos(\theta))^2]} \tag{4.48}$$

Here, the cut-off function is taken as

$$f_c(r) = \frac{1}{2}\begin{cases} 2 & r < R - D \\ 1 - \sin\left(\frac{\pi(r-R)}{2D}\right) & R - D < r < R + D \\ 0 & r > R + D \end{cases} \tag{4.49}$$

where the middle interval function is known as the "skin" of the potential. Note that if the local bond order is ignored, so that $B = 2A = $ constant, and $\lambda_1 = 2\lambda_2$, potential Eqs (4.45) through (4.47), reduces to the Morse models (4.40) and (4.41) at $r < R - D$. In other words, all deviations from a simple pair-wise potential are credited to the dependence of the function B on the local atomic environment. The value of this function is therefore determined by the number of competing bonds, the strength λ of the bonds, and the angles θ between them (θ_{ijk} shows the angle between bond ij and ik). The function ζ in Eq. (4.47) is actually a weighted measure of the number of bonds competing with the bond ij, and the parameter n shows how much the closer neighbors are favored over more distant ones in the competition to form bonds. The potentials proposed by Brenner and coworkers [32, 34] are usually viewed as more accurate but more complex compared to the Tersoff models [26, 27, 30, 31, 36]. The Brenner potentials include more detailed terms V_A, V_R, and B_{ij} to account for different types of chemical bonds that occur in the diamond and graphite phases of the carbon, as well in hydrocarbon molecules.

The *embedded atom method* (EAM) used in metallic systems is another outstanding approach for multibody potential [37]. The most specific point about the EAM is its physical image of metallic bonding, where each atom is embedded in a host electron gas produced by all neighboring atoms. The atom–host interaction is generally more complex than the simple two-body model. This interaction is defined in terms of an experimental embedding energy function. The function incorporates some important many-atom effects by considering the amount of energy required to enter an atom into the electron gas of a given density. The total potential energy U, therefore, comprises the embedding energies G of all the atoms in the system and the electrostatic pair-wise interaction energies V:

$$U = \sum_i G_i \left(\sum_{j \neq i} \rho_j^a \left(r_{ij} \right) \right) + \sum_{i,j>i} V_{ij}(r_{ij}) \tag{4.50}$$

Here, ρ_j^a is the averaged electron density for a host atom j, viewed as a function of the distance between this atom and the embedded atom i. As a result, the host electron density is considered as a linear superposition of contributions from all individual atoms, which are assumed to have spherical symmetry. The EAM has been employed successfully in studying defects and fracture, grain boundaries, interdiffusion in alloys, liquid metals, and other metallic systems and processes.

4.3.2   Link Between Molecular Dynamics and Quantum Mechanics

In the MD technique, the interacting particles are supposed to be either solid spheres with no internal structure, or material points exerting potential forces into their neighborhood. In other words, the interaction of the molecules and atoms does not influence the path of the simulation, and there is no energy exchange between the MD system and the subatomic particles, that is, the nuclei and electrons. However, each of the atoms of the system that is simulated using MD method is a complex sphere, which is capable of changing its internal state by substituting energy with its nearby inductor. In addition, the averaged nature of interatomic forces used in the MD simulations is resolved by its interatomic procedures and state characteristics [8, 14].

The dependence of the potential function U on the distance between molecules and atoms can be obtained fundamentally by QM calculations as well. As a result of using QM calculations within a classical MD simulation, an "energy linkage" between the atomistic and interatomic scales is formed. These main debates are employed in any multiscale method calculated to accurately communicate the MD and QM simulations. Furthermore, due to lack of evidence about the trajectories of particles in a QM model, the energy arguments are exclusively applicable for interchange of the data between the MD and QM subsystems.

To demonstrate the overall idea of MD/QM linkage, imagine a simple system consisting of two interacting hydrogen atoms. A hydrogen molecule H_2 contains of two proton nuclei (+) and two electrons (−). The positions of the electrons with respect to each other and the nuclei are defined by the lengths r_{12} and $r_{\alpha i}$, respectively, where $\alpha = a, b$ and $i = 1, 2$. The distance between electrons and nuclei of the two hydrogen atoms is depicted in Figure 4.5.

The entire energy E of this system includes the energies of two unbound hydrogen atoms, E_a and E_b, plus an atomic binding energy, termed U,

$$E = E_a + E_b + U \tag{4.51}$$

Since the classical MD models assume no energy absorption by the simulated atoms, the values E, E_a, and E_b should be of atomic states with minimum possible energies, that is, the ground states, known from QM calculations. Thus, the total energy of the coupled system

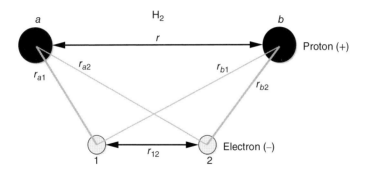

Figure 4.5 Coordination in the hydrogen molecule

is dependent on the values of r, providing $E(r)$, that is, the energy of pair-wise or two-body atomic interactions as a function of r will be

$$U(r) \equiv V_2(r) = E(r) - E_a - E_b \tag{4.52}$$

This function can be interpolated as smooth curve and be employed in the classical MD equations of motion (4.5) or (4.15) and (4.16), which is in fact the main idea of developing a linkage between the MD simulations and quantum mechanics. The energies E_a and E_b can be established by solving the motionless Schrödinger wave equation [38, 39] for each noninteracting hydrogen atom, that is, when they are properly put at the physical infinite separation distance, $r \to \infty$. This equation gives the functional eigen value problem as follows:

$$\hat{H}\psi_\alpha = E_\alpha \psi_\alpha \tag{4.53}$$

$$\hat{H} = -\frac{h^2}{2m}\Delta_\alpha - \frac{e^2}{r_\alpha}; \quad \Delta_\alpha = \frac{\partial^2}{\partial r^2}, \quad \text{and} \quad \alpha = a, b \tag{4.54}$$

Here h, m, and e are Planck's constant, the electron mass, and charge, respectively. Δ is the Laplace operator. $r_a \equiv r_{a1}, r_b \equiv r_{b2}$, and h is the one-electron Hamiltonian operator. This operator resembles the Hamiltonian function of classical dynamics, Eq. (4.25), which represents the total energy of a system in terms of coordinates and momenta of the particles. Similarly, the first term in Eq. (4.54) is the kinetic energy operator, and the second term gives the Coulomb potential of the electrostatic electron–proton interaction. Obviously, $E_a = E_b$ and $\psi_a = \psi_b$ for two identical atoms at the ground state. The mentioned notation is nevertheless preserved for generality, since similar arguments hold for a pair of distinct atoms as well.

Although the wave function solution, ψ_a, provides a complete description of a QM system in each corresponding energy state, it gives no immediate physical insight of itself. Hence, it serves as a mathematical tool only and cannot be determined experimentally. The wave function is therefore only employed for further calculations in order to obtain observable quantities. For example, the product $\psi_a^* \psi_a$, where the star notation means complex conjugate, provides a real-valued probability density function of electrons, and its integration over a spatial domain in the neighborhood of the nucleus of the unbound atom, a, gives the probability of finding the electron in this spatial domain. The one-electron wave function ψ_a is often considered as the hydrogen atomic orbital. All hydrogen orbitals and their corresponding energy levels can be found in quantum mechanics textbooks [40–42].

As mentioned earlier, the total energy E of the bound system H_2 in the ground state can be obtained for at a distance r between the nuclei, by solving the molecular two-electron Schrödinger equation,

$$\hat{H}\Psi = E\Psi \tag{4.55}$$

$$\hat{H} = -\frac{h^2}{2m}\Delta - \sum_{\alpha,i}\frac{e^2}{r_{\alpha i}} + \frac{e^2}{r_{12}} + \frac{e^2}{r} \tag{4.56}$$

where the Laplacian operator Δ involves all the electronic degrees of freedom, and $\hat{H}$ and Ψ are the molecular Hamiltonian and the two-electron wave function, describing the ground state of the coupled system. In all instances, when an N-electron wave function $\hat{\Psi}$ exists, the overall energy of the system can be calculated as an integral on the electronic variables, rather than direct solving of the Schrödinger equation (4.55). For the H_2 molecule, it gives

$$E \approx \frac{\int \cdots \int \hat{\Psi}^*\hat{H}\hat{\Psi}dr_{a1}, \ldots, dr_{b2}dr_{12}}{\int \cdots \int \hat{\Psi}^*\hat{\Psi}dr_{a1}, \ldots, dr_{b2}dr_{12}} - E_a - E_b \tag{4.57}$$

where the star notation is the complex conjugate. Equation (4.57) is obtained by premultiplying the wave equation (4.55) with its complex conjugate solution $\hat{\Psi}^*$ and integrating it over all electronic degrees of freedom. The configuration integrals, such as those in Eq. (4.57), are commonly written in quantum mechanics briefly as

$$E \approx \frac{< \hat{\Psi}(r)|\hat{H}|\hat{\Psi}(r) >}{< \hat{\Psi}(r)|\hat{\Psi}(r) >} - E_a - E_b \tag{4.58}$$

where the polyatomic multielectron wave function $\hat{\Psi}$ depends parametrically on the interatomic distance r. Eventually, the system of coupled MD/QM functions can be shown as

$$\hat{H}\Psi = \left(U + \sum_{\alpha}E_\alpha\right)\Psi, \quad m_i\ddot{r}_i = -\frac{\partial U}{\partial r_i} \tag{4.59}$$

which signifies the simultaneous coupling between the interatomic and atomistic simulations of nanostructured classifications.

4.3.3   Limitations of Molecular Dynamics Simulations

MD simulations have been employed as a powerful tool for solving complex physical phenomena. However, the length and timescales that can be investigated by this means are still limited. The common problems such as nanoscale coatings, ion-beam deposition, nanoindentation, and MEMS devices on the scale of several microns, however, involve models consisting billions of atoms that are too large to be modeled by means of current MD simulation techniques. The major task of the MD simulation is therefore predicting the time-dependent trajectories in a system of interacting particles by time-integration algorithms based on truncated Taylor's expansions. Further detailed descriptions of these simulation algorithms can be found in other references [21, 43]. In general, efficiency and accuracy are the two most important criteria guiding the development of simulation methods, which depend on both the complexity of the employed interatomic potentials and the time-integration algorithms used in the simulation.

With the state-of-the-art computational power, a typical MD simulation can contain several millions of atoms, but the limitations of MD simulation techniques such as effect of boundary conditions and time steps still exist.

4.4 Tight Binding Method

The method of linear combination of atomic orbitals (LCAO) or the tight binding method was first suggested by Bloch [89] and revised later by Slater and Koster [90] in the background of periodic potential obstacles. The main aim of this technique is to create an estimate wave function for an electron in a noncentral field of point origins (nuclei). Such a wave function is referred to as the molecular orbital (MO) and will be employed to obtain the mentioned estimate experimental role for the corresponding multielectron systems in the *Hartree–Fock* and related methods, which will be explained later. The tight binding technique is based on the quantum superposition principle. In quantum mechanics, electron configurations of atoms are described as wavefunctions, and this principle assumes that in the calculation of electronic band structure, wave functions for isolated atoms are located at each atomic site and a linear combination of them can make atomic orbitals using the readily obtainable hydrogen orbital, for the hydrogen ion H_2^+, containing two proton nuclei and one electron, as indicated in Figure 4.6.

Based on tight binding theory, the ground state of a multielectron wave function $\hat{\Psi}$ would be therefore obtainable based on either the exact hydrogen wave functions ψ_a or by an approximate of one-electron MOs ψ as will be explained. The technique enables the following estimated MO [86, 87]:

$$\hat{\psi} = c_a\psi_a + c_b\psi_b \tag{4.60}$$

The physical expression of this estimation is that in the neighborhoods of nuclei a and b, the desirable MO should closely resemble the atomic orbitals ψ_a and ψ_b, respectively. Based on the principle of quantum mechanics, the energy for an assumed (taken) estimated wave function is always bigger than that of a real or a more accurate wave function. Hence, the constants c_a and c_a for (4.60) can be established by minimizing the integral of (4.57). Owing to (4.57) and (4.60), the estimated ground-state energy $\hat{E} \geq E$ gives

$$\hat{E} = \frac{<\hat{\psi}|\hat{H}|\hat{\psi}>}{<\hat{\psi}|\hat{\psi}>} = \frac{c_a^2 H_{aa} + c_b^2 H_{bb} + 2c_a c_b H_{ab}}{c_a^2 + c_b^2 + 2c_a c_b S_{ab}} \tag{4.61}$$

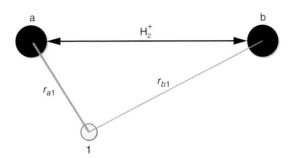

Figure 4.6 Coordination in the hydrogen ion

where

$$S_{ab} = < \psi_a | \psi_b >, \; H_{ab} = < \psi_a | \hat{H} | \psi_b > \tag{4.62}$$

$$\hat{H} = -\frac{h^2}{2m}\Delta - \frac{e^2}{r_{a1}} - \frac{e^2}{r_{b1}} + \frac{e^2}{r} \tag{4.63}$$

Here, $\hat{H}$ is the Hamiltonian operator for a single electron in the field of two protons and r is the separation distance between the two protons. In deriving (4.61), it was also supposed that the atomic orbitals are in the normalization condition, making

$$S_{aa} = < \psi_a | \psi_b >= 1 \tag{4.64}$$

as well as the symmetry conditions, that is, $S_{ab} = S_{ba}$ and $H_{ab} = H_{ba}$ holding for the hydrogen ion. At the variational minimum, we have the conditions $(\partial \hat{E}/\partial c_a) = 0$ and $(\partial \hat{E}/\partial c_b) = 0$. Using these conditions together with the symmetry $H_{aa} = H_{bb}$,

$$c_a^2 - c_b^2 = 0 \tag{4.65}$$

which is only possible when

$$c_a = c_b \; \text{or} \; c_a = -c_b \tag{4.66}$$

Thus, two MOs exist for the H_2^+ ion, one symmetric and one antisymmetric

$$\hat{\psi}^+ = N^+(\psi_a + \psi_b), \; \hat{\psi}^- = N^-(\psi_a + \psi_b) \tag{4.67}$$

where $N^\pm$ are the normalization factors, which can be found in a condition similar to (4.64). Agreeing to (4.66) and (4.60), the molecular states are described by energies as follows:

$$\hat{E}^+ = \frac{H_{aa} + H_{ab}}{1 + S_{ab}} \tag{4.68}$$

$$\hat{E}^- = \frac{H_{aa} - H_{ab}}{1 - S_{ab}} \tag{4.69}$$

4.5 Hartree–Fock and Related Methods

In this section, an outline of a group of numerical methods enabling one to build the ground-state multielectron wave function $\hat{\Psi}$ is provided. The ground-state multielectron wave function is achievable based on either the exact hydrogen wave functions ψ_a or the approximate one-electron MOs ψ. The hydrogen wave functions are employed in multielectron systems in a central field, whereas the MOs are suitable for polyatomic multielectron systems, particularly, H_2. The unified notation ψ_i will be used for both types of input wave functions, where the subscript index i denotes the number of the electron in the system.

In one of the common approximations, electrons are estimated to be independent, leading to elimination of the term (e^2/r^{12}) in the Hamiltonian equation (4.56). Based on this approximation, the two-electron wave function is obtained as the second-order *Slater* determinant,

$$\hat{\Psi} = \frac{1}{\sqrt{2}} \begin{Vmatrix} \psi_1(1) & \psi_2(1) \\ \psi_1(2) & \psi_2(2) \end{Vmatrix} \tag{4.70}$$

where $\psi_i(j)$ is the wave function of electron i written in terms of the variables of electron j. Thus, for a multielectron system, where N is the total number of electrons in the system, N-order determinant would be used as

$$\hat{\Psi} = \frac{1}{\sqrt{2}} \begin{Vmatrix} \psi_1(1) & \cdots & \psi_N(1) \\ \vdots & \ddots & \vdots \\ \psi_1(N) & \cdots & \psi_N(N) \end{Vmatrix} \qquad (4.71)$$

Equation (4.71) satisfies the Pauli rule, stating that two electrons cannot be simultaneously in the same one-electron state. Indeed, if for instance $\psi_i(1) = \psi_j(2)$ at $i \neq j$, then the Slater determinant demonstrates that such a state does not exist. As a result, if the tight binding MOs are used in Eq. (4.71), extra terms will be required for describing nonground states of the hydrogen atom. For that, the electron–electron interaction can be taken into account with the use of the *Hartree* method [91]. Similar to the tight binding technique, it is assumed that the ground-state energy for a chosen approximated wave function is always greater than the accurate wave function. The trial (zero-order) function can therefore be adopted from the independent electron eigen function:

$$\hat{\Psi} = \psi_1^0(1)\psi_2^0(2) \qquad (4.72)$$

where ψ_i^0 is the one-electron hydrogen eigen function, found by solving the tight binding MOs according to Eq. (4.67). The one-electron orbitals obtained from (4.72) are more accurate to be used in order to minimize the integral of Eq. (4.57). This minimization will subsequently reduce to the iteration procedure, where the updated orbitals are found by solving the system of one-electron Schrödinger equations:

$$\left[-\frac{h^2}{2m}\Delta - \sum_\alpha \frac{e^2}{r_{\alpha i}} + \frac{e^2}{r} + \chi_j^0 - E_i \right] \psi_i^1 = 0, \; i = 1, 2, \ldots, N \qquad (4.73)$$

with the energy integrals χ_j involving the trial orbitals of (4.73),

$$\chi_i^0 = \sum_{j \neq i} < \psi_j^0 \left| \frac{e^2}{r_{ij}} \right| \psi_j^0 >, \; i = 1, 2, \ldots, N \qquad (4.74)$$

where r_{ij} are the interelectron distances. For a two-electron system, therefore

$$\chi_1^0 = < \psi_2^0 \left| \frac{e^2}{r_{12}} \right| \psi_2^0 >, \; \chi_1^0 = < \psi_2^0 \left| \frac{e^2}{r_{12}} \right| \psi_2^0 > \qquad (4.75)$$

In the iteration process, after finding the functions ψ_i^1, they are substituted into (4.74) and (4.75), instead of ψ_i^0 to find the new value χ_i^1. The former then is substituted back to (4.73) to obtain ψ_i^2, and so on. If this procedure converges, then the system (4.73) will yield $\psi_i^k \cong \psi_i^{k-1}$, for some k. These functions will represent the desired one-electron orbitals for (4.72). The corresponding value of χ_i^1 will then be known as the *Hartree* self-consistent field. For polyatomic molecules, Eq. (4.73) is solved with the tight binding method at each iteration step; this updates the values of coefficients in the linear representations similar to (4.60). According to (4.73) and (4.75), the physical interpretation of this method is in the assumption that the electron–electron interaction occurs through the averaged field χ_i^k, and for a given electron, this field is created by the rest of electrons in the system.

The *Hartree* method does not comply with the Pauli principle. Therefore, in order to compensate the shortage, by using the trial function forms (4.70) and (4.71) instead of (4.72), a set of equations for the one-electron wave functions will be achieved to constitute a multielectron wave function in a way that it satisfies the Pauli principle as well. This function accounts for some additional characteristics of the quantum system, such as the spin variables of the electrons. Such an approach of solving the multielectron problem is known as the *Hartree–Fock* method [92]. The resultant one-electron equations, as compared to (4.73), include additional terms of the type $\chi_{ij}\psi_i^1$ where χ_{ij} is the exchange integral,

$$\chi_{ij} = \langle \psi_j \left| \frac{e^2}{r_{ij}} \right| \psi_i \rangle \tag{4.76}$$

The successive iteration procedure, similar to the one explained earlier with Eqs (4.73) and (4.75), yields the modified MOs for (4.70) and (4.71), known as the *Hartree–Fock* orbitals. The corresponding values E_i in Eq. (4.73) are called *Hartree–Fock* orbital energies. The *Hartree–Fock* approximation has been used successfully for solving a variety of problems in quantum physics and chemistry [44]. A more detailed description of this method can be found in [93].

4.6 Density Functional Theory

Although the *Hartree–Fock* calculations provide reliable results, they are computationally expensive and are incapable of being employed for large systems. Density functional methods were therefore introduced to provide alternative routes capable of yielding comparable results at lower expenses, hence allowing the simulation of a larger set of molecules and atoms. Within the *Hartree–Fock method*, the consideration begins with an exact Hamiltonian, similar to (4.56), but with an approximate trial wave function. This wave function is written by combining the readily available one-electron orbitals according to Eqs (4.70)–(4.72), and it is further improved by optimizing the one-electron solutions. In contrast, the density functional model starts with a Hamiltonian, relating to an "idealized" many-electron system, for which the exact wave function is immediately available. This solution is updated at each iteration step by optimizing the ideal system closer and closer to the real system [45, 46]. The density functional theory was pioneered by Hohenberg and Kohn [94], who showed that the ground-state energy of multielectron system is a unique functional of the electronic density,

$$E = T[\rho(\mathbf{r})] + V_C[\rho(\mathbf{r})] + E_{XC}(\rho(\mathbf{r})) \tag{4.77}$$

where T and V_C are known functionals corresponding to the kinetic energy of the electrons and the potential energy of electron–nucleus (or electron–ion) and nucleus–nucleus (or ion–ion) Coulomb interactions, respectively. For large multielectron atoms, only the outer layer (valence) electrons are usually considered to contribute to the electronic density of interest, while core electrons and the nucleus are treated together as an ion. In this case, the corresponding electron–ion and ion–ion interaction energies are therefore employed for deriving the V_C term.

An advantage of the density functional theory, according to Eq. (4.77), is that, in contrast with Eq. (4.59), the ground-state energy is obtained without involving a multielectron wave function. However, a proper form of the exchange–correlation functional E_{XC} is employed here, which varies in different modifications of this method. In the simplest approach, called local density functional theory [47], the exchange–correlation energy is determined as an integral of some function of the total electron density as

$$E_{XC} = \int \rho(r)\varepsilon_{XC}[\rho(r)]dr \qquad (4.78)$$

where ε_{XC} is the exchange–correlation energy per electron in a homogeneous electron gas of constant density. For a system of N-electrons, the electron density function is expressed through the modulus of the one electron Kohn–Sham orbital ψ_i,

$$\rho(\boldsymbol{r}) = \sum_{i=1}^{N} |\psi_i(\boldsymbol{r})|^2 = \sum_{i=1}^{N} \psi_i^* \psi_i(\boldsymbol{r}) \qquad (4.79)$$

where the initial value of it is obtained from a set of exact basic functions, that is, the plane wave solutions of the Schrödinger equation for a free electron. In this method, the participation coefficients are optimized in a similar way to that of the Hartree–Fock method, as described later. The updated orbitals are found by solving the Kohn–Sham equation,

$$\left[-\frac{h^2}{2m}\Delta - \sum_\alpha \frac{e^2}{r_{\alpha i}} + \frac{e^2}{r} + \int \frac{\rho(r)}{|r-r'|}dr' + V_{XC} - E_i \right] \psi_i^1(r) = 0, \ i = 1, 2, \dots, N \quad (4.80)$$

Here, the terms E_i are the *Kohn–Sham orbital* energies and V_{XC} is the derivative of the exchange–correlation functional of Eq. (4.77). With respect to the electron density,

$$V_{XC}[\rho] = \frac{\delta E_{XC}[\rho]}{\delta \rho} \qquad (4.81)$$

The improved set of Kohn–Sham orbitals, ψ_i^1, is then used to compute a more accurate density function, according to (4.78). These iterations repeat until the exchange–correlation energy and the density converge.

The local density functional theory provides a very rough approximation for the molecular system, because it assumes a uniform total electron density throughout the molecular system. Nevertheless, nonlocal density functional approaches have also been developed to account for variations in the total density, for example, [48, 49]. This is accomplished by introducing a dependence of ground-state energy on the gradient of the electron density, as well as the density itself.

The original density functional procedure, explained in Eqs (4.78) through (4.81), in general involves N^3 order of computation, for N to be the number of electrons, as compared with N^4 for the *Hartree–Fock* and tight binding method. Hence, significant improvements have been made in the later technique. However, using the Car–Parrinello MD method [50] and the conjugate gradient method [44], even better progresses have been obtained in terms of computation order. The Car–Parrinello method reduces the computation order to N^2, while the conjugate-gradient method can, as shown in [50], be even more efficient. Standard computer packages are currently available to accomplish *ab initio* calculations based on the density functional theory [51].

4.7   Multiscale Simulation Methods

The limitations of atomistic simulations and continuum mechanics, along with practical needs that arise from the heterogeneous nature of engineering materials, have motivated research on multiscale simulations that bridge atomistic simulations and continuum modeling techniques together toward achieving more accurate and less expensive results [52–55]. In order to make the computations tractable, multiscale models generally make use of coarse–fine decomposition. An atomistic simulation method, such as MD, is, on the other hand, used in a small subregion of the domain in which capturing the individual atomistic dynamics accurately is crucial, whereas in all other regions of the domain continuum simulation is used, in which the deformation is considered to be homogeneous and smooth.

Since the continuum region is usually chosen to be much larger than the atomistic region, the overall domain of interest can be considerably large in this part. A purely atomistic solution is normally not affordable on any domain, though the multiscale solution would be presumably capable of providing detailed atomistic information only when and where it is necessary. The key issue is then the coupling between the coarse and fine scales. Depending on the method of information exchange between the coarse and fine regions, multiscale methods can be classified into three groups, namely, hierarchical, concurrent, and multiscale boundary conditions.

4.8   Conclusion

Nanoscale components are generally used in conjunction with other components that are larger in size, and therefore have a mechanical response that is on a much larger length and timescale than that of the nanoscale component. Therefore, in such hybrid systems, typical single-scale simulation methods such as MD or quantum mechanics may not be applicable due to the disparity in length and timescales of the structure. Hence, for such systems, the computer-aided engineering tools must be able to span length scales from nanometers to microns, and timescales from femtoseconds to microseconds. Therefore, these systems cannot be modeled by continuum methods alone, because they are too small, or by molecular methods alone, because they are too large. To support the design and qualification of nanostructured materials, a range of simulation tools must be available to designers just as they are today available at the macroscopic scales in commercial software packages. However, considerable research is still required to establish the foundations for such software and to develop computational capabilities that span the scales from the atomistic to continuum domains. These capabilities should include a variety of tools, from finite elements to MD and QM methods, in order to provide powerful multiscale methodologies.

References

[1] Rafii-Tabar, H. (2008) *Computational physics of carbon nanotubes*, Cambridge University Press, Cambridge.

[2] Shokrieh, M. and Rafiee, R. (2010) A review of the mechanical properties of isolated carbon nanotubes and carbon nanotube composites. *Mechanics of Composite Materials*, **46** (2), 155–172.

[3] Arroyo, M. and Belytschko, T. (2005) Continuum mechanics modeling and simulation of carbon nanotubes. *Meccanica*, **40** (4–6), 455–469.

[4] Lu, Q. and B. Bhattacharya. (2005) The role of atomistic simulations in probing the small–scale aspects of fracture—a case study on a single–walled carbon nanotube. *Engineering Fracture Mechanics*, **72** (13), 2037–2071.

[5] Omata, Y., et al. (2005) Nanotube nanoscience: a molecular–dynamics study. *Physica E: Low–dimensional Systems and Nanostructures*, **29** (3), 454–468.

[6] Dvorkin, E.N. and Bathe, K.–J. (1984) A continuum mechanics based four–node shell element for general non-linear analysis. *Engineering Computations*, **1** (1), 77–88.

[7] Zienkiewicz, O.C. and Morice, P. (1971) *The Finite Element Method in Engineering Science*, Vol. 1977, McGraw–hill, London.

[8] Liu, W.K., S. Jun, and D. Qian. (2008) Computational nanomechanics of materials. *Journal of Computational and Theoretical Nanoscience*, **5** (5), 970–996.

[9] Burkert, U. and Allinger, N.L. (1982) *Molecular Mechanics*, Vol. 177, American Chemical Society, Washington, DC.

[10] Kent, P. (1999) *Trajectories for Learning*, Institute of Education, University of London.

[11] Landau, L.D. and E.M. Lifshitz. (1978) Mecánica. Vol. 1.: Reverté.

[12] Kittel, C. and McEuen, P. (1976) *Introduction to Solid State Physics*. Vol. 8, Wiley, New York.

[13] Goldstein, H., Poole, C.P. Jr., and Safko, J.L. Sr., (2012) *Klassische Mechanik*, John Wiley & Sons.

[14] Liu, W.K., et al. (2004) An introduction to computational nanomechanics and materials. *Computer Methods in Applied Mechanics and Engineering*, **193** (17), 1529–1578.

[15] Li, S. and Liu, W.K. (2004) *Meshfree Particle Methods*, Springer, Heidelberg.

[16] Milgrom, M. (1983) A modification of the Newtonian dynamics as a possible alternative to the hidden mass hypothesis. *The Astrophysical Journal*, **270**, 365–370.

[17] Chandler, D. and P.G. Wolynes. (1981) Exploiting the isomorphism between quantum theory and classical statistical mechanics of polyatomic fluids. *The Journal of Chemical Physics*, **74** (7), 4078–4095.

[18] Haile, J.M. (1992) *Molecular Dynamics Simulation: Elementary Methods*, John Wiley & Sons, Inc.

[19] Allen, M.P. and Tildesley, D.J. (1987) *Computer Simulation of Liquids*, Oxford Science Publications.

[20] Humphrey, W., A. Dalke, and K. Schulten. (1996) VMD: visual molecular dynamics. *Journal of Molecular Graphics*, **14** (1), 33–38.

[21] Hoover, W.G. (1986) *Molecular Dynamics*, Springer-Verlag.

[22] Phillips, J.C. *et al.* (2005) Scalable molecular dynamics with NAMD. *Journal of Computational Chemistry*, **26** (16), 1781–1802.

[23] Mayo, S.L., B.D. Olafson, and W.A. Goddard. (1990) DREIDING: a generic force field for molecular simulations. *Journal of Physical Chemistry*, **94** (26), 8897–8909.

[24] Rappé, A.K., et al. (1992) UFF, a full periodic table force field for molecular mechanics and molecular dynamics simulations. *Journal of the American Chemical Society*, **114** (25), 10024–10035.

[25] Wang, Y., Tomanek, D. and Bertsch, G. (1991) Stiffness of a solid composed of C 60 clusters. *Physical Review B*, **44** (12), 6562.

[26] Tersoff, J. (1989) Modeling solid–state chemistry: Interatomic potentials for multicomponent systems. *Physical Review B*, **39** (8), 5566.

[27] Tersoff, J. (1988) Empirical interatomic potential for silicon with improved elastic properties. *Physical Review B*, **38** (14), 9902.

[28] Mahaffy, R., Bhatia, R. and Garrison, B.J. (1997) Diffusion of a butanethiolate molecule on a Au {111} surface. *The Journal of Physical Chemistry B*, **101** (5), 771–773.

[29] Biswas, R. and D. Hamann. (1985)Interatomic potentials for silicon structural energies. *Physical Review Letters*, **55** (19), 2001.

[30] Tersoff, J. (1988) New empirical approach for the structure and energy of covalent systems. *Physical Review B*, **37** (12), 6991.

[31] Tersoff, J. (1986) New empirical model for the structural properties of silicon. *Physical Review Letters*, **56** (6), 632.

[32] Brenner, D.W. (1990) Empirical potential for hydrocarbons for use in simulating the chemical vapor deposition of diamond films. *Physical Review B*, **42** (15), 9458.

[33] Brenner, D.W. (1992) Erratum: Empirical potential for hydrocarbons for use in simulating the chemical vapor deposition of diamond films. *Physical Review B*, **46** (3), 1948.

[34] Brenner, D.W., et al. (2002), A second–generation reactive empirical bond order (REBO) potential energy expression for hydrocarbons. *Journal of Physics: Condensed Matter*, **14** (4), 783.

[35] Finnis, M. and J. Sinclair. (1984) A simple empirical N–body potential for transition metals. *Philosophical Magazine A*, **50** (1), 45–55.

[36] Tersoff, J. (1992) Energies of fullerenes. *Physical Review B*, **46** (23), 15546.

[37] Daw, M.S., Foiles, S.M. and Baskes, M.I. (1993) The embedded–atom method: a review of theory and applications. *Materials Science Reports*, **9** (7), 251–310.

[38] Park, D. (2012) *Introduction to the Quantum Theory*, Courier Dover Publications.

[39] Rodberg, L.S. and Thaler, R.M. (1967) *Introduction to the Quantum Theory of Scattering*, Academic Press.

[40] Dirac, P.A.M. (1930) *The Principles of Quantum Mechanics*, Oxford University Press.

[41] Feynman, R.P. and Hibbs, A.R. (1965) *Quantum Mechanics and Path Integrals*, McGraw–Hill.

[42] Bethe, H.A. and Salpeter, E.E. (1977) *Quantum mechanics of one–and two–electron atoms*. Vol. 168, Plenum Publishing Corporation, New York.

[43] Rapaport, D.C. (2004) *The art of molecular dynamics simulation*, Cambridge university press.

[44] Payne, M.C., et al. (1992) Iterative minimization techniques for ab initio total–energy calculations: molecular dynamics and conjugate gradients. *Reviews of Modern Physics*, **64** (4), 1045.

[45] Dreizler, R.M. and Engel, E. (2011) *Density functional theory*, Springer.

[46] Parr, R.G. (1983) Density functional theory. *Annual Review of Physical Chemistry*, **34** (1), 631–656.

[47] Kohn, W. and L.J. Sham. (1965) Self–consistent equations including exchange and correlation effects. *Physical Review*, **140** (4A), A1133.

[48] Perdew, J., E. McMullen, and A. Zunger. (1981) Density–functional theory of the correlation energy in atoms and ions: a simple analytic model and a challenge. *Physical Review A*, **23** (6), 2785.

[49] Slater, J., T.M. Wilson, and J. Wood. (1969) Comparison of several exchange potentials for electrons in the Cu+ ion. *Physical Review*, **179** (1), 28.

[50] Car, R. and Parrinello, M. (1985) Unified approach for molecular dynamics and density–functional theory. *Physical Review Letters*, **55** (22), 2471.

[51] Kresse, G. and Hafner, J., (1993) Ab initio molecular dynamics for liquid metals. *Physical Review B*, **47**, 558.

[52] Abraham, F.F. *et al.* (1998) Spanning the continuum to quantum length scales in a dynamic simulation of brittle fracture. *EPL (Europhysics Letters)*, **44** (6), 783.

[53] Rudd, R.E. and J.Q. Broughton. (2000) Concurrent coupling of length scales in solid state systems. *Physica Status Solidi B*, **217** (1), 251–291.

[54] Tadmor, E.B., M. Ortiz, and R. Phillips. (1996) Quasicontinuum analysis of defects in solids. *Philosophical Magazine A*, **73** (6), 1529–1563.

[55] Miller, R., et al. (1998) Quasicontinuum simulation of fracture at the atomic scale. *Modelling and Simulation in Materials Science and Engineering*, **6** (5), 607.

5

Probabilistic Strength Theory of Carbon Nanotubes and Fibers

Xi F. Xu[1] and Irene J. Beyerlein[2]

[1]*School of Civil Engineering, Beijing Jiaotong University, Beijing, China*
[2]*Los Alamos National Laboratory, Los Alamos, NM, USA*

5.1 Introduction

Carbon nanotubes (CNTs) are considered the most promising structural elements for 3D assembling of the next-generation superstrong materials and structures. Reliability and safety of such CNT-based materials critically rely on the quality of the CNTs or, more specifically, on the statistics of the physical properties of the CNTs. Probabilistic analysis of CNT properties not only provides insights into and guidance for design and optimization of CNT-based materials, but also serves as a theoretical foundation for the development of future standards and specifications on industrialized CNTs. This chapter is written with an intention to cover our recent research on understanding and interpretation of the statistical strength of single CNTs and CNT fibers [1–3]. By distinguishing probabilistic modeling from statistical analysis, we elucidate a unique role that a physics-based probabilistic theory plays in the extrapolation of available test data to distribution tails [4]. Recognizing CNTs as the most ideal candidate among all known materials for application of the weakest link model, we present a probabilistic strength theory attempting to explain why testing data of CNT strength seemingly fit well into the classical Weibull distribution, why in many other cases not so, and thereby propose a generalized Weibull distribution for CNTs and all other brittle materials.

In the following sections, Section 5.2 is devoted to a generalized Weibull theory to explain the statistical features of CNT strength. In Section 5.3, a multiscale method is described to show how to upscale strength distribution from nanoscale CNTs to microscale CNT fibers, using a specifically developed local load sharing (LLS) model. Conclusions are made at the end.

Advanced Computational Nanomechanics, First Edition. Edited by Nuno Silvestre.
© 2016 John Wiley & Sons, Ltd. Published 2016 by John Wiley & Sons, Ltd.

5.2 A Probabilistic Strength Theory of CNTs

The nanoscale implied in the term "nanotubes" refers to the CNT diameter typically ranging from 0.8–2 nm for single-walled CNTs (SWCNTs) to 5–20 nm for multiwalled CNTs (MWCNTs). In the other dimension, CNT length reaches the scale of centimeters, which results in an extreme aspect ratio as large as 10^8. The two disparate length scales form a rare, if not unique, geometric feature of CNTs that span from the nanoscale to the macroscale. This extreme geometric feature presents CNTs as an ideal candidate for application of the weakest link model and examination of the classical Weibull distribution.

5.2.1 Asymptotic Strength Distribution of CNTs

Extreme value theory provides asymptotic results to characterize extreme order statistics [5, 6]. Among the three types of extreme value distribution given by the theory, the Weibull distribution is the only one available to characterize the minima of a sequence of nonnegative random variables. The corresponding extreme statistics theorem is given as follows.

Let X_1, X_2, ..., X_n be a sequence of independent and identically distributed (i.d.d.) nonnegative random variables with common distribution $P(X \leq x) = F_0(x)$ and $M_n = \min\{X_1, X_2, \ldots, X_n\}$. If a sequence of real positive numbers a_n exists such that

$$\lim_{n \to \infty} P(M_n a_n \leq x) = F(x) \tag{5.1}$$

or

$$\lim_{n \to \infty} \left(1 - F_0 \left(\frac{x}{a_n}\right)\right)^n = 1 - F(x) \tag{5.2}$$

where F is a nondegenerate distribution function, then the limit function F belongs to the Weibull family. The condition of convergence is to have the lower tail of the parent distribution $F_0(x)$ behave like a power law when x approaches 0; that is,

$$\lim_{x \to 0} \frac{F_0(x)}{x^m} = A > 0, \quad \text{for some } m > 0 \tag{5.3}$$

The limit distribution F for the random variable $M_n a_n$ with $n \to \infty$ is expressed as

$$F(x) = 1 - \exp(-Ax^m) \tag{5.4}$$

which, by letting $A = x_0^{-m}$, becomes exactly the Weibull distribution

$$F(x) = 1 - \exp\left(-\left(\frac{x}{x_0}\right)^m\right) \tag{5.5}$$

where m and x_0 are the shape parameter and scale parameter, respectively. One possible sequence of a_n for the above-mentioned theorem is $a_n = n^{1/m}$. Examples of the parent distribution $F_0(x)$ satisfying the convergence condition (5.3) include lognormal, Gamma, Beta, and Weibull distributions. In other words, the probability distribution for the minimum M_n of a sequence of i.d.d. random variables X_1, X_2, ..., X_n with $n \to \infty$ is ultimately determined by the lower tail of the parent distribution that behaves like a power law, given that a limit distribution does exist that turns out to be of Weibull.

To interpret the above asymptotic result in the case of CNT strength, the following two classical physical models are employed:

1. The first is the weakest link model to describe the failure of brittle materials. Following the convention in fracture mechanics, a flaw size c is defined as the half-length of the "crack" along the cross section of a CNT. Such a flaw size clearly ranges between a, the lattice spacing as the lower bound, and $c_{sup} = \pi d/2$ half the CNT perimeter as the upper limit, with d being the CNT diameter. In the weakest link model, the strength of each weakest link element is independent of one another. A CNT is typically characterized with an extremely low density of point defects with respect to the total number of lattice sites. For instance, the expected equilibrium concentration of point defects in a pristine CNT is reported to be on the order of 10^{-6} or even lower [7]. Given this fact, by neglecting flaw interactions, it is reasonable to approximate strength of a weakest link element as the one fully determined by the largest flaw present within the element. The largest flaw is specifically called the critical flaw. By choosing l, the length of weakest link elements, sufficiently large, the strengths of any two spatially adjacent elements, can safely be ensured independent of one another. A schematic of a CNT structure and weakest link elements is shown in Figure 5.1. With the above justification, it is worth noting that the asymptotic extreme value theory equally applies to dependent weakest link elements, which for simplicity is not considered in this study.

2. The second model is the Griffith failure criterion of fracture mechanics, which asserts that the strength of a specimen has an inverse relation to the square root of the critical crack size c, that is, $\sigma \propto 1/\sqrt{c}$. Using a dimensionless factor ψ to account for the curvature and finite size effect of tubes, the Griffith criterion is extended to a CNT characterized with the critical flaw size c as follows:

$$\sigma = \frac{T}{\psi \sqrt{c}} \tag{5.6}$$

where T serves as the toughness of the CNT. When the critical flaw size c approaches the upper limit c_{sup}, the effect of the factor ψ dominates. By using the stress formula for a finite plate containing a crack [8], the relation between strength and c in this extreme case can be approximated as

$$\sigma \propto \sqrt{\frac{c_{sup} - c}{c_{sup} c}}, \quad \text{when } c \to c_{sup} \tag{5.7}$$

which can be further expanded in Taylor series as

$$\sigma \propto \frac{\sqrt{c_{sup} - c}}{c_{sup}} + O\left(\left(\sqrt{\frac{c_{sup} - c}{c_{sup}}}\right)^3\right), \quad \text{when } c \to c_{sup} \tag{5.8}$$

Denote $F_l(\sigma)$ as the strength distribution of weakest link elements. Given that the limit distribution exists, according to the convergence condition (5.3) the lower tail of $F_l(\sigma)$ can be approximated as a power law, that is,

$$F_l(\sigma) \approx \left(\frac{\sigma}{\sigma_0}\right)^m, \quad \sigma \to 0^+ \tag{5.9}$$

From (5.1), it follows

$$\lim_{n \to \infty} P(M_n \leq x) = F(a_n x) \tag{5.10}$$

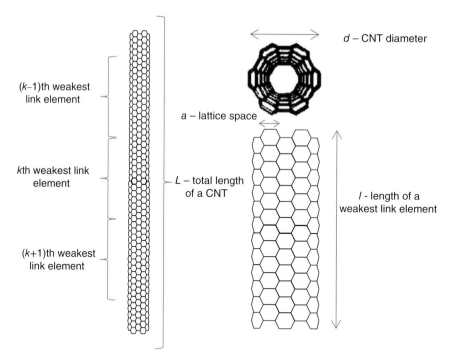

Figure 5.1 Schematic of hexagonal lattice structure and the set-up of weakest link elements in a single-walled CNT

With the sequence $a_n = n^{1/m}$, the Weibull strength distribution $W_L(\sigma)$ is asymptotically given as

$$W_L(\sigma) = F(a_n \sigma)$$

$$= 1 - \exp\left(-n\left(\frac{\sigma}{\sigma_0}\right)^m\right) \tag{5.11}$$

for a long CNT with its length $L = nl$ and the number n being sufficiently large.

Denote $G_l(c) = P(C \leq c)$ as the flaw size distribution in a weakest link element, which certainly depends on the element length l. By using Eq. (5.7), the power law approximation (5.9) for the lower tail of strength distribution can be translated into a power law approximation for the upper tail of $G_l(c)$ as

$$1 - G_l(c) \propto (c_{\text{sup}} - c)^{m/2}, \quad c \to c_{\text{sup}} \tag{5.12}$$

As a closing remark of this section, while the asymptotic result (5.11) is mathematically rigorous, we note it has little value of practical use in that with a sufficiently large n the extreme lower tail of the strength distribution is normally out of engineering interest. Our interests lie in a strength distribution for any finite length of CNTs not just at the lower tail but also at other parts of the distribution curve. Another motive is to address recent debates on the type of distribution suitable for the strength of CNTs [9–14]. Next, we present the formulation of a nonasymptotic strength distribution for CNTs, and thereby explain why the classical Weibull distribution fits testing data of many brittle materials and why in many other experiments it does not work.

5.2.2   Nonasymptotic Strength Distribution of CNTs

To formulate the nonasymptotic strength distribution of CNTs, we need to first take a look at the flaw size distribution $G_l(c)$ distributed between lattice spacing a and the upper limit c_{sup}. Pristine CNTs normally contain point defects occurring in fabrication processing. In these imperfect CNTs and also those even initially free of defects, new point defects continuously accumulate as a function of time and temperature due to thermodynamic effects [15, 16]. These point defects grow into flaws of various sizes subject to physical and chemical forces, such as radiation, cycling of thermal or mechanical loads, oxidation pitting, and so on. The flaw size distribution of CNTs depends on many factors, for example, fabrication method, time exposed to environmental effects, and physical and chemical forces. The information about $G_l(c)$ is hardly quantifiable via experimental measurement. However, we can gain certain valuable insights into the ultimate strength of CNTs by focusing on some qualitative features of the distribution curve $G_l(c)$.

An important point presented in the preceding section is the asymptotic strength distribution of CNTs, which is fully determined by the power law behavior of the tail shape near the upper limit of flaw size. Following this observation, we extend the power law approximation of the upper tail to the full range of the flaw size distribution. In the vicinity of any flaw size $c \in [a, c_{\text{sup}})$, the value $1 - G_l(c)$ can be approximated as a power law

$$1 - G_l(c) \approx B(c)c^{-\alpha(c)} \tag{5.13}$$

where B and α are generally functions of c. The log–log plot of $1 - G_l(c)$ versus flaw size c is schematically shown in Figure 5.2. The slope of the curve is expected to be minimal near the smallest defect a, and to gradually rise with an increase in flaw size until it reaches the maximum near the upper limit c_{sup}. The reason for the curve to have such a trend will be explained shortly in the following section from a physical point of view. To translate the flaw size distribution (5.13) into a strength distribution, we need a certain strength criterion applicable to the discrete nature of point defects similar to the Griffith criterion (5.6) of fracture mechanics. The idea is to generalize the Griffith criterion (5.6) into the following form:

$$\sigma = \frac{T}{\psi c^{1/\beta}} \tag{5.14}$$

where ψ remains a dimensionless factor to account for curvature and finite size effect, and T the toughness but its dimension varies with the value of β. For a large critical flaw size near the upper limit, the flaw can be approximated as a crack in fracture mechanics where the value of β is 2. When the critical flaw decreases from a large size toward the lower bound a, the factor ψ gradually approaches the constant 1, while the exponent β decreases and at the lower bound a reaches the minimum. The value β varies only slightly however. In one atomistic simulation [17], the minimum is found to be approximately 1.8–1.9, close to the maximum

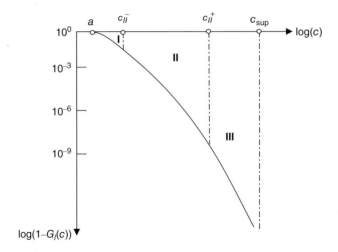

Figure 5.2 Schematic of a log–log plot of the flaw size distribution of weakest link elements. *Source*: adapted from [2, 3]

value 2. By using (5.13) and (5.14), the strength distribution $F_l(\sigma)$ in the vicinity of strength σ is approximated as

$$F_l(\sigma) \approx \left(\frac{\sigma}{\sigma_0(\sigma)}\right)^{m(\sigma)} \tag{5.15}$$

$$m = \alpha\beta, \quad \sigma_0 = \frac{T}{\psi B^{1/\alpha\beta}} \tag{5.16}$$

where m and σ_0 are generally functions of σ. The log–log plot of the strength distribution $F_l(\sigma)$ versus σ is similar to Figure 5.2, which is not drawn here.

Given the power law approximation (5.15) for the strength distribution of weakest link elements, the strength distribution for a CNT with the length $L = nl$ is exactly derived as

$$F_L(\sigma) = 1 - (1 - F_l(\sigma))^n$$

$$= 1 - \left[1 - \left(\frac{\sigma}{\sigma_0(\sigma)}\right)^{m(\sigma)}\right]^n \tag{5.17}$$

which can be well approximated by a Weibull-type function

$$\widetilde{W}_L(\sigma) = 1 - \exp\left(-n\left(\frac{\sigma}{\sigma_0(\sigma)}\right)^{m(\sigma)}\right) \tag{5.18}$$

The function (5.18) is therefore named as a generalized Weibull distribution, in which the shape parameter m is strength dependent, while the scale parameter is additionally dependent on gauge length L or n

$$\sigma_L(\sigma) = \sigma_0(\sigma)n^{-(1/m(\sigma))} \tag{5.19}$$

The curve of $\widetilde{W}_L(\sigma)$ is schematically shown in Figure 5.3, with the details explained in the following section. A special case of (5.18) is the asymptotic Weibull distribution (5.11) for

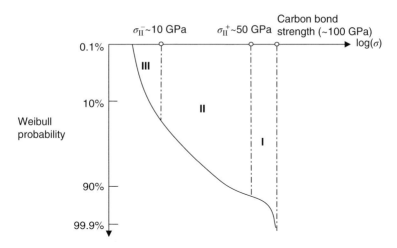

Figure 5.3   Schematic of the Weibull plot for the strength distribution of CNTs. *Source*: adapted from [3]

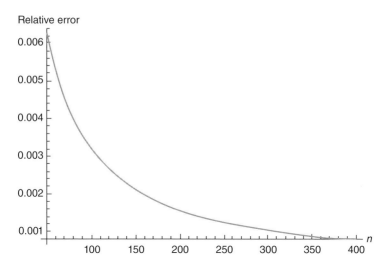

Figure 5.4   Number of weakest link elements n versus the relative error of the generalized Weibull distribution (1.18) when approximating the weakest link distribution (5.17) with $m = 3$ and $\sigma_0 = 150$ GPa [2]

the lower tail when the strength approaches zero. It is worth noting that the Weibull-type approximation of (5.17) is numerically accurate even for a relatively short CNT. A numerical example for error estimate of the approximation is shown in Figure 5.4, where the parameters are taken as $m = 3$ and $\sigma_0 = 150$ GPa for 50-nm-long weakest link elements based on published CNT test data [13, 18]. This figure indicates when the number of elements n reaches 50, corresponding to a short CNT (2.5 μm), the relative error $|\widetilde{W}_L(\sigma) - F_L(\sigma)|/F_L(\sigma)$ is already as small as 0.6%; and the error decreases quickly with a further increase in n. With such a good approximation, we consider the generalized Weibull distribution (5.18) as the nonasymptotic strength distribution applicable for the whole range of strength from the upper tail to the lower tail, and all the lengths of CNTs, both long and short, in energineering applications.

In closing this section, we make the following observations:

1. In the log–log plot, the magnitudes of flaw size and probability are scaled logarithmically, and therefore α the slope of the curve in Figure 5.2 varies much slower than it would on a linear scale. In fact for many brittle materials, a central segment of the distribution curve in Figure 5.2 is close to a straight line, and therefore over the corresponding strength range of this segment, the parameters α and σ_0 are approximately constant. As mentioned earlier, over the whole range of flaw size the exponent β varies only slightly. In such a case, the classical two-parameter Weibull distribution fits well experimental data. In other cases when test data cover a range beyond a straight segment, the two parameters become not constant, and therefore a certain correction of the classical Weibull distribution has to be made, such as the various modified, mixed, or multi-modal Weibull distributions proposed in the past, (for example, see [19]).

2. With the proposal of the generalized Weibull distribution, the duality between Weibull and Gaussian is extended from the asymptotic case to the nonasymptotic case. Owing to the universality of the power law approximation (central limit theorem), the Weibull (Gaussian) distribution arises as the natural outcome of an infinite serial (parallel) system independent of probability distribution of constituent elements in a system. In the case of finite size, Gaussian approximation for a parallel system needs a certain correction in the tails such as Edgeworth series, in a way analogous to Weibull approximation for a serial system that generalizes shape and scale parameters into strength-dependent functions for the whole strength range beyond the lower tail.

3. The above probabilistic strength theory is formulated directly on SWCNTs and MWCNTs subjected to a so-called "sword-in-sheath" type fracture mechanism, that is, the ultimate strength of a CNT is fully determined by a critical flaw. In case of MWCNTs with efficient intershell load transfer, such as those improved by irradiation-induced crosslinking [20], the failure becomes less brittle. By adjusting the exponent β and the "toughness" T in Eq. (5.14), our model can be extended to cover these CNTs as well.

4. The generalized Weibull distribution can be applied to test samples consisting of CNTs with different diameters and lengths, by using surface area in (5.18) and redefining the number n as

$$n = \frac{\text{surface area of a CNT}}{\text{surface area of a weakest link element}} \tag{5.20}$$

   Caution should be taken in the curve fitting in these cases, however, as further noted in the following section.

5. It was first demonstrated in the theory of fiber bundles [21, 22] that variable flaw sizes can lead to a segmented Weibull probability distribution for material strength, with an ever increasing m from the upper tail to the lower tail, which was also confirmed in experiments [23, 24]. As the fracture in such fibrous structures follows a discrete version of fracture mechanics, we might expect very similar behavior with CNTs.

5.2.3 Incorporation of Physical and Virtual Testing Data

Strength of a CNT has been shown to be sensitive to point defects, and a one- or two-atom vacancies may contribute to 20–30% reduction from the ideal strength of perfect CNTs [17], which is similarly shown in the case of WS^2 nanotubes [25]. From a probabilistic point of view, we first use a Poisson model to simplistically explain the chance of experimentally observing the ideal CNT strength. Suppose a CNT has its diameter $d = 20$ nm and length $L = 5$ μm (aspect ratio $= 250$) that is geometrically close to the CNTs tested in [18]. As the CNT contains about 10^7 carbon atoms, with the defect rate $\lambda = 10^{-6}$ a quick estimate for the defect-free probability shows $\exp(-10^{-6} \cdot 10^7) = 4.54 \times 10^{-5}$, that is, with such a defect concentration there is little chance of reaching the theoretical strength in tensile testing [18]. In contrast, if we were to reduce the CNT length to 0.5 μm, that is, reducing the aspect ratio from 250 to 25, the defect-free probability drastically increases nearly 10,000 times to $\exp(-10^{-6} \cdot 10^6) = 0.369$. Therefore it should not be surprising when a tensile test on a CNT with the length $\sim$0.5 μm shows a defect-free strength at the three-digit level [26].

Owing to difficulties in accurate measurement of nanoscale quantities, there are few published reports available on experimental testing of CNT strength, especially on single CNTs. Tensile tests of MWCNTs reported in [18, 12] were statistically analyzed in [13]. The testing data for strengths of 19 CNTs are provided in Table 5.1. In experimental characterization of a strength distribution, a commonly made mistake is to unscrupulously group test data of specimens with different lengths or diameters together in a single curve fitting. In the weakest link model, this mistake is obviously seen in that the weight placed on a long specimen is same as that on a short specimen, while the latter has a much smaller number of weakest link elements. By taking surface area into account, and accordingly dividing the 19 CNTs into three groups [13], the specimens are shown to fit well the classical Weibull distribution. The results graphed in Figure 5.5(a)–(d) are listed in detail in Table 5.2, where the scale parameter σ_L is

Table 5.1 Test data of 19 MWCNTs

No.	Diameter (nm)	Length (μm)	Strength (GPa)
1	28.0	4.10	11
2	28.0	6.40	12
3	19.0	3.03	18
4	31.0	1.10	18
5	28.0	5.70	19
6	19.0	6.50	20
7	18.5	4.61	20
8	33.0	10.99	21
9	28.0	3.60	24
10	36.0	1.80	24
11	29.0	5.70	26
12	13.0	2.92	28
13	40.0	3.50	34
14	22.0	6.67	35
15	24.0	1.04	37
16	24.0	2.33	37
17	22.0	6.04	39
18	20.0	8.20	43
19	20.0	6.87	63

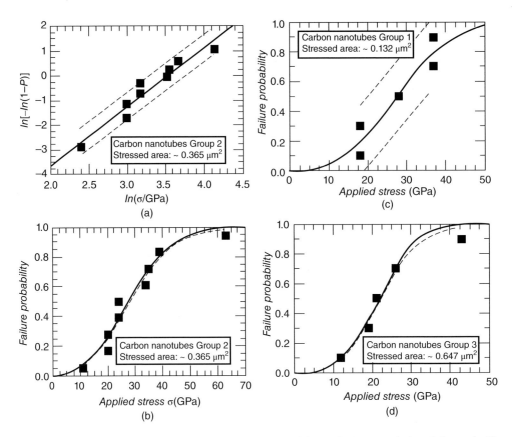

Figure 5.5 (a) Weibull plot of Group 2; (b) normal plot of Group 2; (c) normal plot of Group 1; (d) normal plot of Group 3 (dashed lines delimit 95% confidence band) (Reproduced with permission from [13]. Copyright [2007], AIP Publishing LLC)

characterized with respect to $1\,\mu m^2$ of surface area. It is noted that the data in Group 2 are considered to be more reliable than those in Groups 1 and 3 due to their small sample size.

Based on the above statistical characterization and atomistic simulation [17], we can make an estimate on the flaw size distribution. Given that $T/\psi \approx 40\,\mathrm{GPa}\sqrt{nm}$ [17] and the exponent β is approximately 2, by using Eq. (5.14) the critical flaw size c between 1 and 6 nm is found corresponding to the range of strength between 16 and 40 GPa. This strength range is located at the central part of the strength distribution, apart from the upper and lower tails, which is consistent with the small sample of 19 CNTs. The shape parameter in this core range is found to be 2.5 (Group 2 in Table 5.3). The slope α is calculated as 1.25 since $\beta = 2$. The length of the weakest link element is chosen to be 50 nm. Given the mean strength being about 30 GPa, with $n = 100$ from Eq. (5.17) the parameter σ_0 is found to be 213 GPa. Further from Eq. (5.16), we obtain $B = 0.0153$. Therefore, Eq. (5.13) specifically yields

$$1 - G_\ell(c) \approx 0.0153 c^{-1.25}, \quad c \in [1\,\mathrm{nm}, 6\,\mathrm{nm}] \tag{5.21}$$

With the above result, both the generalized Weibull distribution (5.18) and the true distribution (5.17) yield identically $F_L(18.5\,\mathrm{GPa}) = 0.2$ and $F_L(40.7\,\mathrm{GPa}) = 0.8$, which compare well

Table 5.2 Weibull strength fits for 19 MWCNTs

	Group 1		Group 2		Group 3	
	Area (μm²)	Strength (GPa)	Area (μm²)	Strength (GPa)	Area (μm²)	Strength (GPa)
1	0.0784	37	0.204	24	0.501	19
2	0.107	18	0.268	20	0.515	43
3	0.119	28	0.317	24	0.519	26
4	0.176	37	0.361	11	0.563	12
5	0.181	18	0.388	20	1.139	21
6			0.417	39		
7			0.432	63		
8			0.440	34		
9			0.461	35		
Shape parameter (m)	2.77 ± 2.32		2.48 ± 0.79		3.29 ± 2.43	
Scale parameter (σ_L)	37.6 ± 8.6		25.6 ± 4.8		20.7 ± 3.3	

Note: The errors correspond to the 95% confidence band.
Source: Reproduced with permission from [13]. Copyright [2007], AIP Publishing LLC.

Table 5.3 Three strength regions of CNTs

	Low strength region (Region III)	Intermediate strength region (Region II)	High strength region (Region I)
Strength	0–10 GPa	10–50 GPa	50–100 GPa
Flaw size	Large (≥10 nm)	Intermediate (≥1 nm)	Small (<1 nm)
Shape parameter (m)	Large (e.g., >3)	Intermediate (e.g., 2.5–3)	Small (e.g., <2.5)*

Note: Except for the extreme upper tail.

with the Weibull-fitted results [13]. The estimate (5.21) indicates that 1.53% and 0.163% of the 50-nm-long weakest link elements have a maximum flaw greater than 1 nm and 6 nm, respectively.

In general, a flaw size distribution curve (Figure 5.2) can be categorized into three regions. Region I on the left corresponds to the lower part below nanometer, Region III on the right the upper part above 10 nm size, and Region II the central part between $c_{II}^- \sim 1$ nm and $c_{II}^+ \sim 10$ nm that is most accessible and characterized by physical experiments. For high-quality CNTs, Region I can be subjected to experimental assessment as well. When the three regions of flaw size are translated into their counterparts in the strength distribution $F_l(\sigma)$, Regions I and III exchange their positions, that is, Regions I and III become the upper and lower parts, respectively. Similarly, in the Weibull plot for $\widetilde{W}_L(\sigma)$, Region I, the high strength part lies between the theoretical carbon bond strength, a three-digit value ($\sim$100 GPa) and σ_{II}^+, $\sim$50 GPa. According to numerical simulation [17], Region I corresponds to perfect CNTs or CNTs containing well-separated point defects. Region III is the low strength part lying between 0 and σ_{II}^-, $\sim$10 GPa, that is, corresponding to single-digit strength values.

The sensitivity of CNT strength to flaw size is highest at point defects, which in general reduces with an increase in flaw size. In this light, it is conjectured that the shape parameter m is minimum at the lower bound a, and increases with an increase in flaw size, since in the Weibull distribution the shape parameter m is proportional to the degree of statistical scattering. This trend of m is similarly reflected from the distribution curve of flaw size in Figure 5.2. The exact values of m depend on the quality of the CNTs, while in Table 5.3 the m values between 2.5 and 3 for Region II seem indicative of the present stage of CNT fabrication quality. Note in Figure 5.3, at the extreme upper tail, there is a sudden increase in m due to the limit of the ideal strength value. This phenomenon is confirmed by an experiment on strength of WS^2 nanotubes [25], as shown in Figure 5.6 where the three data points at the upper tail approach the ideal strength limit.

The above categorization of CNT strength regions is summarized in Table 5.3.

With the above formulation, we can gain certain insights into distribution tails that usually are inaccessible to testing data in engineering practice. The tail information, especially of the lower tail, is critical in engineering design with respect to safety and reliability. A simplistic example is provided below to demonstrate the extrapolation of the available data to the tails. As shown in Figure 5.7, a straight line fitting the 9 data points of Group 2 (Table 5.2) yields a shape parameter $m_2 = 2.5$ for Region II. Since in the Weibull plot the slope m continuously decreases from the lower tail to the upper tail, the lower and upper parts of the line (two dashed lines) invading into Regions I and III, respectively, are replaced with two straight segments with different slopes, $m_1 = 3$ and $m_3 = 2$. While the example is for demonstration purposes only, the extrapolation does suggest that the straight extension of test data into the lower and upper tails underestimates and overestimates CNT strength, respectively, and provides a way of evaluating probability bounds.

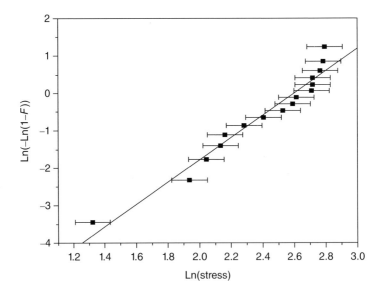

Figure 5.6 Weibull plot of the strength test data showing a rising upper tail, for multiwall WS^2 nanotubes that are 15–30 nm in diameter and 2–5 µm long [25] (Copyright (2006) National Academy of Sciences, U.S.A.)

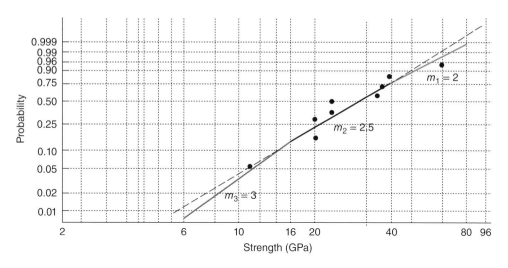

Figure 5.7 Extrapolation of test data in Region II to Regions I and III [3]

5.3 Strength Upscaling from CNTs to CNT Fibers

Microscale CNT fibers are considered the most feasible and crucial intermediate structure to harness high strength of CNTs by bridging nanoscale CNTs and macroscale composites together. CNT fibers are fabricated by spinning millions of nanotubes, for example, 10^5–10^6 MWCNTs into a fiber with a diameter in the range 5–50 μm (Figure 5.8) [28]. To improve the packing density and alignment, CNT fibers can be further post-processed via twisting, and such a twisted fiber is called a CNT yarn (Figure 5.9(a) and (b)), e.g., [27]. In this section, we describe an nth order local load sharing (LLS) model to upscale (or the other way around downscale) the strength distribution from individual CNTs to CNT fibers.

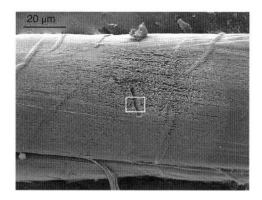
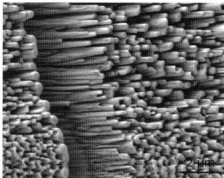

Figure 5.8 Carbon nanotubes appear aligned at the focused ion beam cut cross section of an aerogel-spun fiber (Reproduced with permission from [28]. Copyright [2012], AIP Publishing LLC)

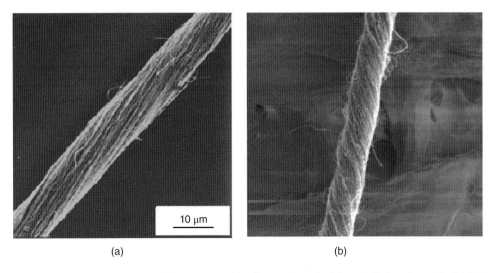

(a) (b)

Figure 5.9 SEM micrographs of CNT yarns with a low-twist angle (a) and a high-twist angle (b) [29] (Copyright Wiley-VCH Verlag GmbH & Co. KGaA. Reproduced with permission)

5.3.1 A Local Load Sharing Model

In a fiber bundle model, the fibers are assumed to be aligned and loaded with the same force along the fiber direction. The strengths of the fibers are treated as independent and identically distributed random variables. As the applied load increases, the weakest fiber breaks first. The load carried by this weakest fiber prior to break is redistributed among the surviving fibers. The nature of this redistribution reflects the interaction of the components in a system, and load sharing rules are specified based on physical insights. Two classical load sharing rules are the equal load sharing (ELS) rule [30], where the load is uniformly distributed among all intact fibers, and the extreme local load sharing (ELLS) rule [31], where the broken fiber affects only the nearest surviving fibers.

Load redistribution in a physical system generally lies in between the two limits of ELS and ELLS. A variety of mechanics-based models have been developed to analyze such a progressive failure process, for example, the Hedgepeth and Van Dyke's shear lag model [32] to evaluate the effect of shear coupling on stress concentration around a local failure. To circumvent complex analysis of load redistribution in progressive failure of random composites, surrogate simple load sharing rules are preferred. In [1], an efficient nth order LLS model is proposed to provide flexibility in spanning the range between ELS and ELLS. The simple LLS rule can be empirically calibrated to account for a variety of complex local interactions.

In the nth-order LLS model, the order n indicates the degree of localization, related to various inelastic mechanisms, such as yielding, softening, and microcracking. An nth order zone is defined to be of size-n centering around a broken bond i. In a 1D system, the zone consists of n sites to the left and n sites to the right of bond i, denoted as $D_i = \{i - n, i - n + 1, \ldots, i - 1, i + 1, \ldots, i + n\}$. When bond i breaks, its nearest surviving bonds within the zone will share a certain ratio ξ ($0 \leq \xi \leq 1$) of the released load, which is calculated based on a simple rule described in an algorithm later. The remaining ratio of the released load $(1 - \xi)$ will be equally shared by all surviving bonds in the system. The ratios ξ and $(1 - \xi)$ represent

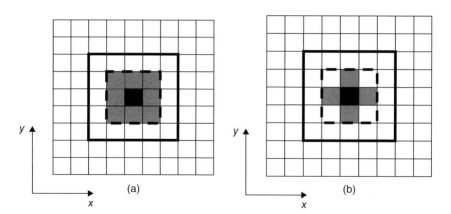

Figure 5.10 Moore's neighborhood (gray area on (a)) and Von Neumann's neighborhood (gray area on (b)) of the broken bond (black square), with the first- and second-order zones indicated with dash and bold lines, respectively [1]

local and global effects due to a local failure, which accordingly are named as localization and globalization parameters, respectively. If there are no surviving bonds within the zone, all the released load will be equally shared by all remaining surviving bonds in the system. When the order $n = 0$, that is, the size of the zone is 0, meaning there is no local interaction, the zeroth-order LLS reduces to ELS, where 100% of the load released by a broken bond is equally shared by all the surviving bonds in the system. At the other end of the spectrum, when the nth order zone is as large as the entire system (i.e., extreme local interactions), the nth order LLS with n sufficiently large becomes ELLS, which corresponds to an extreme localization of load sharing since all the released load is transferred to the nearest surviving bonds only. In this extreme localization case, zero percent of the released load from a broken bond is equally shared by all surviving bonds in the system.

Let $S = \{(i,j)|1 \leq i \leq N_1, 1 \leq j \leq N_2\}$ index a discrete set of sites for the bonds in a 2D rectangular lattice system. In Figure 5.10(a) and (b), Moore's [33] and Von Neumann's [34] nearest neighborhoods are shown consisting of 8 bonds and 4 bonds, respectively. The zone for the first-order LLS rule is outlined by dashed lines, while for the second-order rule by bold lines. Note that an LLS rule is not restricted to a square array; for example, a hexagonal arrangement can be considered. In the following study, Moore's neighborhood is adopted. For simplicity, all bonds in the nearest neighborhood are treated equally. By taking into account stress analysis and physical experiments, a nonuniform distribution scheme can be assigned or calibrated, such as that given by the Hedgepeth and Van Dyke's model [32].

To numerically implement an nth-order LLS rule in a 2D lattice system, a force-controlled quasi-static algorithm [1] is described as a 10-step procedure below. The strengths of bonds are modeled as independent and identically distributed random variables, while the elastic modulus is deterministic and identical for all bonds.

1. Sampling bond strength $s_{kl}, (k, l) \in S$;
2. Initialize forces $f_{kl}^{(0)} = f_0$ with $0 < f_0 < \min\{s_{kl}\}$ for all $(k, l) \in S$;
3. Identify the weakest bond (i,j) to be broken in the current loading step t: $(i,j) = \underset{(k,l) \in S}{\arg\max}(r_{kl} = f_{kl}^{(t)}/s_{kl})$;

4. The total force $F^{(t)}$ is adjusted as $F^{(t)} = \frac{1}{r_{ij}} \sum_{k=1}^{N_1} \sum_{l=1}^{N_2} f_{kl}^{(t)}$;

5. Compute the displacement $u^{(t)} = F^{(t)}/(N_s \kappa)$, κ – elastic modulus, N_s – number of surviving bonds;

6. Update the force on the broken bond $f_{ij}^{(t)} = 0$;

7. Distribute the released load $f_{ij}^{(t)}$ equally to all the bonds (intact or broken) in its neighborhood N_{ij}. For any pre-existing broken bond $(k, l) \in N_{ij}$, the load allocated to it is equally redistributed to the bonds in $N_{kl} \cap D_{ij}$. This redistribution process continues iteratively until the load shared by any pre-existing broken bond in D_{ij} becomes negligibly small;

8. Calculate $\xi f_{ij}^{(t)}$ by summing up all the above redistributed load;

9. Distribute $(1 - \xi) f_{ij}^{(t)}$ equally to all the surviving bonds in the system;

10. Go back to step (3) for the next loading step $(t + 1)$ with all the updated forces.

Five examples are provided in Figure 5.11 to verify the above algorithm, with the calculated distribution factors shown. In Example (a), the sum of the factors $\xi = 1$ for a single bond failure subject to the first-order LLS rule, while all the higher-order LLS rules yield the same results. When two adjacent bonds break simultaneously, Examples (b) and (c) show that $\xi = 0.9524$ and 1 for the first- and second-order LLS rules, respectively, while all the rules with an order

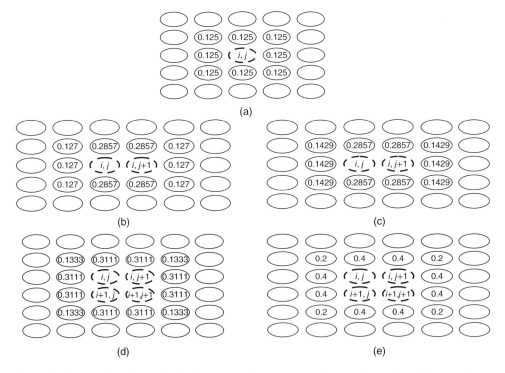

(a)

(b)   (c)

(d)   (e)

Figure 5.11 Calculated load distribution factors for examples: (a) single broken bond; (b) two simultaneously broken bonds using first-order LLS; (c) two simultaneously broken bonds using second-order LLS; (d) four simultaneously broken bonds using first-order LLS; (e) four simultaneously broken bonds using second-order LLS [1]

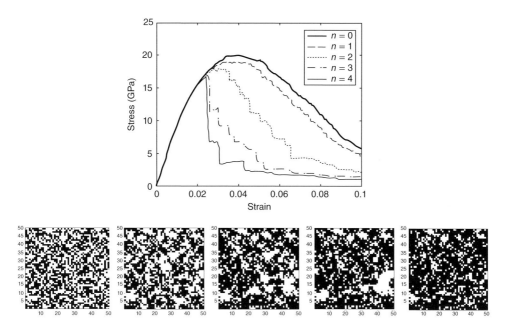

Figure 5.12 A sample set of stress–strain curves and snapshots (white-broken bonds, dark-intact bonds, left to right from zeroth- to fourth-order LLS) at the peak load for a CNT rope consisting of 50×50 CNTs [1]

higher than 2 yield the same results as the second-order LLS rule. When four adjacent bonds break simultaneously, the ξ values for the first- and second-order LLS rules are 0.7555 and 1, respectively, as shown in Examples (d) and (e). Again, all the rules with an order higher than 2 yield the same results as the second-order LLS rule. These examples confirm the specification that for an LLS rule with the order sufficiently high, or more precisely when the nth order zone is larger than the cluster size of breaks, load redistribution and hence failure progression become brittle-like with the localization parameter $\xi = 1$.

The force-controlled quasi-static algorithm is applied to simulate a CNT rope consisting of 50×50 tubes. It is assumed that the strength of CNTs is Weibull distributed with shape and scale parameters 1.5 and 50 GPa, respectively, and the Young's modulus is 1000 GPa. Five stress–strain curves and five snapshots of the rope cross-section at the peak load are provided in Figure 5.12 corresponding to zeroth- to fourth-order LLS rules. The five curves show that with an increase in the LLS order, the peak load converges and the drop of the post-peak load becomes even steeper, indicating a stronger localization effect and a more brittle-type failure.

5.3.2 Interpretation of CNT Bundle Tensile Testing

In this part, we present a lower estimate of individual CNT strength from CNT bundle testing data by using the LLS model. Due to potentially large uncertainties in strength measurements of single CNTs, CNT bundle testing is often conducted as an alternative to single CNT testing. Another issue in single CNT testing is the inadvertent "proof-testing" that similarly occurs in carbon fiber testing since, when extracting specimens, only those strong enough can survive the extracting tension. CNT bundle testing data, however, need to be appropriately interpreted to derive physical properties of individual CNTs. For example, strength of individual CNTs cannot be simply back-calculated as the average stress of the bundle at failure. When statistics or uncertainty is involved, for example, in propagating strength distributions across scales, caution needs to be exercised in such a multiscale analysis [35].

A tensile testing sample of SWCNT bundles was reported in [36], which consists of 15 bundle strengths ranging from 13 GPa to 52 GPa with mean 30 GPa. Weibull fitting of the data shows the shape and scale parameters are 2.7 and 33.9 GPa, respectively [9]. Due to heterogeneity of fracture strength of individual CNTs, the CNTs in a bundle are not broken simultaneously under an identical tensile force, but rather participate in a progressive failure process. The tensile force released from a newly broken CNT is redistributed to its neighbors and others of the bundle. The exact failure process of a CNT bundle depends on many details of experimental set-up, which often are hardly measurable and unavailable from the report. We however can make a certain estimate of individual CNT strength by using the proposed LLS model.

Owing to the specific testing condition, a ring model is considered in which only those CNTs located on the parameter contribute to the ultimate strength of the bundle [36]. Accordingly the LLS model is applied to a 1D ring of bonds. The mean strength formula in the classical equal load sharing model for an infinite bundle system [30] is extended to the zeroth-order LLS of a finite size bundle system as follows:

$$\bar{\sigma} = \varphi \sigma_L \left(\frac{1}{m} \right)^{1/m} \exp \left(-\frac{1}{m} \right) \qquad (5.22)$$

where m and σ_L are the shape and scale parameters for individual CNTs with gauge length L, and ϕ is the factor to account for the effect of finite bundle size (number of fibers). For an infinitely large bundle size, the factor ϕ becomes 1 reducing to Daniels' asymptotic result [30]. The zeroth-order LLS simulation for a ring consisting of a finite number (e.g., dozens) of bonds indicates that the factor ϕ is slightly larger than 1, which is similarly shown in a 2D case later (Table 5.4). Based on the central limit theorem, statistical scattering of CNT bundle strength reduces from that of CNT strength, which is also numerically verified in the following section. Consequently, the shape parameter of CNT strength is smaller than that of CNT bundle strength. Given $m = 2.7$, $\phi = 1$, and $\bar{\sigma} = 30$ GPa, the scale parameter of CNTs, σ_L, is back-calculated from (5.22) to be 63 GPa. Note that $m = 2.7$ is an upper estimate for the scale parameter of CNTs, and the zeroth-order LLS rule results in an upper bound of bundle strength. The two overestimated effects together overwhelm the underestimated effect made in the approximation $\phi = 1$, and therefore the derived $\sigma_L = 63$ GPa is certainly a lower estimate.

Table 5.4 Mean values and c.o.v obtained from ELS and LLS rules on 5-μm-long CNT bundles [1]

	Single CNTs	ELS	Zeroth-order LLS	First-order LLS	Second-order LLS	Third-order LLS	Fourth-order LLS
Mean (GPa)	*45.1*	19.6	20.8	20.2	19.7	19.5	19.5
c.o.v (%)	*67.9*	0	9.1	9.2	9.6	9.9	10.0

Note: All the numbers on the right are obtained from the numbers with italic font.

The corresponding mean strength of individual CNTs is 56 GPa, which is almost double the measured mean value of bundle strength, 30 GPa.

Above we present a simple analysis for the strength relation between a CNT bundle and the constituent CNTs. Below by employing the nth order LLS model, a numerical analysis is described for propagating strength statistics from CNTs to CNT fibers across a wide length-scale gap.

5.3.3   Strength Upscaling Across CNT-Bundle-Fiber Scales

A CNT fiber or yarn consists of 10^5–10^6 CNTs that are made aligned or twisted along a CNT fiber axis. Between nanoscale CNTs and a microscale fiber, CNT bundles can be chosen as an intermediate scale bridging the two scales. Often times, such bundles physically exist inside a CNT fiber due to van der Waals forces, typically consisting of dozens of CNTs. Such a hierarchical structure of a CNT fiber is schematically shown in Figure 5.13, where parameter values are chosen for a benchmark model to simulate. Strength of CNT fibers certainly depends on the quality of constituent CNTs and how fibers are fabricated. Another critical factor is gauge length chosen in fiber strength testing. The reported CNT fiber strengths vary widely, with the highest ranging approximately 3–6 GPa. These values are 1 order of magnitude lower than the strength of individual CNTs as described in Section 5.2. To demonstrate how strength is reduced from CNTs to CNT fibers, we use the LLS model to simulate the strength upscaling process via the benchmark model (Figure 5.13). By identifying the crucial factors that most contribute to the strength reduction, we expect that such a multiscale strength analysis can provide certain insights for continuous improvement of high-strength CNT fibers.

Strength of individual SWCNTs with gauge length 5 µm is assumed to follow the Weibull distribution with the shape and scale parameters being $m = 1.5$ and $\sigma_t = 50$ GPa, respectively. The corresponding mean strength and coefficient of variance (c.o.v) are $\overline{\sigma}_t = 45.1$ GPa and

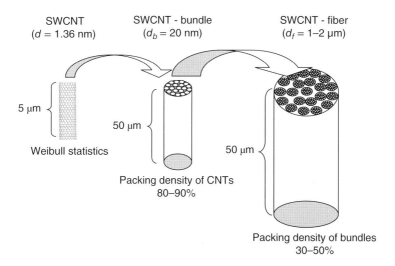

Figure 5.13   Schematic of hierarchical structure of CNT-fibers (the parameter values shown correspond to a benchmark model for simulation) [1]

67.9%, respectively. The diameter and wall thickness of SWCNTs are taken as $d = 1.36$ nm and $t = 0.34$ nm, respectively.

In the benchmark model a bundle consists of 100 SWCNTs aligned in parallel along the bundle axis. Considering a high packing density ρ_t of CNTs in the bundle, about 80–90%, the diameter of the bundle d_b is approximately 20 nm. In applying an nth order LLS rule, the 100 nanotubes are set on a square grid. The characteristic length L_t of CNTs is calculated as two times of the contact length for inter-CNT friction τ to transfer load effectively. By assuming that the friction decreases linearly along the axis from the value τ at the breaking point, a simple estimate of the characteristic CNT length L_t is given as

$$L_t = 2\frac{\eta\bar{\sigma}_t \cdot \pi dt}{\tau/2 \cdot \pi(d+t)} \tag{5.23}$$

where the factor η accounts for the effect of CNT packing density in the bundle. The inter-CNT friction shear stress τ is considered to range between 10 and 20 MPa [37]. For a high packing density the factor η is about 1, and the characteristic CNT length is estimated from (5.23) to be 2.5–4.9 μm. For illustrative purpose, the characteristic length is simply set as 5 μm that is equal to the gauge length of the CNTs. To find the strength distribution of the 5-μm-long bundles, 10,000 Monte Carlo samples are generated for each of the zeroth- to fourth-order LLS rules, and the force-controlled quasi-static algorithm is applied. The normal probability paper plot in Figure 5.14 indicates that the bundle strength obtained for the each LLS rule is approximately Gaussian, except for some deviation in tails. While the Gaussian curve of the zeroth-order LLS

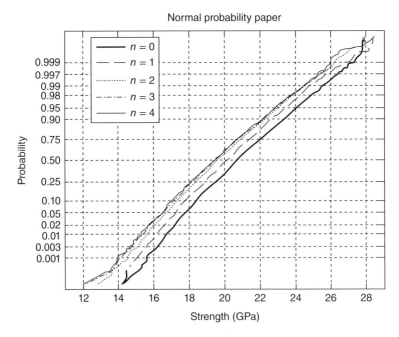

Figure 5.14 Normal probability paper plot for the strengths of CNT bundles using the 2D zeroth to fourth-order LLS rules on 5-μm-long 10×10 CNTs (10,000 Monte Carlo samples or each order of the LLS) [1]

is consistent with the finite size effect on the central limit theorem, Gaussian curves for other LLS rules indicate that in the bundle case the homogenization effect is much stronger than the localization effect. The mean strength and c.o.v of the 5-μm-long bundles using various LLS rules are summarized in Table 5.4, in which both values decrease significantly from those values of single CNTs. With the rise of the LLS order, the mean strength reduces while the c.o.v increases, and the convergence is approximately achieved at the fourth-order LLS. For comparison, the ELS result for an infinitely large bundle is also listed in Table 5.4, which is obtained from (5.22) with $\phi = 1$. An important finding here is that the bundle mean strength drops to less than half CNT mean strength, which compares well with bundle testing data of double-walled CNTs [20]. It should be noted that both the strengths of bundles and nanotubes here are given based on the wall areas of CNTs. The simulated scattering of the bundle strength is certainly narrower than the scatter in the test data since only the statistical strength of individual CNTs is considered and many other uncertainties are not accounted for, such as alignment and entanglement of CNTs, varying bundle size (length and number of CNTs), and so on.

Similar to (5.23), the characteristic bundle length for effective interbundle load transfer is estimated as

$$L_b = 2\frac{\eta N_t \overline{\sigma}_b \cdot \pi dt}{\tau/2 \cdot \pi d_b} \tag{5.24}$$

where N_t the number of CNTs in a bundle is 100. Compared with ρ_t, the packing density of bundles ρ_b in a CNT fiber is relatively low. Suppose $\rho_b = 30$–50%, and the factor η is considered to be around 3–4. The effective bundle length is accordingly estimated from (5.24) to be about 30–80 μm, and we simply choose $L_b = 50$ μm. Given the estimated strength distribution for the 5-μm-long bundles, the strength distribution for 50-μm-long bundles can be calculated by using the weakest link model. The mean strength and c.o.v of the 50-μm-long bundles are listed in Table 5.5, both values of which drop from those of the 5-μm-long bundles. Due to the weakest link effect, the distribution curves for the five LLS rules deviate from Gaussian approximation, as expected.

Next, a CNT fiber containing 2500 bundles is modeled on a 50×50 square grid with each grid representing either a 50-μm-long bundle or a vacancy, taking into account the relatively low packing density ρ_b. Two cases are considered, one with a packing density 50% (1250 CNT bundles randomly distributed over the 2500 grid locations) and the other with a packing density of 30% (750 CNT bundles randomly distributed over the 2500 grid locations). The diameter of such a CNT fiber d_f is approximately 1–2 μm. In each case, 5000 Monte Carlo samples are generated and the zeroth- to fourth-order LLS rules are applied to the force-controlled quasi-static algorithm. The number of the samples and the orders are so chosen that convergence is achieved. The mean values and c.o.v for strength of such a CNT

Table 5.5 Mean values and c.o.v obtained from the weakest link model on 50-μm-long CNT bundles [1]

	Zeroth-order LLS	First-order LLS	Second-order LLS	Third-order LLS	Fourth-order LLS
Mean (GPa)	17.9	17.3	16.7	17.6	17.5
c.o.v (%)	6.2	6.4	6.8	7.1	7.1

Table 5.6 Mean values and c.o.v (%) obtained from LL rules on 50-μm-long CNT fibers

		LLS rule on the 50×50 CNT fibers				
	LLS rule on the 10×10 CNT bundles	Zeroth-order LLS	First-order LLS	Second-order LLS	Third-order LLS	Fourth-order LLS
$\rho_b = 50\%$ (upper estimate)	Zeroth-order ($\rho_t = 90\%$)	4.4/0.55%	4.0/2.1%	4.0/2.5%	4.0/2.5%	4.0/2.5%
$\rho_b = 50\%$ (lower estimate)	Fourth-order ($\rho_t = 80\%$)	3.5/0.61%	3.3/2.1%	3.2/2.7%	3.2/2.7%	3.2/2.7%
$\rho_b = 30\%$ (upper estimate)	Zeroth-order ($\rho_t = 90\%$)	2.6/0.70%	2.5/1.6%	2.4/2.6%	2.4/2.7%	2.4/2.7%
$\rho_b = 30\%$ (lower estimate)	Fourth-order ($\rho_t = 80\%$)	2.1/0.77%	2.0/1.6%	1.9/2.7%	1.9/2.9%	1.9/2.9%

Source: adapted from [1].

fiber are listed in Table 5.6, by further taking into account the packing densities ρ_t and ρ_b, and the hole area within individual CNTs. The hole area is accounted for by multiplying the simulated strength values with the ratio between the wall area and the total cross-sectional area of a CNT, that is, $4dt/(d + t)^2 = 0.64$. The results in Table 5.6 in fact correspond to multiplication of the mean values in Table 1 of reference [1] with the ratio 0.64, since a solid cylinder model of nanotubes is implicitly assumed for CNT strength in [1]. Table 5.6 indicates a decrease in the mean value of the strength as the order of the LLS rule increases, while the opposite trend is observed for the coefficient of variation. The estimated mean strength of the CNT fiber ranges between 3.2 and 4.4 GPa for a 50% packing density, and 1.9–2.6 GPa for a 30% packing density. By using normal probability paper plots, the strength distribution is found close to Gaussian for the zeroth-order LLS rule, and increasingly deviates from normality with the rise of the LLS order.

Based on the above benchmark model results, four concluding remarks of this section are given as follows:

1. The estimated mean strength for CNT fibers, in the range 1.9–4.4 GPa, is approximately 1 order of magnitude lower than the mean strength of individual CNTs, which is consistent with experimental results. Dependent on the types of CNTs and fabrication methods, measured CNT fiber strengths are mostly reported on the single-digit level 1–10 GPa, the range which is about one-tenth of the range for measured CNT strengths (e.g., see Table 5.1).
2. The primary knock-down factors contributing to the above strength reduction are approximated as follows: (1) point defects and nanoscale flaws represented by Weibull statistics reduce the mean strength by a reduction factor f_1 around 0.4. The factor is further reduced with an increase in a fiber length beyond 50 μm; (2) the simple but often forgotten geometric factor, that is, holes and voids within and without of CNTs, reduces strength by a reduction factor f_2 ranging between 0.15 and 0.25 dependent on the packing density. The product of the two factors explains the apparent 1 order of magnitude reduction in strength from individual CNTs to CNT fibers.

3. Given universal thermodynamic effects [15, 16] and rigid geometric constraints posed upon the factors f_1 and f_2 from above, the maximal product of the two reduction factors is considered to be not more than 0.2–0.25 for a CNT fiber with lengths on the millimeter scale. The mean strength of 5-μm-long CNTs is maximally in the high strength region 50–80 GPa (critical flaw <1 nm as shown in Table 5.3), and therefore the upper bound for mean strength of millimeter-long CNT fibers is estimated in the range 10–20 GPa, which noticeably is 1 order of magnitude lower than the theoretical strength of CNTs;

4. There is also a significant reduction in the coefficient of variation of the strength from individual CNTs to a CNT fiber. The estimated c.o.v for the CNT fiber, 0.6–3%, is considered relatively low compared with experimental results due to a number of simplifying assumptions made in our model, for example, bundle packing density constant along the fiber length, fiber packing density constant from fiber to fiber, alignment of nanotubes and bundles, and so on. When these assumptions are further relaxed, the coefficient of variation and mean value of the fiber strength are expected to increase and reduce, respectively.

5.4   Conclusion

The probabilistic strength theory presented in this chapter not only justifies the use of classical Weibull distribution in fitting of CNT testing data but also provides theoretical guidance in the extrapolation of CNT testing data to tails. The theory is considered applicable to the fracture strength of brittle materials. The benchmark model on strength upscaling from CNTs to fibers indicates that the full potential of CNT fibers for exploitation is expected to be in the range between 10 and 20 GPa with respect to mean strength due to universal thermodynamic effects and inherent geometric constraints. Nonetheless, CNTs remain promising structural elements in building up of the next-generation superstrong materials, especially in terms of specific strength.

References

[1] Xu, X.F., Hu, K., Beyerlein, I.J. and Deodatis, G. (2011) Statistical strength of carbon nanotube composites. *International Journal of Uncertainty Quantification*, **1**, 279–295.

[2] Xu, X.F., Jie, Y. and Beyerlein, I.J. (2013) A note on statistical strength of carbon nanotubes. *Computers, Materials & Continua*, **38**, 17–30.

[3] Xu, X.F., Jie, Y. and Beyerlein, I.J. (2014) A probability model for strength of carbon nanotubes. *API Advances*, **4**, 077116.

[4] Freudenthal, A.M. (1968) Statistical approach to brittle fracture, in *Fracture: An Advanced Treatise*, vol. 2 (ed H. Liebowitz), Academic Press, pp. 591–619.

[5] Fisher, R.A. and Tippett, L.H.C. (1928) Limiting forms of the frequency distribution of the largest and smallest member of a sample. *Proc Cambridge Philos Soc*, **24**, 180–190.

[6] Gumbel, E.J. (1958) *Statistics of Extremes*, Columbia University Press.

[7] Collins, P.G. (2010) Defects and disorder in carbon nanotubes, in *Oxford Handbook of Nanoscience and Technology: Frontiers and Advances* (eds A.V. Narlikar and Y.Y. Fu), Oxford University Press.

[8] Rooke, D.P. and Cartwright, D.J. (1976) *Compendium of stress intensity factors*, HMSO Ministry of Defence. Procurement Executive.

[9] Barber, A.H., Andrews, R., Schadler, L.S. and Wagner, H.D. (2005) Stochastic strength of nanotubes: an appraisal of available data. *Appl. Phys. Lett.*, **87**, 203106.

[10] Lu, C. (2005) On the tensile strength distribution of multiwalled carbon nanotubes. *Appl. Phys. Lett.*, **92**, 206101.

[11] Wagner, H.D., Barber, A.H., Andrews, R. and Schadler, L. (2008) Response to "Comment on 'On the tensile strength distribution of multiwalled carbon nanotubes'". *Appl. Phys. Lett.*, **92**, 206102.

[12] Pugno, N.M. and Ruoff, R.S. (2006) Nanoscale Weibull statistics. *J. Appl. Phys.*, **99**, 024301.

[13] Klein, C.A. (2007) Characteristic tensile strength and Weibull shape parameter of carbon nanotubes. *J. Appl. Phys.*, **101**, 124909.

[14] Lu, C., Danzer, R. and Fischer, F.D. (2002) Fracture statistics of brittle materials: Weibull or normal distribution. *Phys. Rev. E*, **65**, 067102.

[15] Dumitrica, T., Hua, M. and Yakobson, B.I. (2006) Symmetry, time, and temperature dependent strength of carbon nanotubes. *Proc. Natl. Acad. Sci. USA*, **103**, 6105–6109.

[16] Ruoff, R.S. (2006) Time, temperature, and load: the flaws of carbon nanotubes. *Proc. Natl. Acad. Sci. USA*, **103**, 6779–6780.

[17] Zhang, S., Mielke, S.L., Khare, R. et al. (2005) Mechanics of defects in carbon nanotubes: atomistic and multiscale simulations. *Phys Rev B*, **71**, 115403.

[18] Yu, M.F., Lourie, O., Dyer, M.J. et al. (2000) Strength and breaking mechanism of multiwalled carbon nanotubes under tensile load. *Science*, **287**, 637–640.

[19] Zinck, P., Gérard, J.F. and Wagner, H.D. (2002) On the significance and description of the size effect in multimodal fracture behavior. Experimental assessment on E-glass fibers. *Engineering Fracture Mechanics*, **69**, 1049–1055.

[20] Filleter, T., Bernal, R., Li, S. and Espinosa, H.D. (2011) Ultrahigh strength and stiffness in cross-linked hierarchical carbon nanotube bundles. *Advanced Materials*, **23**, 2855–2860.

[21] Harlow, D.G. and Phoenix, S.L. (1978) The chain-of-bundles probability model for the strength of fibrous materials: 1. Analysis and conjectures. *J. Comp. Mater.*, **12**, 195.

[22] Beyerlein, I.J. and Phoenix, S.L. (1996) Statistics for the strength and size effects of microcomposites with four carbon fibers in epoxy resin. *Composite Science and Technology*, **56**, 75.

[23] Phoenix, S.L., Schwartz, P. and Robinson, H.H. (1988) Statistics for the strength and lifetime in creep-rupture of model carbon epoxy composites. *Composites Science and Technology*, **32**, 81.

[24] Otani, H., Phoenix, S.L. and Petrina, P. (1991) Matrix effects on lifetime statistics for carbon fibre-epoxy microcomposites in creep rupture. *Journal of Materials Science*, **26**, 1955.

[25] Kaplan-Ashiri, I., Cohen, S.R., Gartsman, K. et al. (2006) On the mechanical behavior of WS2 nanotubes under axial tension and compression. *Proc. Natl. Acad. Sci. USA*, **103**, 523–528.

[26] Demczyk, B.G., Wang, Y.M., Cumings, J. et al. (2003) Direct mechanical measurement of the tensile strength and elastic modulus of multiwalled carbon nanotubes. *Mater Sci Eng A*, **334**, 173–178.

[27] Beyerlein, I.J., Porwal, P.K., Zhu, Y.T. et al. (2009) Scale and twist effects on the strength of nanostructured yarns and reinforced composites. *Nanotechnology*, **20**, 485702.

[28] Wu, A.S., Nie, X., Hudspeth, M.C. et al. (2012) Carbon nanotube fibers as torsion sensors. *Appl Phys Lett*, **100**, 201908.

[29] Zhang, X., Li, Q., Tu, Y. et al. (2007) Strong carbon-nanotube fibers spun from long carbon-nanotube arrays. *Small*, **3**, 244–248.

[30] Daniels, H.E. (1945) The statistical theory of the strength of bundles of threads. *Proc R Soc A*, **183**, 405–435.

[31] Harlow, D.G. and Phoenix, S.L. (1981) Probability distributions for the strength of composite materials I: two-level bounds. *International Journal of Fracture*, **17**, 347–372.

[32] Hedgepeth, J.M. and Van Dyke, P. (1967) Local stress concentration in imperfect filamentary composite materials. *Journal of Composite Material*, **1**, 294–309.

[33] Wolfram, S. (1983) Statistical mechanics of cellular automata. *Rev. Mod. Phys.*, **55**, 601–644.

[34] Van Neumann, J. (1966) Theory of self-reproducing automata, (ed A. Burks), Univ. of Illinois Press, Urbana, IL.

[35] Xu, X.F. (2007) A multiscale stochastic finite element method on elliptic problems involving uncertainties. *Comput. Methods Appl. Mech. Engrg.*, **196**, 2723–2736.

[36] Yu, M.F., Files, B.S., Arepalli, S. and Ruoff, R.S. (2000) Tensile loading of ropes of single wall carbon nanotubes and their mechanical properties. *Physical Review Letters*, **84**, 5552–5555.

[37] Yokobson, B.I., Samsonidze, G. and Samsonidze, G.G. (2000) Atomistic theory of mechanical relaxation in fullerene nanotubes. *Carbon*, **38**, 1675–1680.

6

Numerical Nanomechanics of Perfect and Defective Hetero-junction CNTs

Ali Ghavamian[1], Moones Rahmandoust[1,2,3] and Andreas Öchsner[1]

[1] *School of Engineering, Griffith University, Gold Coast Campus, Southport, Queensland, Australia*
[2] *Protein Research Center, Shahid Beheshti University, G.C., Velenjak, Tehran, Iran*
[3] *Deputy Vice Chancellor Office of Research and Innovation, Universiti Teknologi Malaysia, Johor Bahru, Johor, Malaysia*

6.1 Introduction

Since the discovery of carbon nanotubes (CNTs) by Iijima in 1991, these unique nanostructures which offer exceptional mechanical, electrical, and thermal properties have drawn worldwide attention. Among all the outstanding properties of these carbon-based nanomaterials, CNTs have introduced wide application potentials in many nanoindustry fields as either stand-alone nanomaterials or reinforcement in composite materials because of their prominent mechanical and physical properties, such as strength and lightness, which have encouraged many scholars to investigate their mechanical properties [1, 2]. Over the years, numerous approaches, which are generally divided into experimental and computational approaches, have been employed for the mechanical characterization of these charming nanomaterials, based on which it has been reported that CNTs have a surprisingly high stiffness (a Young's modulus of approximately 1 TPa). Among the computational techniques that are generally divided into the two groups of molecular dynamics (MD) and continuum mechanics (CM) techniques, the finite element method (FEM) has earned substantial popularity among scholars [3].

6.1.1 Literature Review: Mechanical Properties of Homogeneous CNTs

In 1997, Lu [4] evaluated the elastic and shear moduli of CNTs by an experimental technique in which the empirical force constant was considered. Based on his results, the Young's

Advanced Computational Nanomechanics, First Edition. Edited by Nuno Silvestre.
© 2016 John Wiley & Sons, Ltd. Published 2016 by John Wiley & Sons, Ltd.

modulus for single-walled carbon nanotubes (SWCNTs) and for multi-walled carbon nanotubes (MWCNTs) was 0.97 and 1 TPa, respectively. Moreover, he reported a general value of 0.5 TPa for the shear modulus of CNTs. Another experimental investigation on the Young's modulus of CNTs was performed by Wu et al. by combining optical characterization and magnetic actuation techniques. Their results showed that the Young's modulus of individual CNTs and five equally weighted CNT bundles has values of 0.97 and 0.99 TPa, respectively [5].

The linear elastic properties of CNTs was studied by Li and Chou [6] through a structural mechanics approach which pointed out a Young's modulus in the range of 0.89 and 1.033 TPa and a shear modulus between 0.22 and 0.48 TPa while the Young's and shear moduli of CNTs was reported to be about 1.024 and 0.47 TPa, respectively, by To [7] through a finite element (FE) investigation, when the effect of the Poisson's ratio on the CNT's elastic properties was considered.

Nahas and Abd-Rabou [8] also employed the FE simulation based on structural mechanics by which they calculated the elastic modulus to be equal to 1.03 TPa. The mechanical properties of SWCNTs were investigated by two different approaches of analytical and FEMs by Kalamkarov et al. [9], based on CM. Finally, they obtained the Young's and shear moduli of SWCNTs to be equal to 1.71 and 0.32 TPa, respectively, by the analytical calculations, while their FE results expressed the elastic modulus between 0.9 and 1.05 TPa for SWCNTs and 1.32–1.58 TPa for double-walled carbon nanotubes (DWCNTs). Furthermore, the ranges of 0.14–0.47 TPa and 0.44–0.47 TPa were reported for the shear modulus of the SWCNTs and MWCNTs, respectively.

An analytical modeling approach was used by Natsuki et al. to evaluate the elastic and torsional properties of SWCNTs, which yielded the axial modulus of SWCNTs to be in the range of 0.73 and 1.1 TPa and the shear modulus between 0.37 and 0.32 TPa [10], while these quantities were reported in the range of 1.32 and 1.58 TPa for Young's modulus of MWCNTs and 0.37–0.47 TPa for their shear modulus by Rahmandoust and Öchsner [11], based on the FE simulation in which the covalent bonds between atoms were assumed to be beam elements.

Meo and Rossi [12] also created an FE model of an SWCNT and eventually obtained a Young's modulus of about 1 TPa for SWCNTs. Ávila and Lacerda [13] also created FE models of three major SWCNTs' configurations with the ANSYS commercial software and investigated the elastic modulus of the CNT models. They finally pointed out that this quantity varies between 0.97 and 1.30 TPa.

Shokrieh and Rafiee [14] probed the elastic moduli of graphene sheets and CNTs by a linkage between the lattice molecular structure and the equivalent discrete frame structure. According to their results from classical mechanics, the Young's modulus of graphene sheets was found to be about 1.04 TPa, while for CNTs this quantity was obtained in the range of 1.033 and 1.042 TPa. Song et al. [15] investigated the elastic properties of perfect and atomically defective SWCNTs with silicon doping. Finally, they discovered that the Young's modulus of perfect SWCNTs varies in the range of 1.099 ± 0.005 TPa and those of silicon impurities decrease their elastic strength.

An energy-force approach was employed for numerical evaluation of the elastic and torsional properties of SWCNTs by Jin and Yuan [16]. Based on their results, the Young's modulus of SWCNTs varied between 1.350 and 1.238 TPa and the shear modulus of SWCNTs was obtained in the range of 0.546 and 0.492 TPa while the 3D FE model of CNTs, which was developed by Tserpes and Papanikos, demonstrated the Young's modulus between 0.97 and 1.03 TPa and the shear modulus in the range of 0.283 to 0.504 TPa [17].

Yu et al. investigated the torsional properties of SWCNTs by an MD technique under pure torsion that yielded the shear modulus in the range of 0.37 to 0.5 TPa [18]. Liu and Chou obtained the Young's modulus of MWCNTs in the range of 1.05 ± 0.05 TPa and the shear modulus about 0.40 ± 0.05 TPa [19] by the FEM in which each CNT was assumed to be a frame-like structure. FEM was also employed by Fan et al. [20] to evaluate the mechanical properties of MWCNTs. According to their results, the Young's modulus of MWCNTs was found about 1 TPa, and their shear modulus varied between 0.35 and 0.45 TPa. They also pointed out that the increase in the aspect ratio of CNTs leads to a decrease in their critical buckling loads.

The buckling behavior of single- and multi-walled CNTs was investigated by Lu et al. [21] through an MD simulation, in terms of compressive strain. Finally, they found that the buckling behavior of MWCNTs is substantially dominated by the size of their outermost shell, and that smaller nanotubes have higher buckling stability. Liew et al. also used an MD simulation to probe the buckling properties of SWCNTs and MWCNTs by which they eventually obtained an optimum diameter for a maximum buckling load of SWCNTs [22].

An analysis was performed, based on CM, by Ru to use a simple shell model and evaluate the buckling properties of CNTs. Their analysis clearly showed that the van der Waals (vdW) forces between the walls of CNTs are not significantly influential on the critical strain for the infinitesimal buckling load of a DWCNT [23] while the analytical investigation by Chang and Li on the critical buckling strain of axially compressed chiral SWCNTs showed that with a constant diameter, zigzag tubes have more buckling stability than armchair ones. Their results also suggested that the effect of the van der Waals interaction between the layers of DWCNTs is rather negligible [24].

Bending deformation and buckling behavior of SWCNTs and MWCNTs were evaluated by Yao et al. through an FE modeling by which they proposed an explicit relationship between the critical bending buckling curvature and the diameter, length, and chirality of CNTs [25]. An analytical and FE investigation was performed by Rahmandoust and Öchsner to study the buckling behavior of SWCNTs. Eventually, their analytical and FE results, which were in good agreement, revealed that the classical Euler equation and the assumption of SWCNTs as hollow cylinders can successfully predict the buckling behavior of these nanostructures [26].

Buckling behavior of perfect and defective SWCNTs was evaluated by an MD simulation, proposed by Xin et al. under axial compression. Accordingly, it was learnt that buckling and axially compressive properties of SWCNTs noticeably depend on the length, the chirality, the temperature, and the initial structural defects of the tubes. Furthermore, they also reported that the axial critical buckling loads of SWCNTs with a large aspect ratio at normal temperatures can be predicted by the classical Euler formula and, eventually, vacant sites in the CNT structures considerably decrease their buckling strength [27].

A numerical simulation of MWCNTs was performed by Ghavamian et al. [28, 29] to evaluate their elastic and torsional properties in their perfect and atomically defective forms, based on the FEM. Their results illustrated that the Young's modulus of perfect CNTs is about 1 TPa and their shear modulus varies between 0.073 and 0.378 TPa. They also discovered that CNTs have anisotropic behavior and the atomic defects in the structure of CNTs lead to a decrease in their elastic and torsion moduli. It was also seen that the decrease in the value of the elastic modulus of defective CNTs follows a particular trend that could be expressed by mathematical relations in terms of the amount of atomic defects. Later on, the same approach was also employed by Ghavamian and Öchsner [30] for the evaluation of the buckling properties of MWCNTs. Finally, they reported that the atomically defective CNTs have lower critical buckling loads compared to the perfect models, and for the first time, they presented mathematical relations for the prediction of the decrease in buckling strength of defective CNTs, in terms of the amount of atomic defects.

6.1.2   Literature Review: Mechanical Properties of Hetero-junction CNTs

Apart from the mechanical properties of individual CNTs, it has experimentally been observed that it is also possible that two CNTs connect together by a heptagon–pentagon knee and construct hetero-junction (composite) CNTs. The importance of hetero-junction CNTs became highlighted when it was discovered that the junctions in their structure behave as nanoscale metal/semiconductor or semiconductor/semiconductor junctions and thus could be employed as the building blocks of nanoscale electronic devices made entirely of carbon. Moreover, these nanostructures, which are widely employed as solar cells, offer several interesting characteristics, that is, high mobility, excellent air stability, and high conductivity as well as exceptional mechanical characteristics, that is, lightness and high stiffness and large aspect ratios, which are inherited from their fundamental constructive tubes that lead to a higher efficiency and wide application potentials of these solar cells [31–33]. Thus, several scholars devoted their research to the modeling and characterization of such composite CNTs.

Jia et al. combined hetero-junction CNTs and silicon, doped with diluted HNO_3 to produce and achieve an efficient solar cell with an efficiency of 13.8%. They observed that acid infiltration of nanotube networks enhances the efficiency of silicon–carbon nanotube hetero-junction solar cells considerably by reducing the internal resistance. They also reported that the fabrication process of such solar cells is significantly simplified, compared with the conventional silicon cells [33].

Meunier et al. employed several pentagon–heptagon cell insertions to connect different CNTs and build different hetero-junction CNTs. They discovered that such insertions make the kink in the place of CNTs' connection, and as a consequence, it causes a bending angle between two connected CNTs in the structure of hetero-junction CNTs. Finally, they obtained the pentagon–heptagon defect energy in the junction of the tubes of about 6 eV, compared with each of the separate tubes [34]. Saito et al. [35] used a projection method to create a 3D model of hetero-junction CNTs by pentagon and heptagon pairs. Their modeling revealed that the connecting kink between CNTs introduces a 3D dihedral angle. Taking the calculated tunneling conductance for a metal–metal CN junction and a metal–semiconducting CN junction into account, they proclaimed the fact that such junctions can work as the smallest semiconductor devices.

Xosrovashvili and Gorji [36] proposed a numerical simulation of a hybrid hetero-junction SWCNT and GaAs solar cell by the AMPS-1D device simulation tool and probed their physics and performance by junction parameters. Finally, they learnt that the increase in the concentration of a discrete defect density in the absorber layer leads to a reduction in the electrical parameters of the system.

Dunlap [37] employed two different approaches to connect different carbon tubules for the creation of hetero-junction CNTs. First, they assumed that a hetero-junction CNT is formed when a graphene ribbon with infinite length but finite width is rolled into an infinite set of different tubules with different chiralities where the connections between every two CNTs are constructed by locating different obtained configurations one after each other by translation and rotation. Using the second approach, he created hetero-junction CNTs by employing one pentagon and one heptagon cell to connect two halves of perfect infinite tubules into a different tubule. Based on his modeling, it was concluded that the usage of pentagon–heptagon pairs is the best way to connect sets of tubules in a pair-wise way that creates a bending angle in the connection location with an ideal bending angle of 30°.

Rajabpour et al. [38] proposed 3D models of straight hetero-junction CNTs (with the armchair–armchair connection) based on the FEM and investigated the effect of length and chirality on the elastic modulus of hetero-junction CNTs, including two cases of variable length with constant chirality and variable chirality with constant length. According to their results, with the constant length, increase in the chirality of the tubes leads to an increase in the Young's modulus while with a constant chirality, the increase in the length of the tubes results in a slight decrease in the Young's modulus of hetero-junction CNTs. They also concluded that the FEM is a valuable tool for the characterization of hetero-junction CNTs. Hemmatian et al. [39] also used the FEM for the evaluation of the elastic properties of the defected, twisted, elliptic, bended, and hetero-junction CNTs by the creation of 3D FE models. Their results postulated that the torsional angle of twisted CNTs, the cross-sectional aspect ratio of elliptic CNTs, the bending angle of bended CNTs, and the choice of the length and the chirality of hetero-junction CNTs significantly influence their mechanical properties. They also reported the Young's modulus of hetero-junction CNTs to be in the range of 0.96 to 1.27 TPa and finally confirmed that the FEM could be considered a valuable approach for the characterization of CNTs.

Finally, Yengejeh et al. [40, 41] investigated the buckling and torsional properties of nine straight hetero-junction CNTs by numerical simulation, based on the FEM, and compared them with their fundamental tubes. For such a comparison and evaluation, they simulated straight hetero-junction CNTs with armchair–armchair and zigzag–zigzag connections and then obtained their critical buckling load and shear modulus under cantilever and twisting boundary conditions, respectively. Finally, they suggested that both of these quantities for hetero-junction CNTs are in the range of the critical buckling load and shear modulus of their fundamental homogeneous tubes. They also concluded that the buckling and torsional strength of straight hetero-junctions and their fundamental CNTs increased by increasing the chiral number of both armchair and zigzag CNTs.

A general consideration of the literature clearly shows that although many investigations have been done on the properties of CNTs, seldom has a comprehensive research on the mechanical and linear elastic behavior of hetero-junction CNTs been performed and presented. For instance, in most of the investigations, few and insufficient numbers of hetero-junction CNT models have been studied or their focus has mostly been only on straight connections and other possible connections (bent connections) have been neglected. It can also be observed that most researchers are limited to the electrical properties of hetero-junction CNTs, while less attention has been paid to their mechanical properties. Furthermore, all of the investigations on the mechanical properties of hetero-junction CNTs has been focused on the perfect models of these nanomaterials and never has a research have been done on the mechanical properties of atomically defective hetero-junction CNTs. Recently, a comprehensive research has been performed by Ghavamian and Öchsner on a collection of a considerable number of hetero-junction CNTs in their perfect and atomically defective forms, with all possible connection types (straight and bent), made of three homogeneous CNT types (armchair, zigzag, and chiral), based on the FEM in which the Young's and shear moduli and also critical buckling loads of all hetero-junction models have been obtained and compared with the ones for their fundamental and constructive tubes and also the influence of the atomic defects has been quantified. Therefore, in the actual chapter, the focus will be put on the FE modeling of hetero-junction CNTs, and the results from the investigations by Ghavamian and Öchsner will be presented that provide a more realistic insight about the mechanical properties of these nanomaterials for their proper selection and applications in nanoindustry.

6.2 Theory and Simulation

6.2.1 Atomic Geometry and Finite Element Simulation of Homogeneous CNTs

CNTs are assumed many times to be formed by rolling a graphene sheet into a hollow cylinder, with the thickness of a single carbon atom (0.43 nm) [42], diameters ranging from 1 to 50 nm and lengths over 10 μm. As illustrated in Figure 6.1(a)–(e), for the FE simulation of these nanomaterials, the CNTs are constructed of single rings, copied along the CNT axis to create a hollow cylinder, each of which is made of hexagonal unit cells, made of carbon atoms which

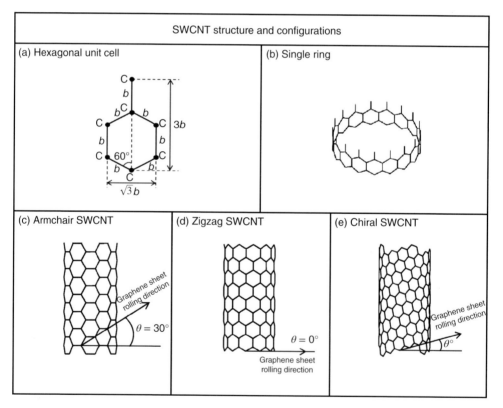

Figure 6.1 (a) Hexagonal unit cell; (b) single ring; (c)–(e) armchair, zigzag, and chiral SWCNTs, respectively. *Source*: Obtained from [47]

Table 6.1   Material and geometric properties of a C–C bond

Material and geometric properties of carbon–carbon covalent bond	
Corresponding force field constants	$k_r = 651.97$ nN/nm $k_\theta = 0.8758$ nN nm/rad^2 $k_\varphi = 0.2780$ nN nm/rad^2
$E = $ Young's modulus $= \dfrac{k_r^2 b}{4\pi k_\theta}$	5.484×10^{-6} N/nm^2
$R_b = $ bond radius $= 2\sqrt{\dfrac{k_\theta}{k_r}}$	0.0733 nm
$I_{xx} = I_{yy} = $ second moments of area $= \dfrac{\pi R_b^4}{4}$	2.2661×10^{-5} nm^4

Source: Obtained from [28].

are connected to each other by covalent carbon–carbon (C–C) bonds that are often modeled as 1D beam elements in a spatial network with six degrees of freedom, that is, three global displacements and three global rotations at their nodes (carbon atoms). Each of these elements also has material and geometric properties, which are obtained by linking solid mechanics and molecular mechanics concepts and calculated, based on an energy approach, by the quantities, called force field constants (k_r, k_θ, and k_φ), as shown in Table 6.1 [28]. In addition, the CNT configurations are defined by the tube chirality or helicity, which is expressed by the chiral vector $\vec{C}_h$ and the chiral or twisting angle θ. The chiral vector itself is defined by two unit vectors $\vec{a}_1$ and $\vec{a}_2$ and two integers m and n (steps along the unit vectors), which is presented by Eq. (6.1) [29],

$$\vec{C}_h = n\vec{a}_1 + m\vec{a}_2 \tag{6.1}$$

Generally, there are three SWCNT configurations, that is, armchair, zigzag, and chiral configurations, which are defined based on the chirality of the tube or the chiral angle by which the graphene sheet is rolled into a cylinder. As illustrated in Figure 6.1(c)–(e), in terms of the chiral vector (m and n) or the chiral angle θ, an armchair SWCNT is formed if ($m = n$) or ($\theta = 30°$). Likewise, if ($\theta = 0°$) or ($m = 0$), a zigzag SWCNT is constructed and eventually, a chiral structure is formed when ($0° < \theta < 30°$) or ($m \neq n \neq 0$). The radius of an SWCNT is also calculated by Eq. (6.2) where $a_0 = \sqrt{3}b$ and $b = 0.142$ nm is the length of the C–C bond [28, 30],

$$R_{CNT} = \frac{a_0 \sqrt{m^2 + mn + n^2}}{2\pi} \tag{6.2}$$

6.2.2   *Atomic Geometry and Finite Element Simulation of Hetero-junction CNTs*

A local atomic view on the structure of hetero-junction CNTs reveals the fact that these structures are constructed of carbon hexagonal unit cells, almost the same as their fundamental CNTs. However, the global view from experimental observations clearly demonstrates that hetero-junction CNTs are constructed of two CNTs with different chiralities, which connect together by a kink with pentagon–heptagon cell pairs (also known as Stone–Wales defects) in the bent locations of the connection [34, 40] (see Figure 6.2(a)). These kinks, whose existence and size seem to be noticeably effective on the mechanical properties of hetero-junction CNTs, are generally divided into two groups, straight and bent connections, as illustrated in Figure 6.2(b)–(d). Straight connections happen when the two connecting fundamental tubes have parallel orientations and their configurations are the same [35, 44]. Bent connections, which are created when the configuration of the fundamental constructive tubes of hetero-junction CNTs are different, are also divided into two groups of large-angle (with

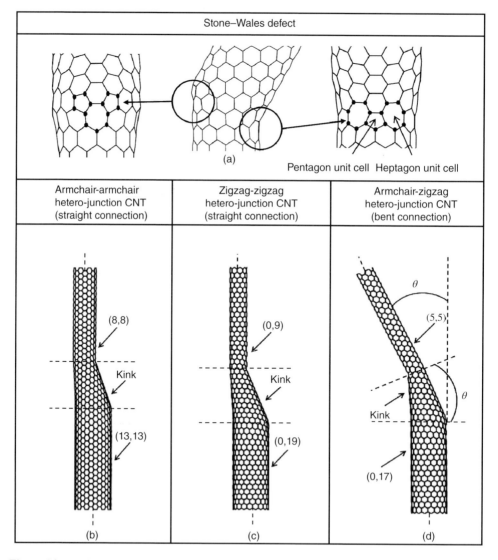

Figure 6.2 (a) Stone–Wales defect; (b) and (c) straight hetero-junction CNT; (d) bent hetero-junction CNT. *Source*: Obtained from [47]

bending angle about 36°) [45] and small-angle connections (with bending angle about 12°) [31, 34]. There is also another classification of the kinks that depends on their corresponding homogeneous tubes' configurations, which build the hetero-junction CNTs. Based on such a categorization, there are six possible hetero-junction CNTs, that is, armchair–armchair, zigzag–zigzag, armchair–zigzag, armchair–chiral, zigzag–chiral, and chiral–chiral hetero-junction CNTs. However, almost all of previous studies focused on hetero-junction CNTs, only with straight kinks (armchair–armchair and zigzag–zigzag connections).

For the FE simulation of the hetero-junction CNTs, first the coordinates of carbon atoms (as model nodes) and their connections (as elements of the models, representing C–C bonds) can be obtained by the CNT simulation specialized software CoNTub V1.0 (an algorithm for connecting two arbitrary CNTs) [46]. Then, this data can be imported into a commercial FE software (MSC. Marc) and modified to model hetero-junction CNTs.

Regarding the fact that the structure of every hetero-junction CNT comprises three parts of thinner tube, thicker tube, and the kink part, and thus it is not such a homogeneous body, for analytical calculations to determine the moduli and critical buckling loads of hetero-junction CNTs, the classical equations from CM were employed to check if these simple relationships could be used to approximately describe the mechanical behavior of these nanostructures. Therefore, the lengths of hetero-junction CNTs are simply calculated by summing up the length of the thinner and wider tubes and the kink part that is expressed by Eqs (6.3) and (6.4) where d is the diameter of the fundamental homogeneous tubes [40], and D_d is the difference between the diameters of the wider and thinner fundamental tubes that is expressed by Eq. (6.5). Finally, each hetero-junction CNT was assumed to be a homogeneous CNT with a constant area and diameter, which were addressed by A_{Hetero} and D_{Hetero}, respectively, whose values were approximated by a general weight percentage relation, presented by Eqs (6.6)–(6.8).

$$L_{\text{Hetero}} = L_{\text{thinner tube}} + L_{\text{kink}} + L_{\text{wider tube}} \tag{6.3}$$

$$L_{\text{Kink}} = \frac{\sqrt{3}}{2}\pi(D_d) \tag{6.4}$$

$$D_d = d_{\text{wider tube}} - d_{\text{thinner tube}} \tag{6.5}$$

$$A_{\text{Hetero}} = \frac{L_{\text{wider tube}} \times A_{\text{wider tube}} + L_{\text{kink}} \times A_{\text{kink}} + L_{\text{thinner tube}} \times A_{\text{thinner tube}}}{L_{\text{Hetero}}} \tag{6.6}$$

$$D_{\text{Hetero}} = \frac{L_{\text{wider tube}} \times D_{\text{wider tube}} + L_{\text{kink}} \times D_{\text{kink}} + L_{\text{thinner tube}} \times D_{\text{thinner tube}}}{L_{\text{Hetero}}} \tag{6.7}$$

$$A_{\text{kink}} = \frac{A_{\text{wider tube}} + A_{\text{thinner tube}}}{2} \tag{6.8}$$

6.2.3   Finite Element Simulation of Atomically Defective Hetero-junction CNTs

According to experimental observations, impurities and vacant sites are the two most likely atomic defects that generally hetero-junction CNTs deal with. For example, silicon is doped intentionally in the structure of CNTs because of their demanded properties, namely, semiconducting behavior for their required applications and improving the efficiency of the

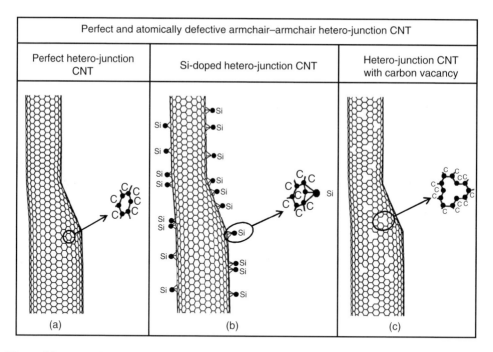

Figure 6.3 (a) Perfect hetero-junction CNT; (b), (c) hetero-junction CNTs with Si-doping and carbon vacancy, respectively. *Source*: Obtained from [47]

corresponding nanodevices. However, the mechanical properties of such structures seem to vary considerably as a result of these atomic defects. Therefore, for simulations of atomically defective hetero-junction CNTs and investigating the influence of the atomic defects on the linear elastic properties of these nanomaterials, finite numbers of randomly chosen carbon atoms are replaced with silicon atoms (nodes) and removed from their structures to introduce different amounts of Si-doping and carbon vacancy to the hetero-junction CNTs and create atomically defective models, as illustrated in Figure 6.3(a)–(c) [47]. It is obvious that in the Si-doped models, the material properties of the Si–C bond ($E_{Si-C} = 7.18327 \times 10^{-6}\,\text{N/nm}^2$) are different from the ones for the C–C bond ($E_{C-C} = 5.484 \times 10^{-6}\,\text{N/nm}^2$). However, the negligible difference in the covalent radius of carbon and silicon, which is 28 pm and only about 3% of their radius, implies the fact that the geometric properties for C–C and Si–C bonds could be assumed to be approximately the same [28].

6.3 Results and Discussion

6.3.1 Linear Elastic Properties of Perfect Hetero-junction CNTs

Ghavamian and Öchsner [43] provided a collection of 43 hetero-junction CNTs simulation models with all possible hetero-connection types, that is, armchair–armchair, zigzag–zigzag, armchair–zigzag, armchair–chiral, zigzag–chiral, and chiral–chiral between 44 straight fundamental homogeneous CNTs of different chiralities and configurations with a length of about 15 nm for their study and eventually investigated and compared the mechanical properties of all these models with the assumption of linear elastic behavior. Considering the anisotropic

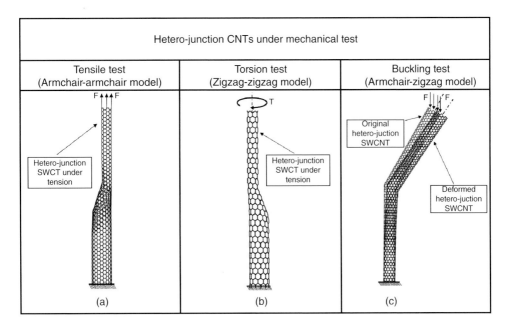

Figure 6.4 (a)–(c) Hetero-junction CNTs under tensile, torsion, and buckling tests, respectively. *Source*: Obtained from [43]

behavior of CNTs, Ghavamian and Öchsner [43] reached the conclusion that classical solid mechanics relations cannot be used for interrelating the elastic, shear, and buckling behavior of CNTs and, therefore, for the evaluation of each of the tensile and shear moduli and also critical buckling loads of homogeneous and hetero-junction CNTs, it is necessary to employ pure tensile, torsion, and buckling loads for the tests, respectively [29] as illustrated in Figure 6.4(a)–(c). Regarding the fact that Ghavavamian and Öchsner have already performed a comprehensive numerical investigation on hetero-junction CNTs, the results from their research will be presented in this chapter.

6.3.1.1   Elastic Modulus of Perfect Hetero-junction CNTs and Their Fundamental Tubes

For evaluating the Young's modulus of hetero-junction CNTs, Ghavamian and Öchsner [43] performed a tensile test by which different types of simulated hetero-junction CNTs accompanied by their fundamental tubes were elongated for an arbitrary displacement, and then the resultant reaction forces of the models were acquired from the FE commercial software to calculate the models' Young's modulus by Eqs (6.9)–(6.11).

$$\sigma = \text{stress} = \frac{P}{A} = \frac{\text{reaction force}}{\text{cross-sectional area}} \tag{6.9}$$

$$\varepsilon_{||} = \text{strain} = \frac{\Delta L}{L} = \frac{\text{axial displacement}}{\text{length of CNT}} \tag{6.10}$$

$$E = \text{Young's modulus} = \frac{\sigma}{\varepsilon_{||}} \tag{6.11}$$

Their results from [43], which were in good agreement with previous studies, showed that the Young's moduli of homogeneous armchair CNTs were about 1.040 TPa. However, zigzag CNTs' elastic moduli was obtained in the range of 0.989 and 1.037 TPa and eventually the Young's moduli of chiral CNTs oscillated between 1.003 and 1.049 TPa. It was also illustrated that the increase in the diameter of zigzag and chiral homogeneous CNTs leads to a very slight increase in their elastic moduli while for armchair tubes, the value of this quantity remained nearly constant. Tensile testing on the hetero-junction CNTs clearly demonstrated the fact that the Stone–Wales defect in the structure of hetero-junction CNTs reduces their tensile stiffness considerably whose amount highly depends on the magnitude of the bending angle in their structure and also the type and size of the kink, which is determined by the diameter of the corresponding wider and thinner fundamental tubes and the difference between them (D_d). According to the results in Figure 6.5(a)–(f), armchair–armchair and zigzag–zigzag hetero-junction CNTs assigned the Young's modulus peaks to themselves (between 0.7 and 1.028 TPa), while the models with armchair–zigzag kinks showed the elastic moduli, in the range of 0.02 and 0.21 TPa, which were the lowest obtained Young's modulus values among the models. The other hetero-junction CNTs' had the elastic modulus varying between 0.05 and 0.57 TPa.

6.3.1.2 Shear Modulus of Hetero-junction CNTs and Their Fundamental Tubes

Torsional properties of hetero-junction CNTs and their corresponding fundamental tubes were reported by Ghavamian and Öchsner [43] who investigated such properties by twisting the models for an arbitrary angle θ under the pure torsion test and then the corresponding reaction torque was obtained to calculate and compare the shear moduli of the models by Eq. (6.12) where θ, T, L, and J represent the twisting angle, the torque, the length, and the polar moment of inertia (calculated by Eq. (6.13)), respectively.

$$G = \frac{TL}{\theta J} \tag{6.12}$$

$$J = \int r^2 dA = 2\pi \int_{r_{in}}^{r_{out}} r^3 dr = \frac{\pi \left(r_{out}^{\,4} - r_{in}^{\,4}\right)}{2} \tag{6.13}$$

In Eq. (6.13), r_{out} and r_{in} are the outer and inner radii of the CNTs, respectively, which are equal to the radius of the CNT (r) plus and minus half of the wall thickness of a CNT. The results, illustrated in Figure 6.6(a)–(f), reveal that the shear modulus of the homogeneous CNTs varies between 0.1 and 0.422 TPa. It was also comprehended that the shear modulus of homogeneous CNTs generally encounters a gradual decrease by increasing the diameter of the CNTs, which confirms the results of previous research works in the literature. Torsion testing on the hetero-junction CNTs yielded shear moduli in the rage of 0.073 and 0.366 TPa for these hybrid nanostructures, which postulates the fact that the Stone–Wales defect causes a considerable decline in the shear modulus of hetero-junction CNTs, compared to their fundamental tubes. It was also perceived that the more the D_d is, the more the value of shear modulus of hetero-junction CNTs eludes the ones for their fundamental tubes. It could obviously be observed that in most cases, the shear modulus of hetero-junction CNTs seemed to be closer to the shear modulus of their corresponding wider tubes. Finally, among all the

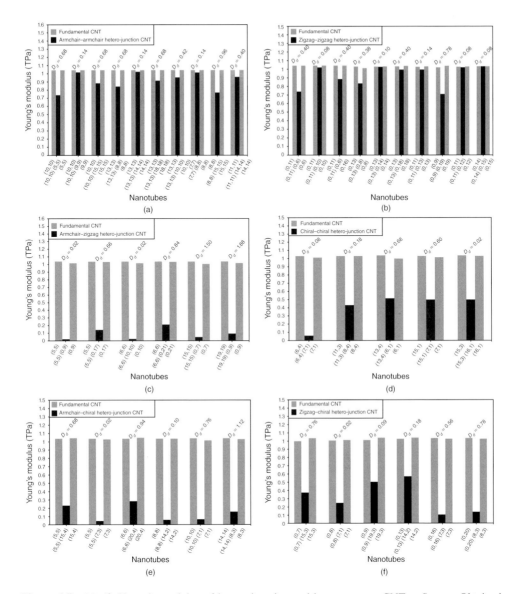

Figure 6.5 (a)–(f) Young's modulus of hetero-junction and homogeneous CNTs. *Source*: Obtained from [43]

hetero-connections CNTs, it seemed that the highest shear modulus (mostly about 0.3 TPa) belongs to zigzag–zigzag hetero-junction models while the shear modulus of armchair–zigzag hetero-junction CNTs had the lowest values (mostly about 0.15 TPa). For the other types of hetero-junction CNTs, the shear modulus was generally obtained to be varying between 0.15 and 0.2 TPa [43].

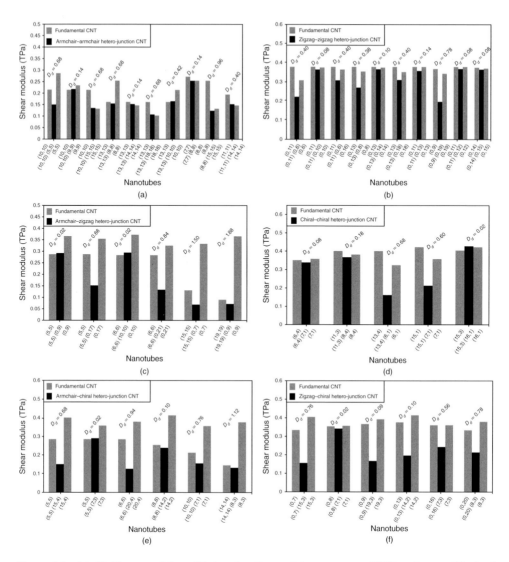

Figure 6.6   (a)–(f) Shear modulus of hetero-junction and homogeneous CNTs. *Source*: Obtained from [43]

6.3.1.3   Critical Buckling Loads of Hetero-junction CNTs and Their Fundamental Tubes

After considering the Young's and shear moduli of hetero-junction CNTs as well as their corresponding fundamental tubes, their buckling strength was evaluated by Ghavamian and Öchsner [43] under a buckling test with the cantilever boundary conditions. For the simulation of this test, first, the critical buckling load of homogeneous CNTs was estimated by the analytical calculation based on the classical Euler formula, which is presented by Eq. (6.14), and then this quantity was obtained through FE operations by a commercial software in which a buckling test was simulated and compared with the figures from analytical calculations. For such calculations, by Eq. (6.14), P_{cr} is the critical buckling load, n is the buckling mode, E is the elastic modulus of the model, and I is the second moment of area that is calculated by Eq. (6.15). K and L are the effective length constant and the length of the tube, respectively. Based on the type of boundary conditions (cantilever) and the number of mode, K is assumed to be equal to 2 and $n = 1$.

$$P_{cr} = \frac{n^2 \pi^2 E I}{(KL)^2} \tag{6.14}$$

$$I = \frac{\pi[(d + t)^4 + (d-t)^4]}{64} \tag{6.15}$$

For the FE simulation of a buckling test, an arbitrary compressive buckling load was exerted to all the models, and their critical buckling loads were obtained as an output of the commercial software. According to the results in Tables 6.2–6.4, the wider homogeneous CNTs have higher critical buckling loads than the thinner ones. As it reveals from the FEM results, which were in a good agreement with the analytical calculations and previous studies, the critical buckling load for CNTs with a length of about 15 nm, varied between 0.58 and 25.42 nN for armchair CNTs with diameters in range of 0.68 to 2.58 nm, and 0.17 to 8.04 nN for zigzag CNTs with diameters between 0.47 and 1.64 nm while the range of 0.23 to 9.51 nN were acquired for chiral CNTs with diameters varying from 0.51 to 1.74 nm. Noticing the figures in Tables 6.2–6.4 shows that although the results from FEM are in good agreement with analytical solution in most cases, the analytical solution does not work in all cases, particularly for CNTs with small diameters, which clearly demonstrates the necessity of a numerical approach for such an investigation [43].

After performing the buckling test on homogeneous CNTs, hetero-junction CNTs that were made of these homogeneous tubes were subjected to such a test and their critical buckling loads were evaluated to reveal their buckling behaviors. Based on the results, illustrated in Figure 6.7(a)–(f), the critical buckling load of hetero-junction CNTs lies in the range of the critical buckling loads of their corresponding fundamental tubes and its magnitude depends on them. However, it can generally be seen that the critical buckling load of hetero-junction CNTs tends to be closer to the one for their corresponding thinner tubes. Moreover, comparing the critical buckling loads of different hetero-junction models clearly demonstrates the fact

Table 6.2 Critical buckling load of armchair homogeneous SWCNTs

Tube chirality	P_{cr} (nN) (analytical)	P_{cr} (nN) (FEM)	Relative difference (%)	Tube chirality	P_{cr} (nN) (analytical)	P_{cr} (nN) (FEM)	Relative difference (%)
(5,5)	0.62	0.58	7.65	(11,11)	5.58	6.01	7.69
(6,6)	1.01	0.99	2.08	(13,13)	9.08	9.85	8.49
(7,7)	1.54	1.57	1.58	(14,14)	11.29	12.26	8.63
(8,8)	2.24	2.33	4.04	(15,15)	13.83	15.02	8.64
(9,9)	3.13	3.31	5.73	(18,18)	23.70	25.42	7.29
(10,10)	4.24	4.53	6.89	(19,19)	28.30	24.61	13.04

Source: Obtained from [43].

Table 6.3 Critical buckling load of zigzag homogeneous SWCNTs

Tube chirality	P_{cr} (nN) (analytical)	P_{cr} (nN) (FEM)	Relative difference (%)	Tube chirality	P_{cr} (nN) (analytical)	P_{cr} (nN) (FEM)	Relative difference (%)
(6,0)	0.24	0.17	28.35	(14,0)	2.31	2.38	2.86
(7,0)	0.36	0.28	20.18	(15,0)	2.82	2.93	4.04
(8,0)	0.50	0.43	13.98	(16,0)	3.39	3.56	5.00
(9,0)	0.68	0.62	9.23	(17,0)	4.04	4.27	5.78
(10,0)	0.91	0.86	5.55	(18,0)	4.77	5.07	6.40
(11,0)	1.18	1.15	2.65	(19,0)	5.58	5.96	6.91
(12,0)	1.50	1.49	0.43	(20,0)	6.48	6.95	7.32
(13,0)	1.88	1.90	1.40	(21,0)	7.47	8.04	7.66

Source: Obtained from [43].

Table 6.4 Critical buckling load of chiral homogeneous SWCNTs

Tube chirality	P_{cr} (nN) (analytical)	P_{cr} (nN) (FEM)	Relative difference (%)	Tube chirality	P_{cr} (nN) (analytical)	P_{cr} (nN) (FEM)	Relative difference (%)
(6,1)	0.30	0.23	22.52	(13,4)	3.00	3.16	5.57
(6,4)	0.63	0.59	7.29	(14,2)	2.82	2.98	5.42
(7,1)	0.43	0.36	15.24	(15,1)	3.07	3.22	4.87
(7,3)	0.66	0.61	7.48	(15,3)	3.78	4.03	6.52
(8,2)	0.71	0.66	6.94	(15,4)	4.21	4.52	7.38
(8,3)	0.87	0.83	4.67	(16,1)	3.67	3.90	6.13
(8,4)	1.05	1.04	1.86	(19,3)	7.02	7.61	8.49
(11,3)	1.76	1.80	2.22	(20,4)	8.67	9.51	9.65

Source: Obtained from [43].

that the magnitude of bending angle, the size of the kink which is determined by D_d and also the diameters of the wider and thinner fundamental tubes are highly determinant in the buckling stability of the resultant hetero-junction CNTs. It can be claimed that the hetero-junction CNTs with lower D_d have closer critical buckling load to the ones of their fundamental tubes. A general overview of the results also expresses the peak values of mostly varying about 5 to 10 nN for critical buckling loads of armchair–armchair hetero-junction models while the minimum critical buckling loads was observed to mostly vary about 1 nN for the models with armchair–zigzag connections. A general range of 1 and 2 nN was also reported for the critical buckling loads of hetero-junction CNTs with the other types of connections; however, a few unexpected high value of about 15 nN and substantially low value of about 0.2 nN were uniquely observed too [43].

6.3.2 Linear Elastic Properties of Atomically Defective Hetero-junction CNTs

After investigating the linear elastic properties of perfect hetero-junction CNTs, atomically defective models that include silicon impurities and vacant sites were created by Ghavamian

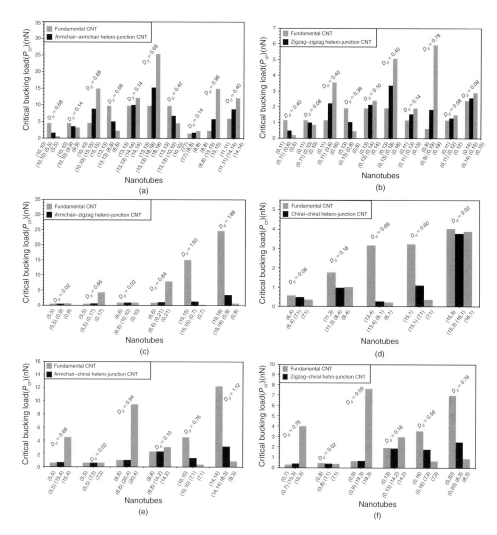

Figure 6.7 (a)–(f) Critical buckling load of hetero-junction and homogeneous CNTs. *Source*: Obtained from [43]

and Öchsner [47, 48] and their Young's and shear moduli, and also critical buckling loads were quantified and compared with the ones for the perfect models to evaluate the influence of the atomic defects on the mechanical properties of hetero-junction CNTs.

6.3.2.1   Linear Elastic Properties of Si-Doped Hetero-junction CNTs

Ghavamian and Öchsner simulated Si-doped hetero-junction CNTs by replacing a finite number of randomly chosen carbon atoms in the whole structure of the models with silicon atoms and then the corresponding material properties were introduced to Si–C bonds as well as C–C bonds. For their research, they exerted 1, 2, and 3% Si-doping to the models to create defective hetero-junction CNTs with controlled amounts of silicon impurity and eventually investigated their Young's and shear moduli, and also critical buckling loads under tensile, torsion, and buckling tests, respectively, to describe their linear elastic behavior and determine the effect of silicon impurity on such behavior. The results, illustrated in Figures 6.8–6.10, clearly show that the Young's and shear moduli and also the critical buckling load of hetero-junction CNTs decrease as a result of silicon impurities in their structure. However, this decrease seems to happen with steeper slope in their Young's modulus and critical buckling loads than in their shear modulus. It was also observed that the decrease in the moduli and critical buckling load of hetero-junction CNTs follows an almost linear trend, which can be approximated by a straight line, expressed by Eq. (6.16) where CMC is the change in the moduli and critical buckling load of the hetero-junction CNT in percent, p represent the percentage of the atomic defect (Si-doping), and a is a coefficient that is equal to -1.59 ± 0.39, -1.10 ± 0.26, and -1.76 ± 0.52 for the change in the Young's and shear modulus and also critical buckling load, respectively [47, 48].

$$CMC = a \times p \tag{6.16}$$

6.3.2.2 Linear Elastic Properties of Hetero-junction CNTs with Carbon Vacancy

For the simulation of hetero-junction CNTs with carbon vacancies, Ghavamian and Öchsner eliminated finite numbers of randomly chosen nodes that represent carbon atoms from the whole structure of the models to introduce controlled amounts of vacant sites and imperfections in their structure and create atomically defective hetero-junction CNTs. For their research, they imposed 0.5, 1, and 1.5% carbon vacancy to the perfect models and finally evaluated the Young's and shear moduli, and also critical buckling loads of the defective models and compared them with the ones for perfect hetero-junction CNTs to investigate the influence of such defects on their tensile, torsion, and buckling behavior. The obtained results from the mechanical tests, which are illustrated in Figures 6.11–6.13, obviously demonstrated that carbon vacancies in the structure of hetero-junction CNTs lead to a considerable linear decreases in their Young's and shear moduli and also critical buckling load which could be predicted by Eq. (6.16) where the coefficient a is equal to -8.17 ± 1.98, -10.03 ± 2.78, and $a = -11.58 \pm 1.98$ for the change in the Young's and shear modulus, and also critical buckling load, respectively [47, 48].

6.4 Conclusion

Overall, it was learned that homogeneous and hetero-junction CNTs are of a great importance due to their exceptional properties, which make these nanomaterials considerably applicable and demanded in the nanoindustry. In this chapter, the research works on a considerable number of different types of perfect and atomically defective hetero-junction CNTs with all possible connection types as well as their constructive homogeneous CNTs of different chiralities and configurations were presented and their elastic, torsional, and buckling properties were numerically investigated based on the FEM with the assumption of linear elastic behavior. The numerical characterization of homogeneous CNTs showed that the Young's modulus of homogeneous CNTs varies between 0.989 and 1.49 TPa while their shear modulus was found in the range of 0.1 and 0.422 TPa. The critical buckling load of homogeneous CNTs

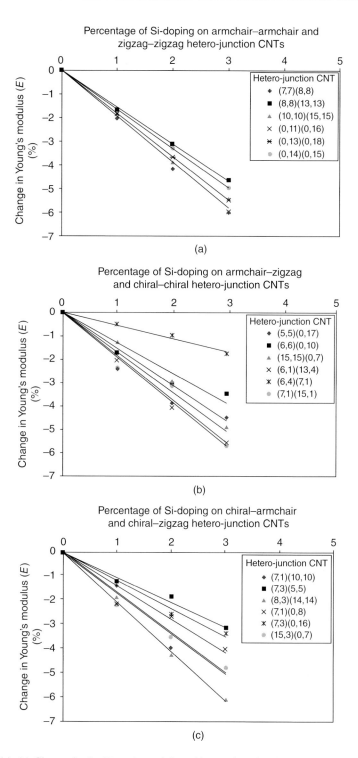

Figure 6.8 (a)–(c) Change in the Young's modulus of hetero-junction CNTs as a result of Si-doping. *Source*: Obtained from [47]

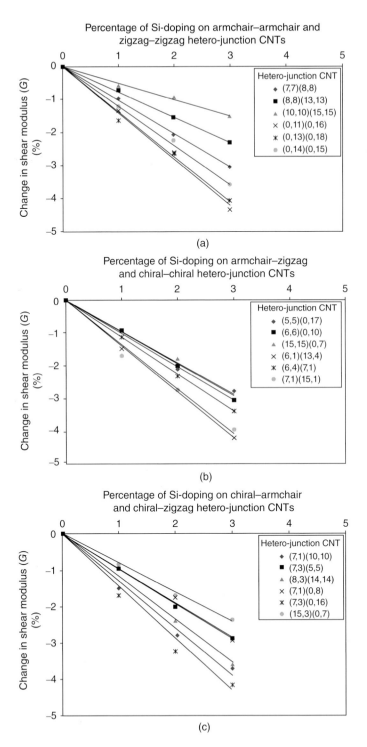

Figure 6.9   (a)–(c) Change in the shear modulus of hetero-junction CNTs as a result of Si-doping.
Source: Obtained from [47]

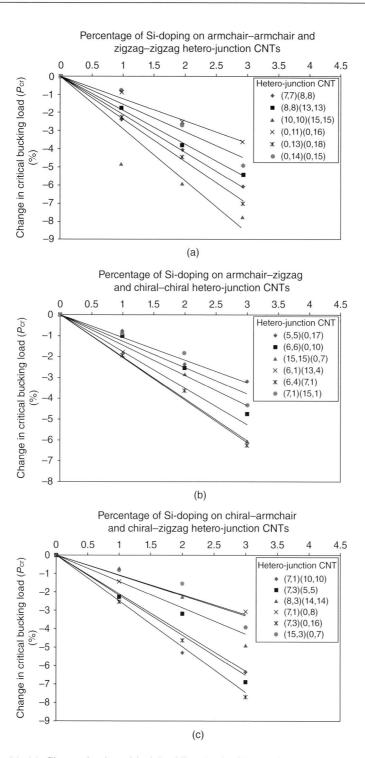

Figure 6.10 (a)–(c) Change in the critical buckling load of hetero-junction CNTs as a result of Si-doping. *Source*: Obtained from [48]

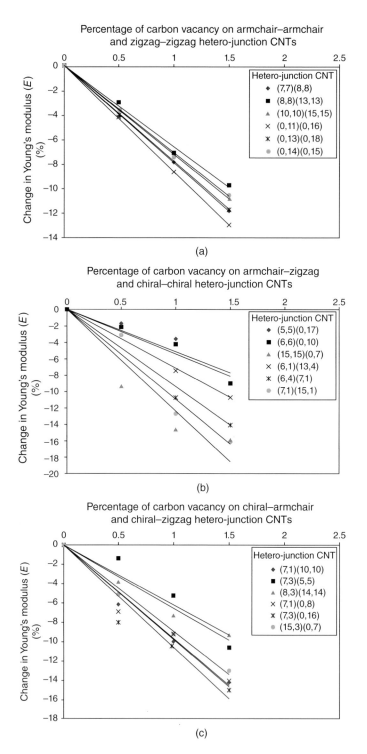

Figure 6.11   (a)–(c) Change in the Young's modulus of hetero-junction CNTs as a result of carbon vacancy. *Source*: Obtained from [47]

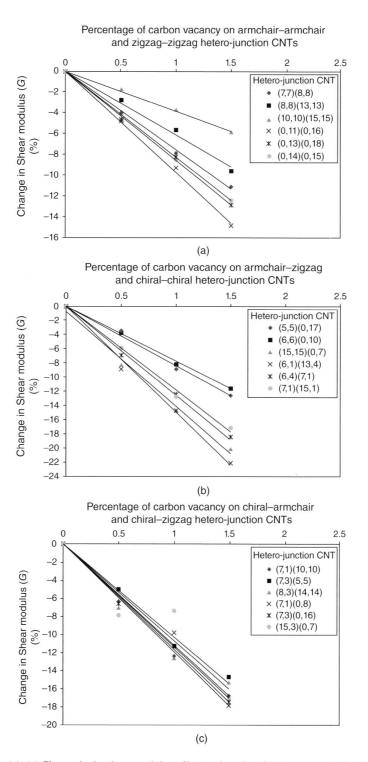

Figure 6.12 (a)–(c) Change in the shear modulus of hetero-junction CNTs as a result of carbon vacancy. *Source*: Obtained from [47]

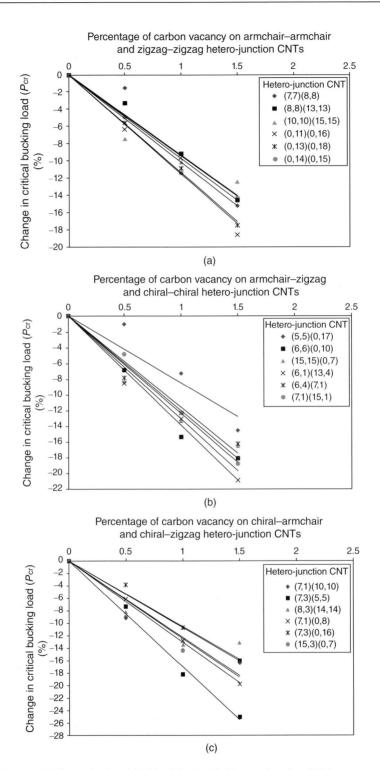

Figure 6.13 (a)–(c) Change in the critical buckling load of hetero-junction CNTs as a result of carbon vacancy. *Source*: Obtained from [48]

was also reported between 0.58 and 0.17 nN. Comparison between the results from mechanical tests on homogeneous CNTs suggests that armchair CNTs have the highest elastic and buckling stabilities while the torsional strength was observed to reach its maximum values in the chiral CNTs than the other CNT configurations. The minimum Young's moduli and critical buckling loads were generally seen in zigzag models, while armchair CNTs demonstrated the lowest shear modulus. An overview of the results from the mechanical tests on the hetero-junction CNTs clearly shows that the magnitude of bending angle and also the size and the type of the kink (Stone–Wales defect) in the structure of hetero-junction CNTs, which depend on the diameters and configurations of their constructive CNTs, are noticeably influential on the mechanical properties of these hetero-junction CNTs and decreases their elastic and torsional strength while the buckling strength of these hetero-junction CNTs lies in the range of the buckling strength of their corresponding fundamental tubes. The shear modulus of hetero-junction CNTs also showed a general tendency to be closer to the one of their wider fundamental CNTs while their critical buckling loads seemed to incline to the ones for their thinner fundamental tubes. The research on the mechanical properties of hetero-junction CNTs revealed the fact that the highest Young's modulus was observed for the hetero-junction CNTs with armchair–armchair and zigzag–zigzag connections with the elastic modulus in the range of 0.7 and 1.028 TPa while this quantity faced its lowest value for the models with armchair–zigzag kinks whose Young's modulus was between 0.02 and 0.21 TPa. On the other hand, zigzag–zigzag hetero-junction CNTs demonstrated the highest shear modulus of mostly about 0.3 TPa while the armchair–zigzag models showed the lowest shear modulus of mostly about 0.15 TPa. Finally, the results from the buckling test on the hetero-junction CNTs yielded the peak values, mostly varying about 5–10 nN for armchair–armchair hetero-junction CNTs while the hetero-junction tubes with armchair–zigzag connection showed the lowest critical buckling loads of mostly varying about 1 nN. Furthermore, after investigating the linear elastic properties of perfect hetero-junction CNTs and their corresponding homogeneous tubes, such properties were evaluated for atomically defective hetero-junction CNTs with silicon impurities and vacant sites of different amounts. It could clearly be observed that the atomic defects in the structure of hetero-junction CNTs lead to an almost linear decrease in the mechanical stability and strength of hetero-junction CNTs, which appeared to be considerably more in the models with carbon vacancy rather than Si-doped models. It can logically be perceived that the offered mathematical relation for expressing and quantifying such a linear decrease in the strength of the defective hetero-junction CNTs in terms of the amount of the atomic defects will be of a great contribution to prediction of the linear elastic properties of these nanomaterials for their proper selection and applications in nanoindustry.

References

[1] Dresselhaus, M.S., Dresselhaus, G. and Saito, R. (1995) Physics of carbon nanotubes. *Carbon*, **33**, 883–891.

[2] Yang, H.K. and Wang, X.T. (2007) Torsional buckling of multi-wall carbon nanotubes embedded in an elastic medium Composite Structures. *Composite Structures*, **77**, 182–192.

[3] Wu, Y., Zhang, X., Leung, A.Y.T. and Zhong, W. (2006) An energy-equivalent model on studying the mechanical properties of single-walled carbon nanotubes. *Thin-Walled Structures*, **44**, 667–676.

[4] Lu, J.P. (1997) Elastic properties of carbon nanotubes and nanoropes. *Physical Review Letters*, **79**, 1297–1300.

[5] Wu, Y., Huang, M., Wang, F. et al. (2008) Determination of the Young's modulus of structurally defined carbon nanotubes. *Nano Letters*, **8**, 4158–4161.

[6] Li, C. and Chou, T.W. (2003) A structural mechanics approach for the analysis of carbon nanotubes. *International Journal of Solids and Structures*, **40**, 2487–2499.

[7] To, C.W.S. (2006) Bending and shear moduli of single-walled carbon nanotubes. *Finite Elements in Analysis and Design*, **42**, 404–413.

[8] Nahas, M.N. and Abd-Rabou, M. (2010) Finite element modeling of carbon nanotubes. *International Journal of Mechanical and Mechatronics*, **10**, 19–24.

[9] Kalamkarov, A.L., Georgiades, A.V., Rokkam, S.K. et al. (2006) Analytical and numerical techniques to predict carbon nanotubes properties. *International Journal of Solids and Structures*, **43**, 6832–6854.

[10] Natsuki, T., Kriengkamol, T. and Morinobu, E. (2004) Prediction of elastic properties for single-walled carbon nanotubes. *Carbon*, **42**, 39–45.

[11] Rahmandoust, M. and Öchsner, A. (2012) On finite element modeling of single and multi-walled carbon nanotubes. *Journal of Nanoscience and Nanotechnology*, **12**, 8129–8136.

[12] Meo, M. and Rossi, M. (2006) Prediction of Young's modulus of single-wall carbon nanotubes by molecular-mechanics based finite element modeling. *Composites Science and Technology*, **66**, 1597–1605.

[13] Ávila, A.F. and Lacerda, G.S.R. (2008) Molecular mechanics applied to single-walled carbon nanotubes. *Materials Research*, **11**, 325–333.

[14] Shokrieh, M.M. and Rafiee, R. (2010) Prediction of Young's modulus of graphene sheets and carbon nanotubes using nanoscale continuum mechanics approach. *Materials and Design*, **31**, 790–795.

[15] Song, H.Y., Sun, H.M. and Zhang, G.X. (2006) Molecular dynamics study of effects of Si-doping upon structure and mechanical properties of carbon nanotube. *Communications in Theoretical Physics*, **45**, 741–744.

[16] Jin, Y. and Yuan, F.G. (2003) Simulation of elastic properties of single-walled carbon nanotubes. *Composites Science and Technology*, **63**, 1507–1515.

[17] Tserpes, K.I. and Papanikos, P. (2005) Finite element modeling of single-walled carbon nanotubes. *Composites: Part B*, **36**, 468–477.

[18] Yu, W., Xi, W.X. and Xianggui, N. (2004) Atomistic simulation of the torsion deformation of carbon nanotubes. *Modelling and Simulation in Materials Science and Engineering*, **12**, 1099–1107.

[19] Li, C. and Chou, T.W. (2003) Elastic moduli of multi-walled carbon nanotubes and the effect of van der Waals forces. *Composites Science and Technology*, **63**, 1517–1524.

[20] Fan, C.W., Liu, Y.Y. and Hwu, C. (2009) Finite element simulation for estimating the mechanical properties of multi-walled carbon nanotubes. *Applied Physics A*, **95**, 819–831.

[21] Lu, J.M., Hwang, C.C., Kuo, Q.Y. and Wang, Y.C. (2008) Mechanical buckling of multi-walled carbon nanotubes: the effects of slenderness ratio. *Physica E*, **40**, 1305–1308.

[22] Liew, K.M., Wong, C.H., He, X.Q. et al. (2004) Nanomechanics of single and multiwalled carbon nanotubes. *Physical Review B*, **69**, 1154291–1154298.

[23] Ru, C.Q. (2000) Effect of van der Waals forces on axial buckling of a double-walled carbon nanotube. *Journal of Applied Physics*, **87**, 7227–7231.

[24] Chang, T., Li, G. and Guo, X. (2005) Elastic axial buckling of carbon nanotubes via a molecular mechanics model. *Carbon*, **43**, 287–294.

[25] Yao, X., Han, Q. and Xin, H. (2008) Bending buckling behaviors of single- and multi-walled carbon nanotubes. *Computational Materials Science*, **43**, 579–590.

[26] Rahmandoust, M. and Öchsner, A. (2011) Buckling behavior and natural frequency of zigzag and armchair single-walled carbon nanotubes. *Journal of Nano Research*, **16**, 153–160.

[27] Xin, H., Han, Q. and Yao, X.H. (2007) Buckling and axially compressive properties of perfect and defective single-walled carbon nanotubes. *Carbon*, **45**, 2486–2495.

[28] Ghavamian, A., Rahmandoust, M. and Öchsner, A. (2012) A numerical evaluation of the influence of defects on the elastic modulus of single and multi-walled carbon nanotubes. *Computational Material Science*, **62**, 110–116.

[29] Ghavamian, A., Rahmandoust, M. and Öchsner, A. (2013) On the determination of the shear modulus of carbon nanotubes. *Composites: Part B*, **44**, 52–59.

[30] Ghavamian, A. and Öchsner, A. (2012) Numerical investigation on the influence of defects on the buckling behavior of single-and multi-walled carbon nanotubes. *Physica E*, **46**, 241–249.

[31] Chico, L., Crespi, V.H., Benedict, L.X. et al. (1996) Pure carbon nanoscale devices: nanotube heterojunctions. *Physical Review Letters*, **76**, 971–794.

[32] Jia, Y., Wei, J., Wang, K. et al. (2008) Nanotube–silicon heterojunction solar cells. *Advanced Materials*, **20**, 4594–4598.

[33] Jia, Y., Cao, A., Bai, X. et al. (2011) Achieving high efficiency silicon-carbon nanotube heterojunction solar cells by acid doping. *Nano Letters*, **11**, 1901–1905.

[34] Meunier, V., Henrard, L. and Lambin, P. (1998) Energetics of bent carbon nanotubes. *Physical Review B*, **57**, 2586–2591.

[35] Saito, R., Dresselhaus, G. and Dresselhaus, M.S. (1996) Tunneling conductance of connected carbon nanotubes. *Physical Review B*, **53**, 2044–2050.

[36] Xosrovashvili, G. and Gorji, N.E. (2014) Numerical simulation of carbon nanotubes/GaAs hybrid PV devices with AMPS-1D. *International Journal of Photoenergy*, **2014**, 1–6.

[37] Dunlap, B.I. (1994) Relating carbon tubules. *Physical Review B*, **49**, 5643–5650.

[38] Rajabpour, M., Hemmatian, H., Fereidoon, A. (27-29 July 2011) *Proceedings of the 2nd international conference on nanotechnology: fundamentals and applications*, Ottawa, Ontario, Canada, paper No. 320, Pages: 320-1–320-6.

[39] Hemmatian, H., Fereidoon, A. and Rajabpour, M. (2014) Mechanical properties investigation of defected, twisted, elliptic, bended and hetero-junction carbon nanotubes based on FEM. *Fullerenes, Nanotubes, and Carbon Nanostructures*, **22**, 528–544.

[40] Yengejeh, S.I., Akbar Zadeh, M. and Öchsner, A. (2014) On the buckling behavior of connected carbon nanotubes with parallel longitudinal axes. *Applied Physics A*, **115**, 1335–1344.

[41] Yengejeh, S.I., Akbar Zadeh, M. and Öchsner, A. (2014) Numerical characterization of the shear behavior of hetero-junction carbon nanotubes. *Journal of Nano Research*, **26**, 143–151.

[42] Gupta, S.S., Bosco, F.G. and Batra, R.C. (2010) Wall thickness and elastic moduli of single-walled carbon nanotubes from frequencies of axial, torsional and inextensional modes of vibration. *Computational Materials Science*, **47**, 1049–1059.

[43] Ghavamian, A. and Öchsner, A. (2015) A comprehensive numerical investigation on the mechanical properties of hetero-junction carbon nanotubes. *Communications in Theoretical Physics*, **64**, 215–230.

[44] Charlier, J.C., Ebbesen, T.W. and Lambin, P. (1996) Structural and electronic properties of pentagon-heptagon pair defects in carbon nanotubes. *Physical Review B*, **53**, 11108–11113.

[45] Lambin, P., Fonseca, A., Vigneron, J.P. et al. (1995) Structural and electronic properties of bent carbon nanotubes. *Chemical Physics Letters*, **245**, 85–89.

[46] Melchor, S. and Dobado, J.A. (2004) CoNTub: an algorithm for connecting two arbitrary carbon nanotubes. *Journal of Chemical Information and Computer Science*, **44**, 1639.

[47] Ghavamian, A., Andriyana, A., Chin, A.B. and Öchsner, A. (2015) Numerical investigation on the influence of the atomic defects on the tensile and torsional behavior of hetero-junction carbon nanotubes. *Materials Chemistry and Physics*, **164**, 122–137.

[48] Ghavamian, A. and Öchsner, A. (2015) Numerical investigation on the influence of the atomic defects on the buckling behavior of hetero-junction carbon nanotubes. *Modelling and Simulation in Materials Science and Engineering*, submitted.

7

A Methodology for the Prediction of Fracture Properties in Polymer Nanocomposites

Samit Roy and Avinash Akepati

Department of Aerospace Engineering and Mechanics, University of Alabama, Tuscaloosa, AL, USA

7.1 Introduction

Fracture can be defined as the separation of a structure into two or more pieces when subjected to mechanical loading. It is associated with the development of a displacement discontinuity in the structure followed by crack growth. When the direction of loading is perpendicular to the direction of crack growth, it is called Mode I fracture (see Figure 7.1). Fracture can be characterized through parameters such as strain energy release rate (G), stress intensity factor (K), and the J-integral (J). The critical values of these parameters represent the fracture strength of a material. Once the critical value is reached, crack initiation occurs in the material.

Although G and K are widely used to characterize fracture, they are applicable only in the linear elastic range of the stress–strain response of the material. On the other hand, J-integral is identical to G in the linear range but has the advantage of being applicable in the nonlinear range as well as being contour path independent. Also, the critical value of J-integral at crack initiation is directly related to fracture toughness of the material, while the value of J as a function of crack length (a) during crack growth provides the resistance of the material to crack growth, or R-curve. Therefore, J-integral could be used as fracture criterion at the nanoscale.

7.2 Literature Review

The use of J-integral [1] as a fracture criterion for inelastic materials can be found extensively in the literature [2, 3]. Researchers have attempted to extend the application of the J-integral to nanostructured materials [4–6], given the need for a suitable metric to quantify debonding

Advanced Computational Nanomechanics, First Edition. Edited by Nuno Silvestre.
© 2016 John Wiley & Sons, Ltd. Published 2016 by John Wiley & Sons, Ltd.

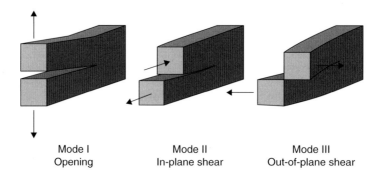

Mode I   Mode II   Mode III
Opening   In-plane shear   Out-of-plane shear

Figure 7.1 Modes of fracture

along material interfaces at the nanoscale (e.g., grain boundary decohesion, dislocation, debonding between CNT and polymer matrix, and so on). Eshelby [7] demonstrated that the J-integral could be interpreted as the divergence of the Eshelby energy–momentum tensor. At finite temperatures, the proper energy potential for computing the J-integral depends on the process or ensemble. Consequently, the Helmholtz free energy is the proper potential for computing J at finite temperatures for an isothermal process as it includes the entropic contribution, whereas the internal energy is the appropriate potential for computing J for an isentropic/adiabatic process [7]. Because free energy is complex to compute directly since it inherently involves computation of entropy density, it is not surprising that most attempts at estimating atomistic J-integral to date were performed at near zero temperatures [8, 4], where differences between the internal energy and Helmholtz free energy are not significant. Recently, Jones and Zimmerman [4] developed a novel molecular dynamics (MD)-based methodology for computing the J-integral in nanostructured materials through the construction of continuum fields from discrete atomic data through the use of Hardy's [9] localization functions. These continuum fields were subsequently used to compute contour-integral expression for J that involves gradients of continuum fields, such as the deformation gradient tensor. Nakatani et al. [10] employed changes in potential energy density in an MD simulation to estimate strain energy density. Because their system cannot be clearly determined to be isothermal or adiabatic, the relevant stress potential metric is unclear. Xu et al. [11] used a system energy release rate approach to compute critical value of J for ductile fracture of a nanosized crystal of nickel. Their analysis computes J-integral using changes in potential energy due to crack growth without entropic effects, even though the process takes place isothermally at a temperature of 300 K. Latapie et al. [12] use a similar MD approach to examine ductile fracture behavior of nanosized crystals of iron (Fe) at elevated temperatures of 100, 300, and 600 K. None of these past attempts at computing fracture energy at finite temperatures have made use of the free energy and entropic contribution, with the exception of a recent work by Jones et al. [5].

7.3   Atomistic J-Integral Evaluation Methodology

This chapter discusses the use of MD simulations to compute atomistic J-integral, in order to quantify the influence of nanofillers (such as graphene platelets) on fracture toughness of a polymer. The concept of J-integral was developed by J. R. Rice [1]. In conventional macroscale fracture mechanics, the total J-integral vector ($\boldsymbol{J}_T$), defined as the divergence of the Eshelby energy–momentum tensor [7] (as shown in Eq. (7.1)), has been used to quantify the crack driving force available from thermomechanical loading as well as material inhomogeneities.

$$J_T = \int_{\partial\Omega} \langle \underline{S} \rangle N dA = \int_{\partial\Omega} (\langle \Psi \rangle N - \langle \underline{H}^T \underline{P} \rangle N) dA = J_U - J_\eta \tag{7.1}$$

In Eq. (7.1), $\underline{S}$ is the Eshelby energy–momentum tensor, Ψ is the free energy density, $\underline{H}$ is the displacement gradient tensor, $\underline{P}$ is the first Piola–Kirchhoff stress tensor, N is the outward normal to the surface $\partial\Omega$ along which contour the J-integral is being evaluated, J_U is the contribution to J-integral due to internal energy (U), and J_η is the entropic contribution (J_U and J_η will be discussed in detail in following sections). Here, it is assumed that ensemble average $\langle * \rangle$ is approximated by the time average of the quantity over a sufficiently long period of time. The critical value of scalar component J_k at crack initiation is related to the fracture toughness of the material, where the subscript k denotes the Cartesian components of the J vector ($k = 1, 2, 3$) in three dimensions and these components are related to, but not the same as, the three primary fracture modes. Specifically, $J_1 = J^I + J^{II} + J^{III}$ and $J_2 = -2\sqrt{(J^I J^{II})}$, where subscripts imply the Cartesian components of the J-vector and the superscripts imply standard fracture modes [13]. In the absence of Mode III, that is, for purely in-plane deformation, the above equations can be solved to obtain the individual fracture modes from the Cartesian components of the J-vector as given by $J^I = \frac{J_1 \pm \sqrt{J_1^2 - J_2^2}}{2}$ and $J^{II} = J_1 - J^I$. Therefore, the Cartesian components of the J-integral can be used as a suitable metric for estimating the crack driving force as well as the fracture toughness of a material as the crack begins to initiate for Mode I, Mode II, and for mixed-mode fracture processes undergoing proportional loading. However, for the conventional macroscale definition of the J-integral to be valid at the nanoscale in terms of the continuum stress and displacement fields (and their spatial derivatives) requires the construction of local continuum fields from discrete atomistic data, using these data in the conventional contour integral expression for J, as given by Eq. (7.1) [4, 5]. One such methodology was proposed by Hardy [9], which allows for the local averaging necessary to obtain the definition of free energy, deformation gradient, and Piola–Kirchhoff stress as fields (and divergence of fields) and not just as total system averages. The formulae used for evaluating each term on the RHS of Eq. (7.1) from MD simulations are given below. The energetic component of total J (J_U) is computed using the formula:

$$J_U = \int_{\partial\Omega} (\langle U \rangle N - \langle \underline{H}^T \underline{P} \rangle N) dA \tag{7.2}$$

In order to facilitate J-integral computation using discrete atomistic data, localization boxes (as shown in Figure 7.2) are constructed along the integral contour, the position of an atom with respect to the box, which determined the value of localization function. The localization function has to satisfy the following conditions: $\psi > 0$ and $\int_\Omega \psi dV = 1$. The localization function used in this work is given by Eq. (7.3).

$$\zeta(X - X^\alpha) = \begin{cases} \dfrac{1}{L_x L_y L_z} & \text{if } \begin{aligned} & X_I - \dfrac{L_x}{2} \le X^\alpha \le X_I + \dfrac{L_x}{2} \\ & Y_I - \dfrac{L_y}{2} \le Y^\alpha \le Y_I + \dfrac{L_y}{2} \\ & Z_I - \dfrac{L_z}{2} \le Z^\alpha \le Z_I + \dfrac{L_z}{2} \end{aligned} \\ 0 \end{cases} \tag{7.3}$$

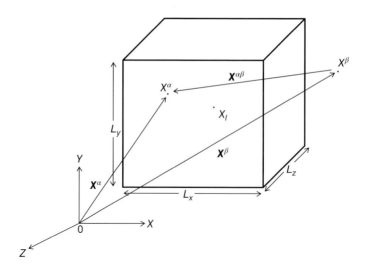

Figure 7.2   Illustration of a localization box with one atom inside and another outside the box

where L_x, L_y, and L_z are the dimensions of the localization box and X_I, Y_I, and Z_I are the coordinates of the centroid, as illustrated in Figure 7.2. The terms required for computation of J_U can now be defined as

$$U(X,t) = \sum_{I=1}^{K} \sum_{\alpha=1}^{M} N_I(X)\zeta_{I\alpha}(X^\alpha - X_I)\phi^\alpha(t) - U_R(X) \tag{7.4}$$

$$\underline{H}(X,t) = \nabla_x u(X,t) = \sum_{I=1}^{K} \sum_{\alpha=1}^{M} \nabla_x N_I(X)\zeta_{I\alpha}(X^\alpha - X_I)u^\alpha(t) \tag{7.5}$$

$$\underline{P}(X,t) = -\frac{1}{2} \sum_{\alpha=1}^{M} \sum_{\beta \neq \alpha}^{M} f^{\alpha\beta} \otimes X^{\alpha\beta} B^{\alpha\beta}(X) \tag{7.6}$$

In Eq. (7.4), ϕ^α is the atomistic potential of atom α, $U_R(X)$ is the reference potential energy density of atom α at 0 K, ζ is the Hardy localization function, M is the number of atoms in the localization box, and $N_I(X)$ are interpolation functions with $I = 1, K$, where K is the total number of nodes. In Eq. (7.5), $u(X,t)$ is the displacement vector. In Eq. (7.6), $f^{\alpha\beta}$ is the vector representing the force between atoms α and β, $X^{\alpha\beta}$ is the vector representing the difference in their positions, and $B^{\alpha\beta}$ is the bond function defined by Hardy [9].

Stored energy density field is given by

$$U(X,t) = \sum_{\alpha=1}^{M} (\phi^\alpha(t) - \phi_X^\alpha)\zeta(X - X^\alpha) \tag{7.7}$$

$$U(X,t) = \sum_{\alpha=1}^{M} \phi^\alpha(t)\zeta(X - X^\alpha) - U_R(X) \tag{7.8}$$

where ϕ^{α} is atomistic potential, ϕ^{α}_{X} is the reference potential energy density of at 0 K, ζ is the localization function, and $U_R(X)$ is a constant. Defining

$$\phi_I(X_I, t) = \sum_{\alpha=1}^{M} \zeta_{I\alpha}(X^{\alpha} - X_I)\phi^{\alpha}(t) \tag{7.9}$$

$$\phi(X, t) = \sum_{I=1}^{K} N_I(X)\phi_I(X_I, t) \tag{7.10}$$

where $N_I(X)$ are the interpolation functions. Substituting (7.9) and (7.10) in (7.7),

$$U(X, t) = \sum_{I=1}^{K} \sum_{\alpha=1}^{M} N_I(X)\zeta_{I\alpha}(X^{\alpha} - X_I)\phi^{\alpha}(t) - U_R(X) \tag{7.11}$$

Displacement gradient is given by

$$\underline{H}(X, t) = \nabla_x u(X, t) = \nabla_x \sum_{I=1}^{K} N_I(X)u_I(t) \tag{7.12}$$

$$\underline{H}(X, t) = \sum_{I=1}^{K} \sum_{\alpha=1}^{M} \nabla_x N_I(X)\zeta_{I\alpha}(X^{\alpha} - X_I)u^{\alpha}(t) \tag{7.13}$$

where $u(X, t)$ is the displacement vector.

The bond function is used to account for the interatomic forces where one atom is inside the localization box and the other one is outside. In such cases, only the fraction of the bond length inside the localization box is used for force computations. The bond function can be defined as

$$B^{\alpha\beta} = \int_0^1 \zeta(\lambda(X_{\alpha} - X) + (1 - \lambda)(X_{\beta} - X))d\lambda \tag{7.14}$$

where λ goes from 0 to 1 if β is inside the box. Otherwise, only till the fraction of $X^{\alpha\beta}$ is inside the box.

Setting $r = \lambda X^{\alpha\beta} + X^{\beta} - X \Rightarrow dr = -dX$ and taking derivative of $\zeta(\lambda X^{\alpha\beta} + X^{\beta} - X)$ with respect to λ,

$$\frac{\partial \zeta}{\partial \lambda} = \left(\frac{\partial \zeta}{\partial r}\right)\left(\frac{\partial r}{\partial \lambda}\right) = \left(-\frac{\partial \zeta}{\partial X}\right)(X^{\alpha\beta}) = -X^{\alpha\beta}.\nabla_x \zeta(\lambda X^{\alpha\beta} + X^{\beta} - X) \tag{7.15}$$

Integrating the above equation with respect to λ from 0 to 1,

$$\int_0^1 \frac{\partial \zeta}{\partial \lambda} d\lambda = -X^{\alpha\beta}.\nabla_x \int_0^1 \zeta(\lambda X^{\alpha\beta} + X^{\beta} - X)d\lambda \tag{7.16}$$

$$\zeta(X^{\alpha} - X) - \zeta(X^{\beta} - X) = -X^{\alpha\beta}.\nabla_x B^{\alpha\beta}(X) \tag{7.17}$$

From the definition of momentum density,

$$\rho_0 \frac{dV}{dt} = \sum_{\alpha=1}^{M} m^{\alpha} \frac{dV^{\alpha}}{dt} \zeta(X^{\alpha} - X) = \sum_{\alpha=1}^{M} (f^{\alpha} + m^{\alpha}b^{\alpha})\psi(X^{\alpha} - X) \tag{7.18}$$

From internal force term,

$$\sum_{\alpha=1}^{M} f^{\alpha} \zeta(X^{\alpha} - X) = \sum_{\alpha=1}^{M} \sum_{\beta \neq \alpha}^{M} f^{\alpha\beta} \zeta(X^{\alpha} - X) = \frac{1}{2} \sum_{\alpha=1}^{M} \sum_{\beta \neq \alpha}^{M} f^{\alpha\beta} [\zeta(X^{\alpha} - X) - \zeta(X^{\beta} - X)] \quad (7.19)$$

Substituting (7.17) and (7.19) in (7.18),

$$\rho_0 \frac{dV}{dt} = \sum_{\alpha=1}^{M} \left\{ \frac{1}{2} \sum_{\beta \neq \alpha}^{M} f^{\alpha\beta} \left[-X^{\alpha\beta} . \nabla_x B^{\alpha\beta}(X) \right] + m^{\alpha} b^{\alpha} \zeta(X^{\alpha} - X) \right\} \quad (7.20)$$

$$\rho_0 \frac{dV}{dt} = \nabla_x . \left\{ -\frac{1}{2} \sum_{\alpha=1}^{M} \sum_{\beta \neq \alpha}^{M} f^{\alpha\beta} \otimes X^{\alpha\beta} B^{\alpha\beta}(X) \right\} + \sum_{\alpha=1}^{M} m^{\alpha} b^{\alpha} \zeta(X^{\alpha} - X) \quad (7.21)$$

Momentum balance:

$$\rho_0 \frac{dV}{dt} = \nabla_x . \underline{P} + \rho_0 b \quad (7.22)$$

Comparing (7.21) and (7.22),

$$\underline{P}(X, t) = -\frac{1}{2} \sum_{\alpha=1}^{M} \sum_{\beta \neq \alpha}^{M} f^{\alpha\beta} \otimes X^{\alpha\beta} B^{\alpha\beta}(X) \quad (7.23)$$

The expression for bond function is defined as follows. Considering a localization box as shown in Figure 7.3 with one atom (α) inside the box and another atom (β) outside the box. In order to scale the influence of the atom outside the box, a new position of atom β is determined as

If $X_1^{\beta} \geq X_1^{I} + \frac{L_X}{2}$, then $X_1^{\beta N} = X_1^{I} + \frac{L_X}{2}$ else $X_1^{\beta N} = X_1^{\beta}$

If $X_1^{\beta} \leq X_1^{I} - \frac{L_X}{2}$, then $X_1^{\beta N} = X_1^{I} - \frac{L_X}{2}$ else $X_1^{\beta N} = X_1^{\beta}$

If $X_2^{\beta} \geq X_2^{I} + \frac{L_Y}{2}$, then $X_2^{\beta N} = X_2^{I} + \frac{L_Y}{2}$ else $X_2^{\beta N} = X_2^{\beta}$

If $X_2^{\beta} \leq X_2^{I} - \frac{L_Y}{2}$, then $X_2^{\beta N} = X_2^{I} - \frac{L_Y}{2}$ else $X_2^{\beta N} = X_2^{\beta}$

If $X_3^{\beta} \geq X_3^{I} + \frac{L_Z}{2}$, then $X_3^{\beta N} = X_3^{I} + \frac{L_Z}{2}$ else $X_3^{\beta N} = X_3^{\beta}$

If $X_3^{\beta} \leq X_3^{I} - \frac{L_Z}{2}$, then $X_3^{\beta N} = X_3^{I} - \frac{L_Z}{2}$ else $X_3^{\beta N} = X_3^{\beta}$

where the superscript "N" denotes the new value of the position and L_x, L_y, L_z are the lengths of the localization box and is the centroid of the box. Now,
If $|X_i^{\beta} - X_i^{\alpha}| \leq 10^{-3}$, then $R_i = 1$ else

$$R_1 = \frac{\left(X_1^{\beta N} - X_1^{\alpha} \right)}{\left(X_1^{\beta} - X_1^{\alpha} \right)}$$

$$R_2 = \frac{\left(X_2^{\beta N} - X_2^{\alpha} \right)}{\left(X_2^{\beta} - X_2^{\alpha} \right)}$$

$$R_3 = \frac{\left(X_3^{\beta N} - X_3^{\alpha} \right)}{\left(X_3^{\beta} - X_3^{\alpha} \right)}$$

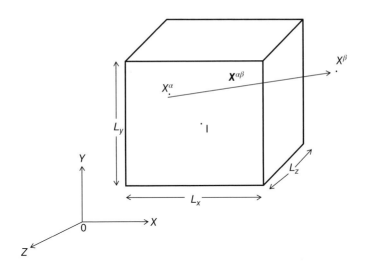

Figure 7.3   Illustration of a localization box with one atom inside and another outside the box

If $R_1 \leq R_2$ then $R = R_1$, else $R = R_2$. If $R_3 \leq R$ then $R = R_3$. This algorithm makes use of the fact that the ratio of the bond length is proportional to the ratio of its projections for a rectangular Cartesian frame of reference. Then, bond function is given by

$$B^{\alpha\beta} = \frac{R}{L_X L_Y L_z} \tag{7.24}$$

7.4   Atomistic *J*-Integral at Finite Temperature

From statistical mechanics, the Helmholtz free energy density is given by Weiner [14],

$$\Psi = U - T\eta = -\frac{k_B T}{V} \text{Log } Z \tag{7.25}$$

where U is the internal energy density, T is the temperature, η is the entropy density, V is the volume of the ensemble, k_B is Boltzmann's constant, and Z is the partition function of the atoms occupying the region. Note that the general definition of free energy density Ψ includes the entropy term and therefore is valid for finite temperature applications of the atomistic *J*-integral. The conventional definition of *J*-integral under isothermal condition does not take into account the entropic contribution to the free energy, and consequently, may lead to significant overestimation of the *J*-integral at the atomistic level at elevated temperature. As alluded to in Eq. (7.1), total *J*-integral at a finite temperature is given by

$$J_T = J_U - J_\eta \tag{7.26}$$

In Eq. (7.26), J_U is the value of *J*-integral without considering entropic contribution and is given by Eq. (7.2), and J_η is the entropic contribution. It is evident that at 0 K, the $T\eta$ in Eq. (7.8) is absent, and hence, the equation for J_U can be used to compute J_T. However, at higher temperatures, $T\eta$ term can have significant contribution due to thermal excitation of the atoms leading to a reduction in J_T. Entropic contribution J_η can be quantified for a defect-free crystalline material using the local harmonic (LH) approximation and is given by Eq. (7.27) [5].

$$J_\eta = \int_{\partial\Omega} \left\langle \frac{k_B T}{V_\alpha} \log\left(\left(\frac{\hbar}{k_B T}\right)^3 \sqrt{\det D_{LH}} \right) \right\rangle N dA \tag{7.27}$$

In Eq. (7.27), k_B is the Boltzmann's constant, $\hbar$ is the Planck's constant, V_α is the volume of the atom, T is the absolute temperature, and D_{LH} is the dynamical matrix based on the LH approximation of atoms vibrating within a defect-free crystal lattice. The LH approximation is essentially an Einstein model of the vibrational frequencies and has been used extensively in MD–continuum coupling. As shown in Eq. (7.27), computation of the dynamical matrix is necessary to quantify the entropic contribution to J-integral. At elevated temperatures, there is a significant random thermal vibration of the atoms about their mean position (not to be confused with the motion of atoms due to propagating stress waves), and the vibrational modes of the lattice are given by the eigenvalues of the dynamical matrix. The procedure to compute dynamical matrix is discussed next.

From statistical mechanics, the dynamical matrix required for quantifying entropic contribution to J-integral is given by Jones et al. [5],

$$D_{LH} = \frac{1}{m}\frac{\partial^2 \Phi}{\partial u_0^2} = \frac{1}{m}\frac{\partial}{\partial u_0} \sum_{\beta \neq 0} \phi'(r_{\beta 0}) \frac{\overline{r_{\beta 0}}}{r_{\beta 0}} \tag{7.28}$$

where $\Phi = \frac{1}{2V} \sum_\alpha^N \sum_{\beta \neq 0}^N \phi(r_{\alpha\beta})$, in which $r_{\alpha\beta}$ is the interatomic distance between atoms α and β, u_0 is the displacement vector, m is the atomic mass, $\phi(r)$ is the pairwise potential, V is the volume, $r_{\beta 0}$ is the position vector of the atom measured from the origin of the MD box. Using the chain rule of differentiation, Eq. (7.28) can be expressed as

$$D_{LH} = \frac{1}{m} \sum_{\beta \neq 0} \left(\phi''\left(r_{\beta 0}\right) \frac{\overline{r_{\beta 0}}}{r_{\beta 0}} \otimes \frac{\overline{r_{\beta 0}} + \phi'(r_{\beta 0})\frac{1}{r_{\beta 0}}\left[I - \frac{\overline{r_{\beta 0}}}{r_{\beta 0}} \otimes \frac{\overline{r_{\beta 0}}}{r_{\beta 0}} \right]}{r_{\beta 0}} \right) \tag{7.29}$$

where I is the identity tensor. Using data from MD simulations in Eq. (7.29), the dynamical matrix can be evaluated which can then be used to compute the entropic contribution to J-integral using Eq. (7.27). For example, consider a pairwise potential given by

$$\phi(r) = k(r - r_0)^2 \tag{7.30}$$

where k is the energy constant and r_0 is the equilibrium interatomic distance. Position vector of an atom β given by

$$\overline{r_\beta} = x_\beta \hat{i} + y_\beta \hat{j} + z_\beta \hat{k} \tag{7.31}$$

$$r_\beta^2 = x_\beta^2 + y_\beta^2 + z_\beta^2 \tag{7.32}$$

The terms required for evaluation of D_{LH} can then be written as

$$\frac{\overline{r_\beta}}{r_\beta} \otimes \frac{\overline{r_\beta}}{r_\beta} = \begin{bmatrix} \dfrac{x_\beta^2}{r_\beta^2} & \dfrac{x_\beta y_\beta}{r_\beta^2} & \dfrac{x_\beta z_\beta}{r_\beta^2} \\[2ex] \dfrac{x_\beta y_\beta}{r_\beta^2} & \dfrac{y_\beta^2}{r_\beta^2} & \dfrac{y_\beta z_\beta}{r_\beta^2} \\[2ex] \dfrac{x_\beta z_\beta}{r_\beta^2} & \dfrac{y_\beta z_\beta}{r_\beta^2} & \dfrac{z_\beta^2}{r_\beta^2} \end{bmatrix} \tag{7.33}$$

$$I - \frac{\overline{r_\beta}}{r_\beta} \otimes \frac{\overline{r_\beta}}{r_\beta} = \begin{bmatrix} 1 - \dfrac{x_\beta^2}{r_\beta^2} & -\dfrac{x_\beta y_\beta}{r_\beta^2} & -\dfrac{x_\beta z_\beta}{r_\beta^2} \\[2ex] -\dfrac{x_\beta y_\beta}{r_\beta^2} & 1 - \dfrac{y_\beta^2}{r_\beta^2} & -\dfrac{y_\beta z_\beta}{r_\beta^2} \\[2ex] -\dfrac{x_\beta z_\beta}{r_\beta^2} & -\dfrac{y_\beta z_\beta}{r_\beta^2} & 1 - \dfrac{z_\beta^2}{r_\beta^2} \end{bmatrix} \tag{7.34}$$

$$\phi'(r)\frac{1}{r_\beta}\left[I - \frac{\overline{r_\beta}}{r_\beta} \otimes \frac{\overline{r_\beta}}{r_\beta} \right] = \frac{2k(r_\beta - r_0)}{r_\beta} \begin{bmatrix} 1 - \dfrac{x_\beta^2}{r_\beta^2} & -\dfrac{x_\beta y_\beta}{r_\beta^2} & -\dfrac{x_\beta z_\beta}{r_\beta^2} \\[2ex] -\dfrac{x_\beta y_\beta}{r_\beta^2} & 1 - \dfrac{y_\beta^2}{r_\beta^2} & -\dfrac{y_\beta z_\beta}{r_\beta^2} \\[2ex] -\dfrac{x_\beta z_\beta}{r_\beta^2} & -\dfrac{y_\beta z_\beta}{r_\beta^2} & 1 - \dfrac{z_\beta^2}{r_\beta^2} \end{bmatrix} \tag{7.35}$$

$$\phi''(r)\frac{\overline{r_\beta}}{r_\beta} \otimes \frac{\overline{r_\beta}}{r_\beta} = \begin{bmatrix} \dfrac{2kx_\beta^2}{r_\beta^2} & \dfrac{2kx_\beta y_\beta}{r_\beta^2} & \dfrac{2kx_\beta z_\beta}{r_\beta^2} \\[2ex] \dfrac{2kx_\beta y_\beta}{r_\beta^2} & \dfrac{2ky_\beta^2}{r_\beta^2} & \dfrac{2ky_\beta z_\beta}{r_\beta^2} \\[2ex] \dfrac{2kx_\beta z_\beta}{r_\beta^2} & \dfrac{2ky_\beta z_\beta}{r_\beta^2} & \dfrac{2kz_\beta^2}{r_\beta^2} \end{bmatrix} \tag{7.36}$$

Substituting Eqs (7.35) and (7.36) in Eq. (7.29), the final evaluation of D_{LH} can then be written as

$$
D_{LH} = \frac{1}{m} \sum_{\beta \neq 0} \left(\begin{bmatrix} \dfrac{2kx_\beta^2}{r_\beta^2} & \dfrac{2kx_\beta y_\beta}{r_\beta^2} & \dfrac{2kx_\beta z_\beta}{r_\beta^2} \\[2mm] \dfrac{2kx_\beta y_\beta}{r_\beta^2} & \dfrac{2ky_\beta^2}{r_\beta^2} & \dfrac{2ky_\beta z_\beta}{r_\beta^2} \\[2mm] \dfrac{2kx_\beta z_\beta}{r_\beta^2} & \dfrac{2ky_\beta z_\beta}{r_\beta^2} & \dfrac{2kz_\beta^2}{r_\beta^2} \end{bmatrix} + \frac{2k(r_\beta - r_0)}{r_\beta} \begin{bmatrix} 1 - \dfrac{x_\beta^2}{r_\beta^2} & -\dfrac{x_\beta y_\beta}{r_\beta^2} & -\dfrac{x_\beta z_\beta}{r_\beta^2} \\[2mm] -\dfrac{x_\beta y_\beta}{r_\beta^2} & 1 - \dfrac{y_\beta^2}{r_\beta^2} & -\dfrac{y_\beta z_\beta}{r_\beta^2} \\[2mm] -\dfrac{x_\beta z_\beta}{r_\beta^2} & -\dfrac{y_\beta z_\beta}{r_\beta^2} & 1 - \dfrac{z_\beta^2}{r_\beta^2} \end{bmatrix} \right) \tag{7.37}
$$

Defining $k = 469$ (kcal/(mole Å^2)), $r_0 = 1.40 \, \text{Å}$, valid for optimized potentials for liquid simulations (OPLS) potential for a single graphene sheet at 300 K, a sample numerical value for the computation of D_{LH} is shown below.

$$
D_{LH} = \begin{bmatrix} 4.5956 & 4.3837 & 0 \\ 4.3837 & 4.2034 & 0 \\ 0 & 0 & 0.0114 \end{bmatrix} \times 10^7 \tag{7.38}
$$

7.5 Cohesive Contour-based Approach for *J*-Integral

As mentioned earlier, entropic contribution (due to $T\eta$ term in Eq. (7.25)) can have a significant effect on the computed J-integral value because that portion of the free energy goes into thermal expansion and, therefore, is no longer available for isothermal crack growth at finite temperature ($T > 0$ K). As defined in Eq. (7.26), total atomistic J-integral at finite temperature (J_T) is given by $J_T = J_U - J_\eta$, where J_U is the contribution to J-integral due to internal energy (U) computed from Eq. (7.2), and J_η is the entropic contribution (Eq. (7.27)). As discussed in Jones et al. [5], this methodology is only applicable to a defect-free crystalline structure and hence cannot be applied to amorphous polymers.

However, the path independence of J-integral facilitates the use of a cohesive contour-based technique to compute J-integral [15]. A J-integral contour in the shape of a narrow strip encompassing the thin cohesive contour ahead of the crack tip is constructed, as illustrated in Figure 7.4. Because of the way the contour is designed, it can be shown that by making the contour width (h) sufficiently small, the Helmholtz free energy term can be ignored from the J-integral equation for Mode I as well as for mixed-mode loadings. If we consider the scalar (indicial) form of the expression for total J-integral,

$$
(J_T)_k = \int_\Gamma \langle \Psi \delta_{kj} - P_{ij} H_{ik} \rangle N_j d\Gamma \tag{7.39}
$$

where Ψ is the Helmholtz free energy density, δ_{kj} is the Kronecker delta, and repeated indices imply summation. If we set $k = 1$ to obtain $(J_T)_1$, and $j = 2$ to define the normal N_2 along the narrow cohesive contour in Figure 7.4, the first term on the RHS of Eq. (7.39) vanishes because $\delta_{12} = 0$ by definition. Therefore, the only terms that need to be evaluated along the narrow cohesive contour are the Cartesian components of the first Piola–Kirchhoff stress tensor

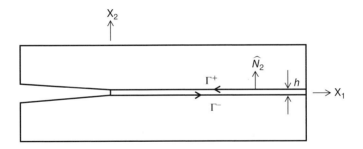

Figure 7.4　*J*-Integral contour for the cohesive contour-based technique

(P_{i2}) and components of the displacement gradient tensor (H_{i1}), using Eqs (7.5) and (7.6), respectively, giving

$$(J_T)_1 = \int_\Gamma \langle -P_{i2}H_{i1}\rangle N_2 d\Gamma \tag{7.40}$$

This approach is especially useful for thermoset polymers since the LH approximation-based approach in Eq. (7.27) cannot be used for quantifying entropic contribution for amorphous (disordered) thermoset polymers. It should be noted that the expression in Eq. (7.40) is an approximation to the analytical value for J_T, and convergence studies will need to be carried out to determine how narrow the width of the cohesive contour, h, needs to be to provide a sufficiently accurate value of J for an amorphous polymer with and without nanographene.

7.6　Numerical Evaluation of Atomistic *J*-Integral

Numerical integration through Gaussian quadrature was employed to evaluate atomistic *J*-integral using the equations given in previous sections. The *J*-integral contour around the crack is divided into segments, and each segment is further divided into localization boxes, as shown in Figure 7.4. The discrete atomistic values of potentials and displacement gradients obtained from MD simulations were converted into field quantities using the finite element-type interpolation functions for a nine-node element (given by Eq. (7.41), in terms of local coordinates ξ and χ) (i.e., set $K = 9$ in Eqs (7.5) and (7.6)).

$$N_1 = \frac{1}{4}(1 - \xi)(1 - \chi)(\xi\chi)$$

$$N_2 = -\frac{1}{4}(1 + \xi)(1 - \chi)(\xi\chi)$$

$$N_3 = \frac{1}{4}(1 + \xi)(1 + \chi)(\xi\chi)$$

$$N_4 = -\frac{1}{4}(1 - \xi)(1 + \chi)(\xi\chi)$$

$$N_5 = -\frac{1}{2}(1 - \xi^2)(1 - \chi)(\chi)$$

$$N_6 = \frac{1}{2}(1 + \xi)(1 - \chi^2)(\xi)$$

$$N_7 = \frac{1}{2}(1 - \xi^2)(1 + \chi)(\chi)$$

$$N_8 = -\frac{1}{2}(1 - \xi)(1 - \chi^2)(\xi)$$

$$N_9 = (1 - \xi^2)(1 - \chi^2) \tag{7.41}$$

J-integral is given by

$$\boldsymbol{J}_{AB} = h \int_a^b (U\widehat{N} - \underline{H}^T \, \underline{P}\widehat{N})dY \tag{7.42}$$

Referring to Figure 7.6, assuming $Y = C_1\xi + C_2$ and applying boundary conditions $a = C_1(-1) + C_2$ and $b = C_1(1) + C_2$ leads to $C_1 = (b - a)/2$ and $C_2 = (b + a)/2$

$$dY = C_1 d\xi = \left(\frac{b - a}{2}\right) d\xi$$

$$\boldsymbol{J}_{AB} = h \int_{-1}^1 (U\widehat{N} - \underline{H}^T \, \underline{P}\widehat{N})C_1 d\xi$$

$$\boldsymbol{J}_{AB} = C_1 h \sum_{k=1}^{N_G} [U(\xi_k)\widehat{N} - \underline{H}^T(\xi_k)\underline{P}(\xi_k)\widehat{N}]w_k$$

$$\boldsymbol{J}_{AB} = C_1 h \sum_{k=1}^{N_G} [U(\xi_k)\widehat{N} - \underline{H}^T(\xi_k)\underline{P}(\xi_k)\widehat{N}]w_k$$

$$\boldsymbol{J}_{AB} = C_1 h \sum_{k=1}^{N_G} [U(Y_k)\widehat{N} - \underline{H}^T(Y_k)\underline{P}(Y_k)\widehat{N}]w_k$$

where $Y_k = C_1\xi_k + C_2$.

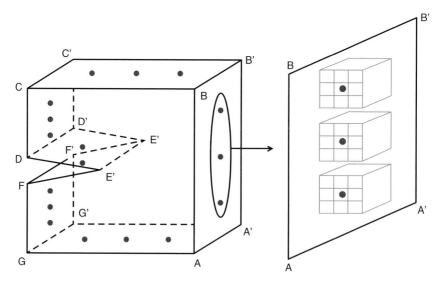

Figure 7.5 Gaussian quadrature points along the integration contour-depicting localization boxes on one side

Table 7.1 Sample values of J-integral computed along one path for graphene at 0 K with OPLS potential

K (MPa m$^{1/2}$)	J_I (J/m^2)				
	AB	CD	FG	BC	GA
0	0	0	0	0	0
1.22	1.2369	0.6075	0.778	−0.4306	−0.4305
2.44	5.0363	2.526	2.7425	−1.8107	−2.1275
3.67	11.0652	5.7634	6.087	−4.2981	−4.3981

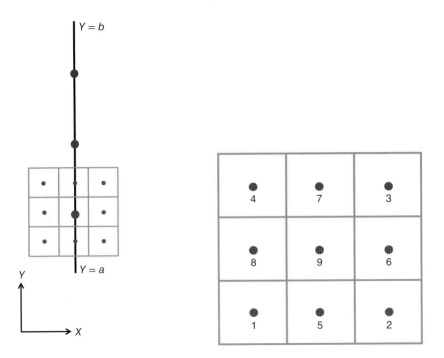

Figure 7.6   Close-up view of the computation box and interpolation grid

J-integral can be calculated using this procedure at all Gaussian quadrature points. Referring to Figures 7.5 and 7.6, sample values of J-integral computed for a graphene sheet at 0 K for each section of the path are shown in the Table 7.1. Details regarding the graphene sheet used, simulation parameters, and more results are presented in later sections.

7.7   Atomistic J-Integral Calculation for a Center-Cracked Nanographene Platelet

A single nanographene platelet with a center-crack was modeled using MD (large-scale atomic/molecular massively parallel simulator (LAMMPS) software) and was subjected to a far-field uniaxial stress in the Y-direction (pure Mode I loading). The atomistic J-integral

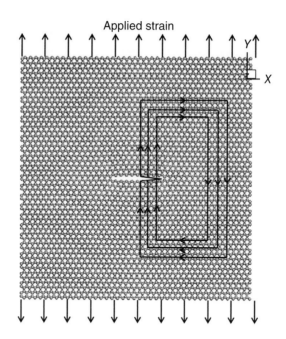

Figure 7.7   Graphene sheet with nanoscale crack and contours for J-Integral computation

was computed on concentric contours as shown in Figure 7.7. In this example, the graphene platelet was modeled using harmonic style for bonds and angles, and OPLS style for dihedrals. The graphene platelet, measuring $10.5\,\text{nm} \times 10\,\text{nm} \times 0.34\,\text{nm}$, was made up of 4420 carbon atoms in armchair configuration. The initial length of the pre-existing (starter) crack was 2.2 nm in the graphene platelet as shown in Figure 7.7. The coarse timestep for this simulation (Δt) was set equal to 0.1 fs and the applied strain rate was $2 \times 10^{11}\,\text{s}^{-1}$.

In order to establish proof of concept for atomistic J-integral computation, the initial MD simulations were carried out at temperature $T = 0\,\text{K}$, that is, when there is no entropic contribution to the free energy due to thermal motion in accordance with Eq. (7.25). Pressure barostatting was not used in these simulations. A convergence study was conducted on the effect of localization box size on the computed values of stress in the graphene sheet away from the crack tip. The computed value of stress becomes box size independent of box sizes greater than 2 nm ($\sim$20 Å), as shown in Figure 7.8. The atomistic J-integral results are shown as a function of applied Mode I stress intensity factor (K_I) in Figure 7.9 for a graphene platelet and compared with linear elastic fracture mechanics (LEFM) predictions. The plot shows computed values for the three paths around the crack for a localization box size of 27 Å. The figure shows that the quadratic dependence of the atomistic J-integral on Mode I stress intensity factor (K_I) is in good agreement with plane-stress LEFM predictions for pure Mode I loading, that is, $J_I = K_I^2/E$, where E is the Young's modulus ($E = 0.85\,\text{TPa}$) of nanographene at 0 K obtained from our MD simulations. The stress intensity factor can be computed from the virial stress values obtained from MD output. The equation for computation of stress intensity factor (K) can be written in terms of virial stress (σ), half crack length (a), and geometry factor (f) by $K_I = \sigma f \sqrt{\pi a}$.

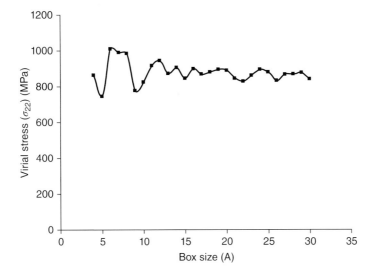

Figure 7.8   Influence of localization box size on computed stress values for graphene

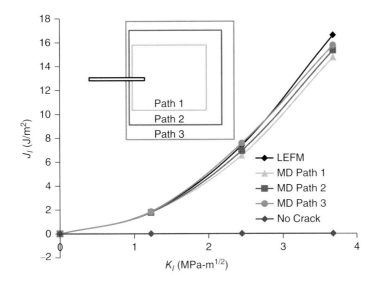

Figure 7.9   Atomistic J_I versus K_I for a graphene platelet at 0 K compared with LEFM

The figure also verifies that J is identically zero in the absence of a singularity within the integration contour, as depicted by the "no crack" line in the figure. More importantly, it is clear from Figure 7.9 that the calculated value of J_I is reasonably path-independent for the graphene platelet even at the nanoscale. This result facilitates the use of calculating J-integral at finite temperatures using a cohesive contour technique discussed earlier.

7.8   Atomistic *J*-Integral Calculation for a Center-Cracked Nanographene Platelet at Finite Temperature ($T = 300$ K)

Atomistic *J*-integral was computed for a center-cracked graphene platelet at temperatures of 0.1, 100, 200, and 300 K to verify the model at finite temperatures and quantify the entropic contribution to *J*. The center-cracked graphene platelet used is the same as the one used for 0 K computations as shown in Figure 7.7. The system was elevated to the desired temperature and was allowed to equilibrate at that temperature and at a pressure of zero atmospheres using NPT fix in LAMMPS. Poisson's effect was considered in this simulation and a value of 0.22 was used for the Poisson's ratio, which is in good agreement with the literature [16]. The graphene platelet was stretched under Mode I tension, as was the case for 0 K, and the resulting data from the MD simulation were used to compute atomistic *J*-integral. Three different box sizes of 22, 27, and 32 Å were used to compute *J*-integral at 0.1, 100, 200, and 300 K (averaged over five concentric paths) using the techniques discussed earlier, and the results were found to be independent of the box size. The results of one of these computations are shown in Figure 7.10, with the symbols representing the mean value and the error bar representing the standard deviation based on averaging over five concentric contours. J_U is the portion of the atomistic *J*-integral that includes only internal energy contribution (i.e., without

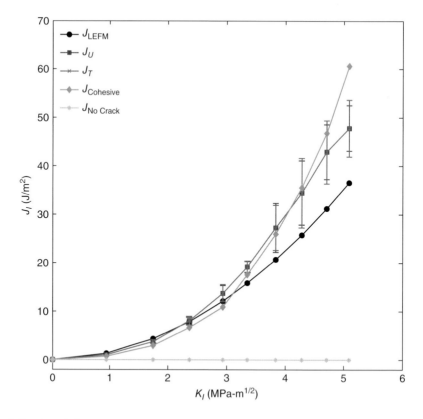

Figure 7.10   Atomistic J_I versus K_I for a graphene platelet at 0.1 K with harmonic potential and 27 Å localization box size

entropic contribution) averaged over five concentric paths. As one might expect, it shows a good agreement with strain energy-based LEFM results except at low temperatures. J_T is the total computed J-integral including both energetic and entropic contributions (again, averaged over five paths). This figure also shows J_{cohesive} computed using the novel cohesive contour technique discussed earlier, and it is in good agreement with J_T.

Since most experiments are conducted at room temperature, results for 300 K case will be discussed in detail. Figure 7.11 shows the atomistic J-integral comparison compared with LEFM prediction using a localization box size of (27 Å). As earlier, J_U is the portion of the atomistic J-integral that includes only internal energy contribution (i.e., without entropic contribution), averaged over five concentric paths. As one might expect, it shows good agreement with strain energy-based LEFM results. J_T is the total computed J-integral including both energetic and entropic contributions (again averaged over five paths), and it is evident that it is significantly ($\sim$50%) lower than LEFM predictions in accordance with Eq. (7.25). Figure 7.11 also shows J_{cohesive} computed using the novel cohesive contour technique discussed in previous sections, and it is in good agreement with J_T. This is an important benchmark result, as it enables potentially the use of cohesive contour-based technique for computing J-integral at the nanoscale for the case of amorphous polymers, where direct computation of entropic contribution using the LH approximation approach is not feasible.

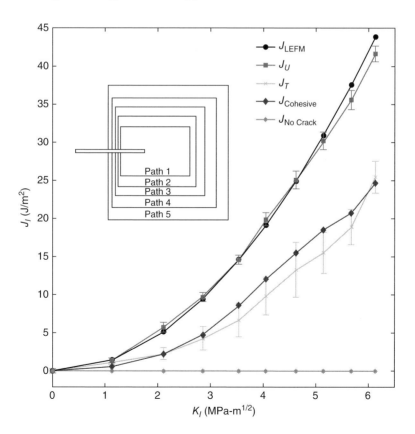

Figure 7.11 Atomistic J_I versus K_I for a graphene platelet at 300 K with harmonic potential with 27 Å localization box size

7.9 Atomistic *J*-Integral Calculation for a Center-Cracked Nanographene Platelet Using ReaxFF

Optimized potentials for liquid simulations (OPLS) based on harmonic force field was employed in all of the MD simulations presented in the previous section. While the OPLS force field works well for evaluating mechanical properties, it cannot simulate the phenomena such as bond forming and breakage. In order to better understand the fracture behavior of the material, it is important to be able to simulate crack initiation and propagation. For that purpose, a new force field, ReaxFF, was necessary. ReaxFF is a bond-order-based force field that allows for continuous bond formation/breaking [17]. Bonded interactions are generated on-the-fly, based on distance-dependent bond-order functions.

As a benchmark case study, MD simulations were carried out on a center-cracked nanographene sheet using ReaxFF force field. The potential parameters used in this simulation were obtained from Chenoweth et al. [18]. The nanographene sheet consists of 4398 atoms and is similar in dimensions to the one used in OPLS simulations, that is, $105\,\text{Å} \times 100\,\text{Å}$, in armchair configuration. An initial central crack of approximately $20\,\text{Å}$ was generated by deleting the atoms at the center of the graphene sheet. All the finite temperature simulations were done using NPT fix in order to fully account for Poisson's effect. A timestep of $0.1\,\text{fs}$ and a strain rate of $3 \times 10^{11}\,\text{s}^{-1}$ were used for the simulations. A total tensile strain of approximately 25% was gradually applied on the MD simulation box in the direction perpendicular to the crack to simulate Mode I loading. The ReaxFF force field was able to successfully simulate the initiation and propagation of the center crack in the nanographene sheet, as depicted in Figure 7.12(a) and (b). For preliminary benchmarking, atomistic *J*-integral was evaluated at 0.1 and 300 K. Based on the results obtained for harmonic OPLS simulations for nanographene, it was established that for a computation box size above $20\,\text{Å}$, the *J*-integral computations are fairly independent of localization box size. Hence, a box size of $27\,\text{Å}$ was selected for consistency between OPLS and ReaxFF computations.

Figure 7.13 shows the comparison of atomistic *J*-integral values computed at 0.1 K using various methods (discussed in previous sections) compared to the values from LEFM for a localization box size of $27\,\text{Å}$ (for one contour). Figure 7.14 shows the comparison of atomistic

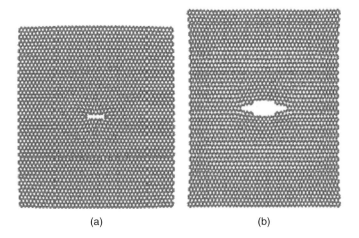

(a)   (b)

Figure 7.12   Center-cracked graphene at the (a) beginning and (b) end of the simulation using ReaxFF

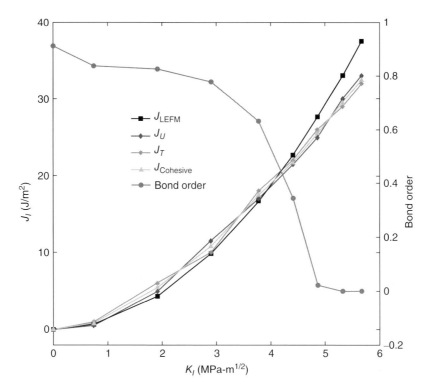

Figure 7.13 *J*-integral computation for graphene as a function of K_I using ReaxFF at 0.1 K; also shown is change in bond-order with loading (27 Å localization box size)

J-integral values computed at 300 K using different methods compared to the values from LEFM for a localization box size of 27 Å (averaged over three contours). In these plots, J_U is the portion of the atomistic *J*-integral that includes only internal energy contribution (i.e., without entropic contribution), and it shows a good agreement with purely strain energy-based LEFM results. J_T is the total computed *J*-integral including both energetic and entropic contributions, calculated using the partition function approach, as discussed in Jones et al. [5]. $J_{cohesive}$ is the *J*-integral value computed using the narrow cohesive contour approach discussed earlier. There is a good agreement between LEFM, the energetic *J*-integral (J_U), and the total *J*-integral (J_T) at 0.1 K as depicted in Figure 7.13, because at 0.1 K the entropic contribution is insignificant as described in Eq. (7.25). However, using the classical definition of *J* (i.e., $J_T = J_U$) does not account for the entropic contribution that becomes significant at higher temperatures, as underscored by the difference between the total *J*-integral (J_T) and the energetic *J*-integral (J_U) computed at 300 K as shown in Figure 7.14. As the temperature increases, the entropic contribution becomes more significant leading to greater than 50% error in the predicted value of total *J* (J_T), as illustrated in Figure 7.14. The classical *J*-integral evaluation scheme cannot account for this discrepancy, but the "exact" partition function method (J_T) and the "approximate" cohesive contour-based approach ($J_{cohesive}$) are able to account for entropic contribution and are in good agreement with each other as can be seen from Figure 7.14. The y-axis on the right-hand side of the plots also depicts the variation

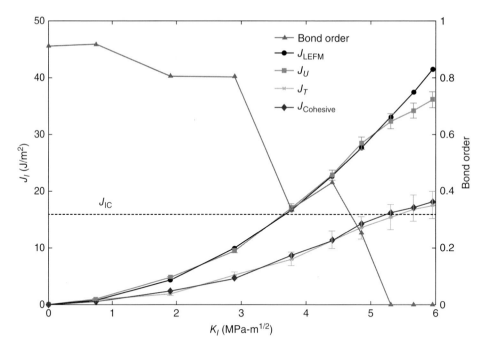

Figure 7.14 J-integral computation in graphene as a function of K_I using ReaxFF at 300 K; also shown is change in bond-order with loading (27 Å localization box size)

in bond order between the atoms just ahead of the crack tip. It can be observed that as the sheet is stretched in tension, the atomic bond gets stretched, and accordingly the bond order decreases as a function of loading (K_I). When the bond order becomes zero the bond breaks, providing a critical value of J_T (J_{Ic}) at crack initiation. Interestingly, as shown in Figure 7.14, the slopes of the J_T and $J_{cohesive}$ curves change as the bond order goes to zero, and this critical value ($J_{Ic} = \sim 18 \, J/m^2$) agrees reasonably well with its value published in the literature ($J_{Ic} = \sim 16 \, J/m^2$ [19]) for nanographene at 300 K. The oscillations in the bond-order value at 300 K are possibly because of thermal excitation of atoms at elevated temperature. The results indicate that the event of bond breakage corresponds fairly well with when the value of computed J-integral deviates from LEFM predictions. This change of slope of J versus K could be considered as the point of crack initiation and could directly be related to the fracture toughness of the material system in a predictive sense.

7.10   Atomistic J-Integral Calculation for a Center-Cracked EPON 862 Model

The atomistic J-integral computation methodology discussed earlier was successfully verified for a nanographene platelet in Section 7.9. Since it is now established that ReaxFF is more effective in determining crack initiation compared to OPLS, all the MD simulations of the polymer were conducted using ReaxFF potential. A ReaxFF-based EPON 862 model was developed with around 4200 atoms (35 Å × 35 Å × 35 Å MD box size) and 85% crosslinking. The ReaxFF parameters used for this model are available in the literature [20]. The initial

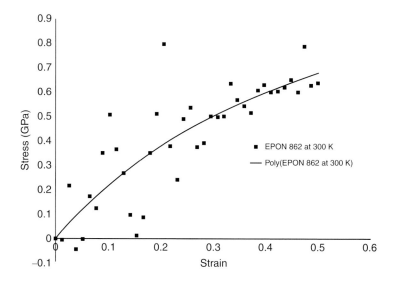

Figure 7.15   Stress versus strain response of EPON 862 at 300 K

4200 atoms system was found to be too small for simulating crack propagation and computing J-integral. Hence, a larger system was constructed by fourfold replicating the original system. The new EPON model consists of 16,806 atoms with dimensions of $70\,\text{Å} \times 35\,\text{Å} \times 70\,\text{Å}$. Before introducing a crack, the system was subjected to uniform tension and the stress–strain response along with a cubic polynomial fit, as shown in Figure 7.15, was found to be in agreement with the Young's modulus of EPON 862 presented in the literature of $E = 3\,\text{GPa}$ [21]. The density of the model was verified to be around 1.17 gm/cc, which is also in agreement with the literature [21].

In LAMMPS, regions of specified shape and dimensions can be created. Using that functionality, four rhombus-shaped "prism" regions were created in the model so that the intersection of those four regions formed a region shaped like a sharp crack. By deleting the atoms present in the intersecting region, a central crack of approximately 26 Å length was introduced in the epoxy model as shown in Figure 7.16, and was subjected to Mode I tension. A total strain of 50% was gradually applied to the epoxy polymer block at a temperature of 300 K. Although the applied strain was very high, significant crack propagation was not observed initially. In order to overcome this problem, the top half atoms and bottom half atoms of the model were grouped as "top" and "bottom" and a force was applied on those atoms in opposite direction, similar to a compact tension test. At higher timestep values, the model breaks into two at the crack. Hence, in order to observe a proper crack propagation, a timestep of 0.005 fs was used, which successfully simulated crack propagation. The resultant strain rate was $25 \times 10^{12}\,\text{s}^{-1}$. The simulations were conducted using NVT ensemble.

Figure 7.16(a) and (b) shows the EPON model at the beginning and at the end of the simulation, respectively. Some crack extension can be observed in Figure 7.16(b). Similar to nanographene computations, influence of localization box size on the computed values of stress was studied and is presented in Figure 7.17. From the figure, it is evident that the computed stress values monotonically converge and become independent of the localization box size for box size above 25 Å.

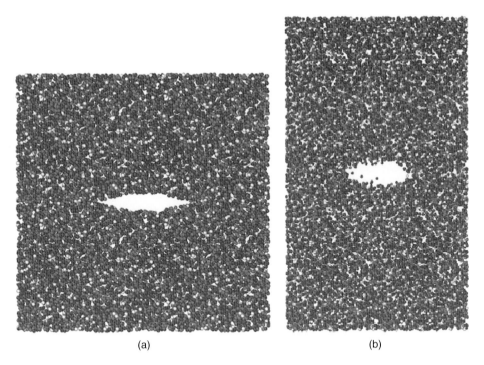

(a)   (b)

Figure 7.16 Center-cracked EPON 862 at the (a) beginning and (b) end of the simulation using ReaxFF at 300 K

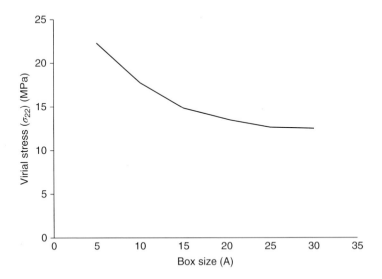

Figure 7.17 Influence of localization box size on computed stress values at 300 K for neat EPON 862

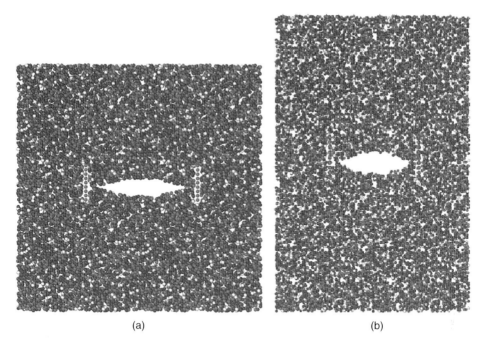

(a)   (b)

Figure 7.18   Center-cracked EPON 862 reinforced with 3 wt% nanographene at the (a) beginning and (b) end of the simulation using ReaxFF at 300 K

In order to study the effect of nanoscale reinforcement on fracture behavior in polymers, an EPON 862 model with 3 wt% nanographene was generated and subjected to Mode I tension. Although considered high for nanographene composites, 3 wt% was selected so that results from MD simulations could be compared with experimental fracture test data available for EPON 862 from previous experiments. Simulating 0.1 or 0.5 wt% nanographene would require a really large-scale MD model, which is computationally prohibitive. Figure 7.18(a) and (b) shows the polymer model with nanographene reinforcement perpendicular to the crack front at the beginning and at end of the MD simulation, respectively. Similar to the neat EPON 862 case, 50% strain was applied at a strain rate of 25×10^{12} s^{-1}. The timestep was 0.005 fs using NVT ensemble. It is evident from Figure 7.18(b) that the presence of the nanographene platelet inhibits crack propagation in the reinforced EPON model. Furthermore, there is some evidence of crack deflection (i.e., crack tilting and/or twisting) and nanographene warping, analogous to what was observed in fracture experiments [22]. The crack has to propagate around the nanoparticle in a mixed-mode manner, which manifests as increased fracture toughness.

7.11   Conclusions and Future Work

A methodology for computing atomistic J-integral was derived, the critical value of which could be used as a metric for fracture toughness of a material. A novel cohesive contour technique to capture the effect of entropy at elevated temperatures was also developed. The methodology was verified by applying it to a single nanographene sheet using OPLS and ReaxFF potentials. Due to its inherent ability to model bond breakage, ReaxFF potential

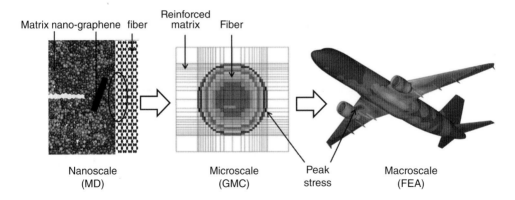

Figure 7.19 Schematic of a hierarchical multiscale model for analysis of structures

was used to successfully model crack propagation in a graphene sheet. The J-integral values obtained from the methodology discussed in this chapter were found to be in good agreement with the values available in the literature. After verifying the applicability of J-integral to predict fracture toughness of a graphene sheet, the methodology was then applied to a neat EPON 862 polymer model and a polymer model with 3 wt% nanographene reinforcement. Although ReaxFF was able to model crack propagation in the polymer, only a small crack extension was observed, even at 50% strain. One reason could be that the ReaxFF parameter set has been validated at lower strain levels. Hence, a better parameter set for ReaxFF is required for EPON 862 that is valid for large deformations. The ability of nanographene to prevent crack extension was qualitatively observed using MD simulations. Work is currently underway to quantify any enhancement in fracture toughness in this system using the atomistic J-integral. The fracture toughness prediction methodology discussed in this chapter could be applied to other polymer systems with appropriate potential functions. The change in values of J-integral with crack propagation could be used to develop resistance curves (R-curve) for polymers with nanoscale reinforcements. The data obtained from these nanoscale studies can be used in multiscale analysis of advanced materials. One such hierarchical multiscale modeling scheme is illustrated in Figure 7.19. Here, data from nanoscale MD simulations can be input into microscale analysis using techniques such as generalized method of cells (GMC) including the influence of nanoparticle on the fracture toughness of the polymer matrix. The microscale model is, in turn, called by a macroscale finite element analysis (FEA) model at each element integration point, and failure of fiber and/or matrix is determined at that point. If there is failure at the microscale, then load redistribution takes place at the macroscale through element stiffness reduction. The process is repeated iteratively. In this manner, the hierarchical multiscale technique could be employed to predict progressive failure in nanoparticle reinforced composite material at the structural level.

Acknowledgments

The authors would like to acknowledge sponsorship of this research by NASA Fundamentals Aeronautics Research Program, Contract No. NNX11AI32A, with Dr Brett Bednarcyk as Technical Monitor. The authors would also like to thank Dr Gregory Odegard at Michigan

Technological University in Houghton, for his assistance with the EPON 862 polymer modeling, and Dr Vinu Unnikrishnan at University of Alabama for his assistance with MD simulations.

References

[1] Rice, J.R. (1968) A path independent integral and the approximate analysis of strain concentration by notches and cracks. *Journal of Applied Mechanics*, **35** (2), 379–386.

[2] Begley, J. and Landes, J. (1972) J-integral as fracture criterion. *American Society for Testing of Materials*, **514**, 1–20.

[3] Mindess, S., Lawrence, F.V. and Kesler, C.E. (1977) The J-integral as a fracture criterion for fiber reinforced concrete. *Cement and Concrete Research*, **7** (6), 731–742.

[4] Jones, R.E. and Zimmerman, J.A. (2010) The construction and application of an atomistic J-integral via Hardy estimates of continuum fields. *Journal of the Mechanics and Physics of Solids*, **58** (9), 1318–1337.

[5] Jones, R.E., Zimmerman, J.A., Oswald, J. and Belytschko, T. (2011) An atomistic J-integral at finite temperature based on Hardy estimates of continuum fields. *Journal of Physics: Condensed Matter*, **23**, 015002.

[6] Zimmerman, J.A. and Jones, R.E. (2013) The application of an atomistic J-integral to a ductile crack. *Journal of Physics: Condensed Matter*, **25**, 155402.

[7] Eshelby, J.D. (1975) The elastic energy-momentum tensor. *Journal of Elasticity*, **5**, 321–335.

[8] Inoue, H., Akahoshi, Y. and Harada, S. (1994) A fracture parameter for Molecular Dynamics method. *International Journal of Fracture*, **66** (4), R77–R81.

[9] Hardy, R.J. (1982) Formulas for determining local properties in molecular dynamics simulations: Shock waves. *The Journal of Chemical Physics*, **76** (1), 622–628.

[10] Nakatani, K., Akihiro, N., Yoshihiko, S. and Kitagawa, H. (2000) Molecular dynamics study on mechanical properties and fracture in amorphous metal. *American Institute of Aeronautics and Astronautics*, **38** (4), 695–701.

[11] Xu, Y.G., Behdinan, K. and Fawaz, Z. (2004) Molecular dynamics calculation of the J-integral fracture criterion for nano-sized crystals. *International Journal of Fracture*, **130** (2), 571–583.

[12] Latapie, A. and Farkas, D. (2004) Molecular dynamics investigation of the fracture behavior of nanocrystalline α-Fe. *Physical Review B*, **69** (13), 134110.

[13] Rigby, R.H. and Aliabadi, M.H. (1993) Mixed-mode J-integral method for analysis of 3D fracture problems using BEM. *Engineering Analysis with Boundary Elements*, **11** (3), 239–256.

[14] Weiner, J.H. (1983) *Statistical Mechanics in Elasticity*, John Wiley & Sons.

[15] Klein, P. and Gao, H. (1998) Crack nucleation and growth as strain localization in a virtual-bond continuum. *Engineering Fracture Mechanics*, **61** (1), 21–48.

[16] Zhao, P. and Shi, G. (2011) Study of poisson's ratios of graphene and single-walled carbon nanotubes based on an improved molecular structural mechanics model. *Tech Science Press*, SL, **5** (1), 49–58.

[17] Van Duin, A.C.T., Dasgupta, S., Lorant, F. and Goddard, W.A. (2001) ReaxFF: a reactive force field for hydrocarbons. *Journal of Physical Chemistry A*, **105**, 9396–9409.

[18] Chenoweth, K., van Duin, A.C.T. and Goddard, W.A. (2008) ReaxFF reactive force field for molecular dynamics simulations of hydrocarbon oxidation. *The Journal of Physical Chemistry. A*, **112**, 1040–1053.

[19] Zhang, P., Ma, L., Fan, F. et al. (2014) Fracture Toughness of Graphene. *Nature Communications*, **5**, 3782.

[20] Liu, L., Liu, Y., Zybin, S.V. et al. (2011) ReaxFF-lg: correction of the ReaxFF reactive force field for London dispersion, with applications to the equations of state for energetic materials. *The Journal of Physical Chemistry. A*, **115** (40), 11016–11022.

[21] Odegard, G.M., Jensen, B.D., Gowtham, S. et al. (2014) Predicting mechanical response of crosslinked epoxy using ReaxFF. *Chemical Physics Letters*, **591**, 175–178.

[22] Rafiee, M., Rafiee, J., Srivastava, I. et al. (2010) Fracture and fatigue in graphene nanocomposites. *Small*, **6** (2), 179–83.

8

Mechanical Characterization of 2D Nanomaterials and Composites

Ruth E. Roman[1], Nicola M. Pugno[2,3,4] and Steven W. Cranford[1]

[1]*Department of Civil and Environmental Engineering, Northeastern University, Boston, MA, USA*
[2]*Department of Civil, Environmental and Mechanical Engineering, University of Trento, Trento, Italy*
[3]*Centre of Materials and Microsystems, Bruno Kessler Foundation, Trento, Italy*
[4]*School of Engineering and Materials Science, Queen Mary University, London, UK*

8.1 Discovering 2D in a 3D World

The remarkable emergence of nanotechnology has made flourish new possibilities for the characterization of nanomaterials. These small-sized structures have at least one dimension in the nanoscale (<100 nm) and can be categorized according to their dimensionality as zero-dimensional (0D), one-dimensional (1D), two-dimensional (2D), and three-dimensional (3D) materials (Figure 8.1(a)–(d)). Herein, we focus in 2D nanostructures.

Intuitively, one can easily discern the difference between 2D and 3D objects: restrict the size of a material volume to its length and width alone and reduce to zero its height. From a materials science or engineering perspective, a reduction in dimensions necessarily changes how a system is both described and characterized. Yet our own minds are accustomed to visualization in 3D – what consequences occur when a material is constrained to two dimensions poses intriguing questions. Over 100 years ago, a simple thought experiment considered life restricted to two dimensions – a hypothetical "flatland." The novel *Flatland: A Romance of Many Dimensions* is an 1884 satire by the English author Edwin Abbott Abbott. The book used the fictional (and limited) 2D world of "Flatland" to comment on the hierarchy of Victorian culture, but the novel's more enduring contribution is its examination of *dimensions*. The story describes a 2D world occupied by geometric figures, while the narrator (himself a lowly square) guides the readers through some of the implications of planer life. The square

Advanced Computational Nanomechanics, First Edition. Edited by Nuno Silvestre.
© 2016 John Wiley & Sons, Ltd. Published 2016 by John Wiley & Sons, Ltd.

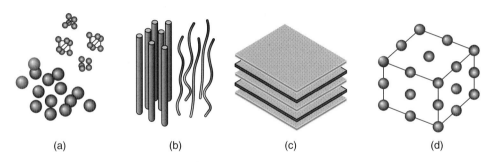

Figure 8.1 Dimensional classification of nanomaterials. (a) 0D with all dimensions in the three directions are in the nanoscale ($\leq 10^{-9}$ m). Examples of this kind are nanoparticles, quantum dots, and clusters. (b) 1D where one dimension of the nanostructure will be outside the nanometer range. Nanorods, nanotubes, and nanofibers are included in this classification. (c) 2D with two dimensions outside the nanometer range. These include different types of thin films and plates. (d) 3D with all dimensions outside the nanoscale. Bulk amorphous materials and materials with a nanocrystalline structure are included in this category

is visited by a 3D sphere, which he cannot comprehend, and only observes in discrete circles of varying sizes, for example, 2D cross sections of the sphere. The 2D materials we discuss herein would be very comfortable in Flatland – having length and width prescribed by a 2D atomic crystallinity, but constrained to a plane (with minor deviations). Moreover, like the square observing the sphere in Flatland, three dimensions of such materials can be imagined (and ultimately experimentally achieved) by grouping discrete "slices" of the 2D structures in multilayer composite systems. However, this typically does not change the 2D nature of the materials. Thus, in order to understand such materials, we are limited to 2D perspectives – a *molecular flatland*. Unlike the poor square in Abbott's novel, however, 2D materials exist in the 3D world, but have only been a recent discovery.

In 2004, a one-atom thickness monolayer of carbon atoms was isolated through relatively simple mechanical tape exfoliation of graphite [1], quietly heralding an era of 2D materials. Novoselov and Geim successfully produced the first single-layered 2D material, graphene, an achievement that granted both of them the Noble prize in Physics in 2010. Its exceptional electrical properties, potential in electronic, structural and thermal applications, and mechanical and chemical stability has suggested graphene to be one of the most remarkable structure in existence. Of course, all these superlatives mean that it has plethora of promising applications. Since then, the potential of other atomistically 2D materials has created a new paradigm of Materials Science. As graphene itself has shown, the properties of 2D structures can be astounding, and monolayers of crystals such as hexagonal boron nitride (hBN) or molybdenum disulfide could be just as mind blowing as graphene. They possess a high degree of anisotropy with nanoscale thickness and infinite length in other dimensions, exhibit great potential in applications such as energy storage and conversion systems, and hold enormous promise due to its unique functionalities and properties, which are not found when the material is in its bulk form. In order to be successful in future applications, the behavior of such materials subject to load must be fully understood, including limits in deformation and strength, and ultimate failure response. In addition, nanocomposites reinforced with 2D nanolayers have become of great interest due to their potential applications. These composite materials, in which the matrix material is reinforced by one or more nanolayers in order to improve performance properties, commonly use polymers, ceramics, and metals as matrix [2], and form a 3D arrangement.

This chapter reviews 2D nanomaterials and composites from the standpoint of their mechanical properties. We discuss first the state of the art for graphene-related nanomaterials, and then we focus in some of the more important experimental and analytical tools used for the characterization of nanomaterials, and the different approaches used to characterize the mechanical properties and behavior of an atomistic system, such as strength, stiffness, and failure. Finally, we focus in multilayers and nanocomposites and the challenges of mixing different materials in the context of mechanical properties.

8.2   2D Nanostructures

The study of 2D nanostructures has witnessed an increasing development over the past several years due to the success of graphene and its remarkable characteristics. These novel materials have extraordinary properties and promising applications. Recently, several works have been published on graphene-like materials, including silicene, germanene, and molybdenum disulfide, which have been studied and designed from both experimental and computational sides, allowing to predict interesting properties and establishing new fabrication methods. Here, we briefly review some of these fascinating 2D nanostructures that represent a new breakthrough in materials science.

8.2.1   *Graphene*

The most prominent 2D nanostructure is graphene, which is a honeycomb arrangement of sp^2-hybridized carbon atoms that has almost the same per-atom crystal energy as diamond (sp^3 structure) (Figure 8.2(a) and (b)). As each graphene carbon has only three bonds instead of four for diamond, the graphene C–C bonds are about 25% stronger [3]. Thus, in plane, it is the strongest material ever measured, and is both chemically stable and inert. Graphene has a particular electronic structure, due to the presence of a Dirac cone in its electronic band structure, which makes graphene conducts electricity better than any other known material at room temperature [4]. In addition, it does not show a band gap, but the density of states is zero at the Fermi level [3]. Today, graphene can be produced by several methods that depend on the quality and dimension of the material to be obtained. Furthermore, we can list other exceptional properties of graphene, such as high specific surface area, high carrier mobility, high thermal conductivity, half-integer quantum Hall effect, and ambipolar electric field effect.

Several researchers have determined the mechanical properties of a single layer of graphene, and it has been reported to have the highest elastic modulus and strength, compared with other materials. The Young's modulus of a graphene sheet is reported on the order of 1 TPa (335 N/m) using different techniques including Raman spectroscopy and AFM [5, 6]. Similarly, computational and experimental works report that the ultimate stress for single layer of graphene is 130 GPa, for an approximate ultimate strain of 20%–25% strain [5]. The bending modulus for graphene has been calculated as well, based on *ab initio*, molecular dynamics (MD) and other techniques resulted in values in the order of 1.5 to 2.1 eV for monolayer graphene and approximately 130 eV for bilayer graphene sheets [7].

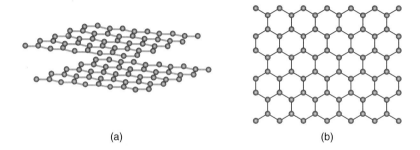

(a)   (b)

Figure 8.2   Atomic structures of (a) graphite and (b) graphene

8.2.2   Graphynes and Graphene Allotropes

The recent advent of graphene has motivated further investigation of similar 2D systems, including all-carbon allotropes of graphene, as well as other 2D crystals. In this group, we highlight the family of graphynes. The graphyne family is an all-carbon allotrope of graphene, in which the framework in general consists of characteristic hexagonal carbon rings connected by acetylenic linkages (single- and triple-bond, Figure 8.3(a) and (b)), which is made up of a variation of the sp^2 carbon motif-forming graphene. In this group, we highlight graphdiyne, a member of the graphyne family proposed by Haley et al. in 1997 [8] with characteristic diacetylenic links. Presently, thin films of graphdiyne have been successfully fabricated on a copper (Cu) substrate by a cross-coupling reaction using hexaethynylbenzene [9], suggesting the future feasibility of extended graphyne structures. In the process, the Cu foil serves as both the catalyst and substrate for growing the graphdiyne films.

Although the mechanical performance of graphdiyne is noticeably inferior to graphene, the graphdiyne structure is predicted to be the most synthetically approachable, and the most stable diacetylenic non-natural carbon allotrope [9]. It has been found that graphdiyne exhibits semiconductive characteristics and, advantageously, the natural band gap in graphdiyne, which can vary as a function of directional anisotropy, is similar to that of silicon and makes it the possible supplant material to silicon electronic devices. The possible applications of graphdiyne sheets and nanoribbons due to its promising properties include nanoelectronics, energy storage, anode materials in batteries, for hydrogen storage, or as membranes for gases separation [10].

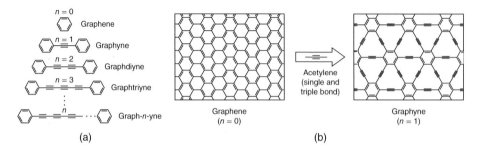

(a)   (b)

Figure 8.3   (a) Classification of graphyne family. (b) Schematic of graphene to graphyne

Table 8.1   Mechanical properties for extended graphynes [11]

Structure	n	Reclined chair direction			Zigzag direction		
		E (GPa)	σ_{ult} (GPa)	ε_{ult} (%)	E (GPa)	σ_{ult} (GPa)	ε_{ult}
Graphyne	1	532.5	48.2	8.2	700.0	107.5	13.2
Graphdiyne	2	469.5	36.0	6.3	578.6	45.5	8.0
Graphtriyne	3	365.0	26.8	7.7	476.7	43.7	9.9
Graphtetrayne	4	370.4	24.8	7.0	453.3	32.5	9.7

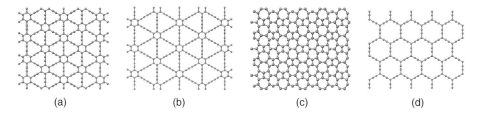

    (a)   (b)   (c)   (d)

Figure 8.4   Atomic structures of graphene allotropes. (a) Graphyne, (b) graphdiyne, (c) graphene allotrope with Stone–Wales defects, (d) supergraphene

The mechanical properties of the extended graphyne family in both armchair (or reclined chair) and zigzag directions have been reported in previous studies, and are summarized in Table 8.1. A consistent degradation of the properties is observed, with the addition of acetylene linkages to the structure.

With the purpose of defining further approaches to achieve controllable manipulation of the atomic, electronic, and mechanical properties of all-carbon graphene, different graphene allotropes have been developed (see Figure 8.4(a) and (b)). An alternative approach is the introduction of variation in structure and topology of graphene, such as atomic-scale defects (Stone–Wales defects, see Figure 8.4(c)), or the substitution of the original carbon–carbon bonds in graphene by acetylene carbyne-like chains (Figure 8.4(d)) [12]. The introduction of acetylene links introduces an effective reduction in stiffness, strength, and stability, enabling the properties to predict the nanostructures properties as a function of acetylene repeats [11].

The 2D materials family is of course not limited to carbonic crystals, although similar problems are faced when attempts are made to synthesize other 2D materials.

8.2.3   *Silicene*

Among the various 2D crystalline structures similar to graphene is also all-silicon-based silicene. Owing to its current use in semiconductor electronics combined with a similar hexagonal graphene-like lattice, this monolayer allotrope of silicon (Si) has attracted increasing attention due to its potential compatibility with Si-based electronics. It was first mentioned in a theoretical study by Takeda and Shiraishi [13] in 1994 and then investigated by Guzman-Verri et al. in 2007, who labeled it silicene [14]. Structurally, silicene is a 2D sheet of hexagonally arranged silicon atom, analogous to graphene, but with consecutive bonds buckled out-of-plane. Unlike graphene, silicene is not stable as a perfectly planar sheet, with

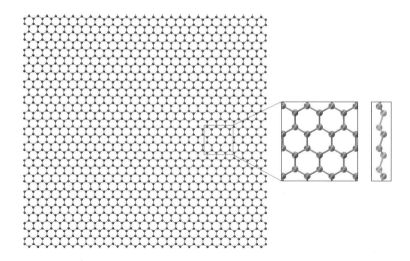

Figure 8.5   Schematic of silicene

chair-like distortions in the rings leading to ordered surface ripples and enhancing reactivity with other material surfaces. Being said, as sp^3 hybridization is more stable in silicon than sp^2 hybridization, silicene is energetically favorable as a low-buckled structure, as depicted in Figure 8.5. This buckling confers advantages on silicene over graphene because it should, in principle, generate both a band gap and polarized spin-states that can be controlled with a perpendicular electrical field [15].

The mechanical properties (in-plane stiffness and ultimate strength) of silicene in both armchair and zigzag directions, as well as the bending rigidity have been previously deduced theoretically in the literature using techniques such as MD and DFT. The in-plane stiffness in zigzag direction is reported in the range of 50–65 N/m, whereas in armchair direction the reported values of in-plane stiffness range from 59 to 65 N/m [16, 17]. The ultimate stress is reported approximately 5–8 N/m for both directions, with the corresponding value of maximum strain between 10% and 20% [16, 17]. The bending modulus has been obtained by curve fitting with energy versus curvature data points and is calculated to be approximately 38 eV. Compared with monolayer and bilayer graphene, it can be noted that this value falls between these two values. This could be due to the out of buckling structure of silicene, which creates an effective thickness and bending inertia, compared to flat graphene, which does not have this effect [16].

8.2.4   Boron Nitride

Boron nitride (BN) is another 2D nanostructure with graphene-like structure that consist of alternating boron and nitrogen atoms in a hexagonal arrangement, where boron and nitrogen atoms are bonded with strong covalent sp^2 bonds (Figure 8.6), while for multiple layers they are stacked together by electrostatic interaction and weak van der Waals forces, as in graphite. Therefore, BN layers could be peeled off from bulk BN crystal by micromechanical cleavage and used as a dielectric layer [18]. For this reason, bulk BN is commonly called white graphite, and in analogy the monolayer hBN is known as white graphene [19].

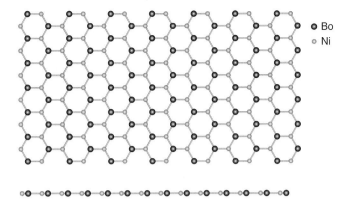

Figure 8.6   Atomic structure of boron nitride (BN)

BN, unlike graphene, is an insulator due to the difference in electronegativity, where the narrowing of the sp² p-bands is responsible for the loss of conductivity [18]. BN is well known for a variety of crystalline structures (cubic, rhombohedral, amorphous); however, its hexagonal layered structure is the most stable. Due to a combination of exceptional properties, from which we can highlight its high oxidation resistance, large thermal conductivity, good electrical insulation, chemical inertness, nontoxicity, apart from being environmentally friendly, this versatile inorganic compound can be used for a wide range of industrial applications, such as surface coatings, ceramic composites, lubricants, or insulators [20].

The mechanical parameters of boron nitrate have been reported in different studies using theoretical calculations such as density functional theory (DFT). The Young's modulus of single-layered hBN is about 780 GPa and is nearly independent of loading orientation. The bending rigidity is also isotropic and is about 0.95 eV. The ultimate tensile stresses are 102, 88, and 108 GPa for the zigzag, armchair, and biaxial strains, respectively [21].

8.2.5   Molybdenum Disulfide

Molybdenum disulfide (MoS_2) is an inorganic 2D material, semiconductor with hexagonal structure, and the most prominent member of the family of transition metal dichalcogenides (TMD). TMDs are representing by the formula MX_2, where M = Mo, W, V, Nb, Ta, Ti, Zr, Hf, and X = S, Se, Te, constituting a fascinating group of materials with an extensible range of remarkable mechanical, optical, and electronic properties and applications [19]. Structurally, it consists of three-atom-thick layers of a monatomic molybdenum plane between two monatomic sulfur planes, such as a sandwich structure (Figure 8.7), in which the atoms are bonded with strong covalent bonding, and the layers are held together by van der Waals interactions [22]. MoS_2 can be obtained easily from the bulk material using exfoliation techniques, such as graphene and BN [23].

The presence of a direct band gap in monolayer MoS_2 makes it promising for nanoelectronics and photoelectronics applications. Another special feature of this material is that the electronic structure can be tuned by varying the number of layer due to perpendicular quantum confinement effect [19]. Additionally, its functional and structural properties can

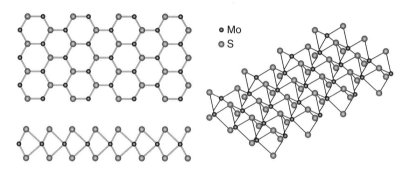

Figure 8.7   Atomic structure of molybdenum disulfide (MoS$_2$)

be modulated by the application of mechanical strain, electric field, surface adsorption, and defects [19, 24].

Due to the remarkable mechanical properties of MoS$_2$, such as large ultimate strains, resulting in high flexibility, and high strength and stiffness, the material is presented as a promising candidate for elastic energy storage for clean energy, as reinforcement in nanocomposites, and for fabrication of flexible electronic devices. Furthermore, it is environmentally friendly. Several studies report the mechanical properties of MoS$_2$, using DFT calculations. For a single layer of MoS$_2$, the Young's modulus is reported in the range of 184–270 GPa, while the ultimate strength is reported between 12 and 24 GPa [19, 24, 25].

8.2.6 Germanene, Stanene, and Phosphorene

The remarkable properties and extensive applications of carbon-based graphene have lead to the research of layered structures of the other elements in main group IV (Si, Ge, Sn, and Pb), resulting in the development of new and innovative atomic 2D nanostructures: germanene and stanene. In their stable state, these elements form a buckled hexagonal structure, unlike graphene. This is related to the nature of graphene possessing an sp^2 hybridization, whereas silicene, germanene, and stanene exhibit the sp^2–sp^3 hybridization due to their preferential state of sp^3 bonding [26].

One of these promising materials is germanene, which is the germanium analog of silicene. Based in first principled DFT calculations, Cahangirov et al. reported that the low-buckled honeycomb structure of germanium analog of graphene can be stable [27]. Trying to produce germanene, the hydrogen-terminated germanene was synthesized in 2013 from the topochemical deintercalation of calcium digermanide (CaGe2) [28]. Afterwards, Li et al. [29] have reported the successful fabrication of germanene by an annealing process after having deposited germanium onto a Pt(111) surface. As in free-standing silicene, germanene presents no band gap at the Fermi level, indicating metallic properties [3].

Two-dimensional tin, called stanene, is another versatile Group IV elemental 2D material. Stanene presents an intrinsic buckling that allows for functionalization by an out-of-plane electric field, and almost full sp^3 hybridized bonding. It also owns a slightly metallic band alignment, with massless Dirac fermions, analogous to silicene and germanene, which are excellent properties for potential nanoelectric applications [30]. Although there is a lack of knowledge on the fundamental properties of some of these Group IV 2D allotropes, the notable correspondence in their crystal and electronic structures enables the properties of germanene and stanene to be extrapolated through relative estimations with graphene and silicene [26].

Recent theoretical studies on the in-plane stiffness of Group IV elemental layers reported an observed reduction in the stiffness with increasing atomic weights (from Si to Sn). This observation is related to the tendency of metallic bonding with increasing atomic weight that is evidenced by the increasing bond length [31].

One of the latest 2D nanostructure developed is phosphorene, which was isolated by liquid exfoliation of black phosphorus [32]. Similar to graphite, black phosphorous is made of stacked layers of phosphorene held together by weak van der Waals forces [33]. Phosphorene, such as graphene, is the elementary 2D nanolayer that composes black phosphorus crystals, but unlike graphene, it has a unique, vertically skewed/wrinkled hexagonal structure (Figure 8.8). Monolayer phosphorene is a semiconductor with a predicted direct band gap, which combined with tunability of its properties makes it an ideal candidate for nanoelectronic, optoelectronic, and photovoltaic applications [26].

It should also be noted that chemical functionalization, especially hydrogenation, is also being used to develop new 2D nanomaterials, such as germanane. The hydrogenated counterpart of graphene is graphane, a fully saturated 2D hydrocarbon with sp^3 hybridized bonds that opens up. Graphane is an insulator and lacks the Dirac cone of graphene [34]. Hydrogenation process can lead to the deterioration of graphene mechanical properties, such as Young's modulus and shear modulus due to membrane shrinkage and extensive membrane corrugation caused by functionalization [35]. Also under study is silicane, the fully hydrogenated version of silicene, which is a wide-band-gap semiconductor that is projected as a promising potential candidate not only for nanoelectronic applications but also for hydrogen storage [36]. For hydrogenated BN, the hybridized states of boron or nitrogen atoms change from sp^2 into sp^3. Hydrogenation opens a band gap in graphene, but it reduces the band gap of BN sheets. Silicane, the fully hydrogenated version of silicene, is a wide-band-gap semiconductor that is projected as a promising potential candidate not only for nanoelectronic applications but also for hydrogen storage [19].

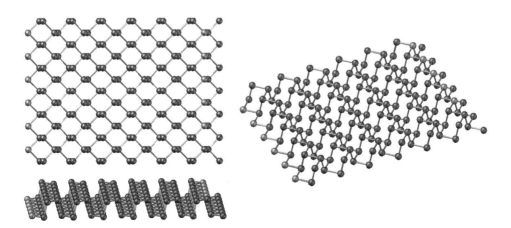

Figure 8.8 Atomic structure of phosphorene

8.3   Mechanical Assays

There are two basic approaches to investigate the mechanical performance of a nanomaterial: experimental and computational modeling. The difference between these two methodologies is that modeling presents how materials *should* behave, experimentation presents how materials *do* behave. Both methodologies face challenges when applying to study nanostructures. On the one hand, experiments should be capable to identify, measure, and manage the different variables on the system; on the other hand, modeling methods must deal with assumptions, constraints and boundary conditions, unknowns, and the complexity of real material structure and environments. While experimental approaches are necessary to characterize emerging 2D materials, we concentrate primarily on computational methods herein. Due to the precise control of material geometries and boundary conditions, computational methods are advantageous when exploring theoretical mechanistic models, requiring less assumptions and conditions than experimental techniques. Ultimately, both models and experiments broadly contribute to the understanding of the nanostructures behavior.

8.3.1   Experimental

Mechanical characterization through experimental methods represents a successful approach to study the behavior and properties of 2D nanostructures. Various experimental techniques are being used to image single layer or multilayers of 2D structures such as optical microscope, atomic force microscopy (AFM), scanning electron microscopy (SEM), and transmission electron microscopy (TEM). Probing the atomic structure of 2D materials is a demanding task due to the small sample size and the choice of technique on the measurement requirements. Often, a combination of two or more techniques completes the characterization of nanomaterials, allowing the fast estimation of parameters as thickness distribution and facilitating their observation and identification [19].

One of the most used methods for carrying out experimental measurements of mechanical properties in 2D nanostructures is scanning probe microscopy (SPM), which is a family of methods that use a sharp probe of nanometer dimension to detect changes in the material surface structure on the atomic scale [37]. SPM provides 3D real space images and enables spatially localized measurements of structure and properties. AFM is one of the prominent members of the SPM family, which can produce topographic images of a surface with atomic resolution in all three dimensions, moreover, combining with appropriately designed attachments, such as nanoindentation, and allowing its use in measuring the mechanical properties such as Young's modulus and hardness of various types of materials [38].

AFM has been used with multilayered graphene sheets in static deflection tests to measure the effective spring constants [39]. Another approach was employed to probe graphene sheets [5, 40], hBN [41], and monolayer MoS_2 suspended over circular holes [42] (Figure 8.9), which yielded the Young's modulus, bending rigidity, and breaking strength by comparing the experimental data to a continuum plate model [43]. The requirement that the material to be suspended translates into a significant limitation because of the use of nanoindentation method with some 2D nanostructures. The presence of a substrate, over which the nanostructure may be either deposited or directly grown epitaxially (e.g., silicene over silver [44], graphdiyne over copper [9]), makes it hard to separate the intrinsic mechanical properties of the nanolayer from that of the substrate from the nanoindentation measurements [40].

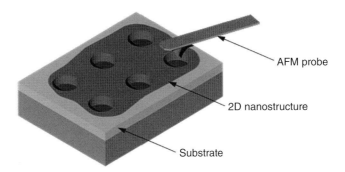

Figure 8.9   Schematic of an indentation experimental setup showing a nanostructure on substrate, suspended over open holes. The mechanical properties are probed by deforming and breaking the resulting suspended free-standing sheet with an atomic force microscope (AFM)

Raman spectroscopy is a vibrational technique that provides access to information relating to the lengths, strengths, and arrangements of chemical bonds in a material, but the method is not highly sensitive to information such as chemical composition [40]. Raman technique uses the interrogation of bond vibrations by optical spectroscopy and provides information about crystallite size, the presence of sp^2–sp^3 hybridization, the introduction of chemical impurities, the optical energy gap, elastic constants, defects, edge structure, strain, number of layers, and mechanical and structural properties for single or multilayer structures, both freestanding or on a substrate [45, 46]. Raman spectroscopy has been widely used in mechanical characterization of graphene layers [40].

Other useful experimental technique with broad applications with 2D nanostructures is TEM, in which a beam of electrons is transmitted through the sample, interacting with the specimen as it passes through [47]. The scattering processes experienced by electrons during their passage through the specimen determine the kind of information obtained, such as the layer sizes, the elemental composition, the nature of crystallinity, and interlayer stacking relationships [38, 48].

8.3.2   Computational

The advancement of computational and processing power in recent decades has rapidly paced the theoretical investigations on various materials ranging from bulk to nanomaterials. Computational techniques such as MD, DFT, and Monte Carlo simulations provide advantageous understanding to elucidate and study the nanoscale phenomena. Moreover, computational methods have become an indispensable tool to the investigation of material systems, complementing the experimental analysis for conducting materials design and property prediction.

MD is a simulation technique that consists of numerically solving the classical equations of motion for a group of atoms, in which the motion of each atom in the material is characterized by its position, velocity, and acceleration. The molecular dynamic method encompasses five basic steps: definition of initial conditions (initial temperature, number of particles, density, time steps, etc.), identify initial positions and velocities of the system particles, integration of Newton's equations of motion, calculation of the force acting on every particle, and computation of the average of measured quantities. Each atom is considered as a classical particle that obeys Newton's laws of mechanics. The only physical law that is used to simulate the dynamical behavior of the atoms is Newton's law, along with a definition of how atoms interact with each other. Those interactions are the so-called interatomic potentials or force fields that describe attractive and repulsive forces in between pairs or larger groups of atoms [49].

DFT, based on quantum theories of electronic structure, is currently the most commonly employed quantum mechanics method, which has evolved into a powerful tool for computing electronic ground-state properties of a large number of nanomaterials. The entire field of DFT method relies on the theorem that the ground-state energy of a many-electron system is a unique and variational functional of the electron density, and this conceptual proposal is implemented in a mathematical form to solve the Kohn–Sham (KS) equations [19]. We note that, due to the system size constraints of DFT, in this chapter we focus on the MD approaches to mechanical characterization of 2D materials, which are more versatile in implementation. Being said, DFT and other quantum-level approaches have a degree of accuracy unattainable by the potential formulations of MD. Moreover, when analytical potentials are available, the effect of various types of interactions can be investigated by switching off, or modifying, the corresponding potential energy terms, being able to probe our understanding of the studied problems. Being said, prudence is necessary when selecting any approach to characterize such systems, with understanding of both the advantages and limitations of any method.

8.4 Mechanical Properties and Characterization

Atomistic simulations have proved to be a unique and powerful way to investigate the mechanical properties of 2D systems at a very fundamental level, and provide a means to investigate complex systems with unparalleled control and accuracy. MD has evolved as a suitable tool for elucidating the atomistic mechanisms that control deformation at the nanoscale and for relating this information to macroscopic material properties. Mechanistic behavior, function, and failure are directly linked to distinct atomistic mechanics and require atomistic and molecular level modeling as an indispensable tool for studying 2D materials [50]. The implementation of full atomistic calculations of mechanical test cases by classical MD has been used successfully to derive a simplified set of parameters to mechanically characterize monolayer systems such as graphene, simple graphyne, and silicene, which are in good agreement with the experimental data as far as available. The atomistic-level characterization techniques described herein are equally applicable to different structures and can be immediately applied to various 2D material geometries.

One of the most logical approaches to characterize the mechanical properties and the behavior of an atomistic system is to emulate the testing procedures and measurement of the macroscale material, including tensile and compression tests, three-point bending tests, torsion, and so on. For example, a small model system can be constructed as a computational "test specimen" with direct control over the loading and boundary conditions, from which the specimen can be subject to various testing procedures and mechanical properties directly determined. The dominance of specific mechanisms is controlled by geometrical parameters, the chemical nature of the molecular interactions, as well as the structural arrangement. Such response-based approaches are advantageous when there is inherent difficulty in defining traditional elastic "properties" such as bending stiffness or Young's modulus.

8.4.1   Defining Stress

To obtain the maximum strength or ultimate stress (along with corresponding extensibility or ultimate strain) for different 2D nanostructures, the most direct approach is to apply a uniaxial tensile strain along the representative directions in the material, analogous to a traditional macroscopic tensile coupon test. One just need to extract the corresponding stress though atomistic measures – this is typically accomplished using *virial stress* (following the *virial theorem* of Classius, 1870, which relates the average over time of the total kinetic energy of a stable system of constant particles bound by potential forces). The virial stress is commonly used to find the macroscopic (continuum) stress in MD computations [51, 52].

Mechanical stress is the measure of the internal forces acting within a deformable body and is inherently a continuum concept. In the molecular interpretation there is no continuous deformable body, but a set of discrete atoms. To reconcile this fact, we fundamentally define material stress as the change in free energy density is a function of material strain, or

$$\sigma_{ij} = \frac{\partial u}{\partial \varepsilon_{ij}} \tag{8.1}$$

where σ_{ij} is the component of stress induced by a strain, ε_{ij}, and the free energy density, u, is defined as the energy per unit volume (the indices i and j refer to the basic vectors of an arbitrary coordinate system). The formulation of stress in terms of energy landscape is convenient, as it is also a common description of atomistic systems in terms of potential energy functions or force fields, where

$$u = \frac{1}{\Omega} \sum_{a \in \Omega} \phi_a(\chi) \tag{8.2}$$

in which Ω is the considered system volume, a the atoms within that volume, and ϕ_a the potential energy function dependent on some state variable χ (the total strain energy $U = u\Omega$). The continuum stress interpretation of atomic force fields is important, allowing the intensity and nature of internal interactions in materials to be measured and equated with macroscopic metrics (such as Young's modulus).

In order to obtain the atomistic–molecular counterpart of the stress tensor (as defined in Eq. (8.1)), we consider a small sample volume, Ω, of an atomistic system.[1] This sample volume contains N atoms, described by positions $r_i^{(a)}$ for $a \in 1, \ldots, N$. The components of the position vector can be denoted by $i = 1, 2, 3$ in 3D space (e.g., $r_1^{(a)}, r_2^{(a)}, r_3^{(a)}$). The number of atoms is large enough to allow an adequate definition of elastic field (stress and strain) in the region. The virial stress approach allows us to determine the components of the macroscopic stress tensor by considering both the kinetic energy of the atoms and the interatomistic forces through the virial components, S_{ij}, in the representative volume Ω, where

$$S_{ij} = \sum_{a \in \Omega} \left[-m^{(a)} v_i^{(a)} v_j^{(a)} + \frac{1}{2} \sum_{b \in \Omega} \left(\left(r_i^{(a)} - r_i^{(b)} \right) F_j^{(ab)} \right) \right] \tag{8.3}$$

which generates the six components of the symmetric stress–volume tensor, S_{ij}, where $m^{(a)}$ is the mass of particle "a," $v_i^{(a)}$ and $v_j^{(a)}$ are the velocities in the ith and jth vector component basis, $r_i^{(a)} - r_i^{(b)}$ denotes the distance between particle "a" and atom "b" along the ith vector component, while $F_j^{(ab)}$ is the force on particle "a" exerted by particle "b" along the jth vector

---

[1] Assumed elastic and following the derived continuum laws as $\Omega \to \infty$.

component (defined by the interatomistic potential, $\phi_a(\chi)$). We observe that the velocity v_i of each atom is composed by a term corresponding to an effective macroscopic drift. As a sum of both the kinetic energy and directional force of the potential, the virial can be (colloquially) considered a kind of "directional strain energy," akin to the strain energy, U.

To reduce random fluctuations, in addition to averaging over the representative spatial volume, Ω, it is recommended to average further over small time interval around the desired time of the stress. We introduce the time average:

$$\langle f(t) \rangle = \lim_{\tau \to \infty} \left[\frac{1}{\tau} \int_0^\tau f(t)\, dt \right] \tag{8.4}$$

The total stress can be calculated as

$$\sigma_{ij} = \langle S_{ij} \rangle \Omega^{-1} \tag{8.5}$$

We note that the virial stress defined as in Eq. (8.5) is equivalent to Cauchy stress only in the framework of linear approximation. However, the virial stress can easily be calculated regardless of nonlinearity, large strains, and/or failure/fracture response, dependent on the fidelity of the atomistic potentials.

One primary advantage of the virial stress approach is that it can shed insight into the coupling of deformation modes and stress distributions in a molecular system, by simultaneously and directly evaluating all components of stress. By calculating the virial stress, any direct force–extension data (e.g., from an applied spring load) can be directly compared to the virial stress and strain along the axis of extension. If there is limited coupling (e.g., no shear modes under pure tension), the results of the virial stress–strain and the applied load–displacement should agree both qualitatively and quantitatively. Disagreement between the axial stress and "force over area" calculations can indicate either the dominance or coupling with a second-order mode of deformation, such as shearing or twist. Moreover, while per atom virial stresses may be inappropriate (in terms of an equivalence to the continuum interpretation of stress), they can be used to map a virial stress field and indicate potential localizations of high force and failure events. This can shed great insight into heterogeneous material systems to indicate any potential "weak links" [50].

Virial stress combined with explicitly applied loads and displacements are powerful tools in the determination of the mechanical properties of complex molecular systems. Ultimately, they rely on fundamental stress–strain relationships formulated by continuum theory, and effectively "translate" simulation results into representative mechanical characterization. However, some systems require a more fundamental interpretation of mechanical behavior when the system is unsuitable for applied loads or prescribed boundary conditions. For such systems, fundamental energy methods can be prescribed [50].

8.4.2 Uniaxial Stress, Plane Stress, and Plane Strain

The exemplary "strength" of 2D materials is an attractive and ubiquitous indicator of their potential in nanotechnology. Indeed, defect-free graphene is said to have an intrinsic strength high than any other material. As such, we first focus on the characterization of ultimate stress. Similar to macroscale counterparts, material strength is most commonly characterized by simply loading a representative sample until failure. Simulated uniaxial tensile tests can be applied to a 2D material sheet by fixing the boundaries in the desired direction and deforming a unit cell by stretching along tested direction at a uniform rate (thereby inducing a uniform strain rate). It is noted that the fixed edges are free to move orthogonal to the applied strain, and measurement of transverse deformation/contraction can thus be used to determine Poisson's ratio, v. Free edge and nonperiodic boundary conditions result in a sheet of finite size, enabling size-dependent studies if desired. Due to the relatively high ratio of edge atoms to bulk atoms in 2D materials, it has been shown that limiting the width of the system (e.g., nanosheets vs nanoribbons) can greatly affect the mechanical response [53].

As stress is fundamentally a 3D concept (undergraduates are often introduced to stress in terms of a force (1D vector) acting normal to a plane (2D surface), necessitating 3D), what is the stress interpretation of an ideal 2D system? In continuum mechanics, a material is said to be under *plane stress* if the stress vector is zero across a particular surface. When that situation occurs over an entire element of a structure (as is often the case for thin plates), the stress analysis is considerably simplified, as the stress state can be represented by a tensor of dimension two. A related notion, *plane strain*, is often applicable to very thick members, whereby strain is limited to zero in a direction, but stress is allowed. Two-dimensional materials represent a combination of both plane stress and plane strain, as neither stress nor strain can possibly be induced out of plane. To illustrate, first, we consider the uniaxial extension of an elastic system. If the extension is in the x-direction, and the material is presumed isotropic and homogeneous and within its elastic limit, then Hooke's law applies

$$\sigma_x = E\varepsilon_x \tag{8.6}$$

where σ_x is the stress, ε_x the strain, and E is the Young's modulus (e.g., uniaxial stiffness). Ultimate stress or strength can easily be taken as the maximum stress experienced prior to system failure during MD simulation.

Realizing, however, that extensions/strains in other directions are proportional to Poisson's ratio, $v_{ij} = -\varepsilon_i/\varepsilon_j$, the strains in three dimensions can be represented as

$$\begin{bmatrix} \varepsilon_x \\ \varepsilon_y \\ \varepsilon_z \end{bmatrix} = \frac{1}{E} \begin{bmatrix} 1 \\ -v_{yx} \\ -v_{zx} \end{bmatrix} \sigma_x \tag{8.7}$$

Similar relationships apply for stress in the other two directions, so we obtain the equations that apply to any 3D stress problem where axial strains are only considered:

$$\begin{bmatrix} \varepsilon_x \\ \varepsilon_y \\ \varepsilon_z \end{bmatrix} = \frac{1}{E} \begin{bmatrix} 1 & -v_{xy} & -v_{xz} \\ -v_{yx} & 1 & -v_{yz} \\ -v_{zx} & -v_{zy} & 1 \end{bmatrix} \begin{bmatrix} \sigma_x \\ \sigma_y \\ \sigma_z \end{bmatrix} \tag{8.8a}$$

Typically, for the case of plane stress, the condition that $\sigma_z = 0$ would be applied, whereas for plane strain, the condition that $\varepsilon_z = 0$ would be applied. These conditions result in the plane stress and plane strain formulations that can be found in any traditional elasticity text. However, for 2D materials, we can apply both conditions simultaneously, such that

$$\begin{bmatrix} \varepsilon_x \\ \varepsilon_y \\ 0 \end{bmatrix} = \frac{1}{E} \begin{bmatrix} 1 & -v_{xy} & -v_{xz} \\ -v_{yx} & 1 & -v_{yz} \\ -v_{zx} & -v_{zy} & 1 \end{bmatrix} \begin{bmatrix} \sigma_x \\ \sigma_y \\ 0 \end{bmatrix} \tag{8.8b}$$

which results in the equations:

$$\varepsilon_x = \frac{1}{E}(\sigma_x - \nu_{xy}\sigma_y) \tag{8.9a}$$

$$\varepsilon_y = \frac{1}{E}(\sigma_y - \nu_{yx}\sigma_x) \tag{8.9b}$$

$$0 = \frac{1}{E}(-\nu_{zx}\sigma_x - \nu_{zy}\sigma_y) \tag{8.9c}$$

The first two equations are identical to the equations of plane stress, which can be rearranged such that:

$$\begin{bmatrix} \sigma_x \\ \sigma_y \end{bmatrix} = \frac{E}{1-\nu^2} \begin{bmatrix} 1 & \nu_{xy} \\ \nu_{yx} & 1 \end{bmatrix} \begin{bmatrix} \varepsilon_x \\ \varepsilon_y \end{bmatrix} \tag{8.10}$$

The final equation above seems to imply that $\sigma_x = -\sigma_y$, which does not seem reasonable (e.g., stretching in one direction does not cause graphene or similar 2D material to undergo equivalent compression in the other direction). However, we realize that both Poisson ratios are in terms of the component "z." Since $\nu_{ij} = -\varepsilon_i/\varepsilon_j$ where $\varepsilon_z = 0$, then $\nu_{zx} = \nu_{zy} = 0$, and

$$0 = \frac{1}{E}(-\nu_{zx}\sigma_x - \nu_{zy}\sigma_y) = \frac{1}{E}(-(0)\sigma_x - (0)\sigma_y) \tag{8.11}$$

So 2D materials can be represented by classical plane stress conditions while neglecting and strain in the third direction. Introducing shear contributions and assuming, in plane, that $\nu_{xy} = \nu_{yx} = \nu$, results in

$$\begin{bmatrix} \sigma_x \\ \sigma_y \\ \tau_{xy} \end{bmatrix} = \frac{E}{1-\nu^2} \begin{bmatrix} 1 & \nu & 0 \\ \nu & 1 & 0 \\ 0 & 0 & \frac{1}{2}(1-\nu) \end{bmatrix} \begin{bmatrix} \varepsilon_x \\ \varepsilon_y \\ \gamma_{xy} \end{bmatrix} \tag{8.12}$$

While it may be simple to state that 2D materials can be considered plane stress, it is critical to understand that the underlying assumptions are different – the resulting formulation is just a convenient interpretation.

With these formulations, multiaxial stress/strain conditions can be applied to extract a number of critical properties via the calculation of virial stress as discussed in Section 8.4.1. In terms of uniaxial strength (ultimate stress) or even biaxial strength (when $\sigma_x = \sigma_y$), the maximum value is taken from the test simulation.

8.4.3 Stiffness

Beyond strength, planar stiffness or axial rigidity is a key mechanical property of 2D materials. Quite different approaches have been employed to derive and measure the stiffness of 2D materials [16]. From a 3D continuum mechanics approach, the stiffness coefficients are defined by

$$C_{ijkl} = \frac{\partial^2 u}{\partial \varepsilon_{ij} \partial \varepsilon_{kl}} \tag{8.13}$$

where u is the potential energy density of the system, and ε_{ij} and ε_{kl} are strains along different orientations. Again, for details, the reader is directed to any classical elasticity text. We note the relationship with Eq. (8.1) resulting in

$$C_{ijkl} = \frac{\partial}{\partial \varepsilon_{kl}} \sigma_{ij} \tag{8.14}$$

which leads to the colloquial (but useful) definition that the stiffness coefficients are simply the derivative of the stress–strain relation. For uniaxial conditions, as described earlier, the Young's modulus (E) can be defined as

$$E = C_{xxxx} = \frac{\partial^2 u}{\partial \varepsilon_{xx}^2} = \frac{\partial^2 u}{\partial \varepsilon_x^2} \tag{8.15}$$

where again u is the strain energy density and ε_x the strain in the uniaxial direction. Note that, in Eq. (8.15), there is no need to calculate any stress quantities – merely the total strain energy. This is beneficial in energetic or *ab initio* approaches, such as the aforementioned DFT approach, which does neither applies potentials nor derives forces, negating the virial stress.

Typically, the strain energy density u is defined as the total energy U per volume V, or $u = U/V$. For 2D materials, however, the strain energy can be defined per unit area, as there is no definitive "thickness." As such, the 2D modulus can be defined as

$$E_{2D} = \frac{1}{A_0} \frac{\partial^2 U(\varepsilon)}{\partial \varepsilon_2^2} \tag{8.16}$$

where U is the total energy of a 2D system of finite area A_0. The ambiguity for the thickness of monoatomic crystal structures such as graphene and other 2D nanomaterials has been discussed in a previous study [54], and due to the buckled structure of some 2D nanomaterials such as silicene, it is difficult to theoretically assign. This situation has been discussed in previous studies [16], suggesting the stress and elastic moduli of monolayer systems be reported in force per unit length (N/m) rather than force per unit area (N/m^2 or Pa), consistent with Eq. (8.15). As such, the in-plane stiffness and in-plane stress (maximum force per unit length) are more accurately used to represent the stiffness and strength of the structure, for a more apt comparison to other 2D crystals where σ and ε are the stress and strain [54, 55]. Note that the 2D modulus can be derived for any thin 2D-like system, by scaling the traditional modulus by the height of the system, or $E_{2D} = Eh$.

Since, if using MD approaches, stress can be calculated, we can alternatively use the stress–strain relation to determine uniaxial moduli. The modulus/stiffness can be then calculated by

$$E = \frac{\partial \sigma_x}{\partial \varepsilon_x} \tag{8.17}$$

where σ_x and ε_x are the stress and strain, and the subscript "x" indicates the direction of loading. Thus, a tangential fitting of the stress–strain response of a 2D material will indicate the stiffness at a particular strain value. A linear fit of the initial small deformation regime is typically associated with the Young's modulus E.

A careful scrutiny between Eqs (8.14) and (8.17) indicates that they may not result in the same quantity. If using viral stress, the definition of σ_x only considers components of force in the x-direction, thus energetic components in other directions may be omitted from the calculation. Coupling of deformation mechanisms may increase or decrease the strain energy density, u, without emerging in the σ_x calculation. Thus, both approaches are necessary to reveal any atomistic mechanisms that may not be apparent *a priori* when calculating stiffness.

Following the simple derivation of Young's modulus as indicated by Eq. (8.17), other classical mechanical properties such as the shear modulus and bulk modulus can be calculated in a similar manner. We first consider Hooke's law in shear:

$$\tau = G\gamma \tag{8.18}$$

where the shear stress τ is related to the shear strain γ via the shear modulus G. The continuum interpretation of the shear modulus can be stated as

$$G = C_{xyxy} = \frac{\partial^2 u}{\partial \varepsilon_{xy}^2} = \frac{\partial}{\partial \varepsilon_{xy}} \sigma_{xy} = \frac{\partial \tau_{xy}}{\partial \gamma_{xy}} \tag{8.19}$$

where $\sigma_{xy} = \tau_{xy}$. Subjecting a system to simple shear strain, γ_{xy}, we can again use tangential or linear fits to attained data to derive a shear modulus. Of note, subjecting shear to 2D system commonly results in out-of-plane buckling due to the compressive response, limiting the applied shear strain range. In effect, the shear stiffness is relatively easy to determine in the small deformation regime, but the shear strength is more difficult to achieve due to nonlinear geometric effects.

Finally, the bulk modulus can be determined via a biaxial straining of a 2D system. While bulk modulus K typically relates hydrostatic stress to volumetric change for 2D systems, the equivalent K_{2D} related biaxial stress to areal change defined by

$$\sigma_{\text{biaxial}} = K_{2D} \frac{\Delta A}{A_0} \tag{8.20}$$

where ΔA is the change in area, A_0 the initial area, and σ_{biaxial} is the stress simultaneously applied in the x- and y-directions.

8.4.4   Effect of Bond Density

The remarkable strength and stiffness are commonly considered a function of their limited "thickness," as all chemical bonds are limited in-plane, and thus appear "stronger" in that direction. Thus, the strength is merely a result of a higher, planar bond density due to 2D structures. This idea can be explored if one could vary the bond density of a structure, while the geometry remains the same. Theoretically, this is possible in extended graphynes.

We have established that the strain energy and deformation of graphene or silicene sheets can be described by continuum elasticity theory [56–58], and as such the same assumptions are applied to graphynes. Likewise, for the mechanical characterization, it is assumed that under small deformation, graphyne can be approximated as a linear elastic material. Previous studies have reported a continuous degradation of modulus with the introduction of acetylene links (n) from graphyne ($n = 1$) to graphtetrayne ($n = 4$). The relatively smooth degradation as a function of bond density through n can be extended and predicted via a simple spring model formulation. The springs model can be used to predict the mechanical properties of graphynes with arbitrarily long acetylenic linkages, considering the graphyne system as a network of springs (see Figure 8.3). Such an elastic approximation (similar to a truss model) has been introduced for graphene [59] and carbon nanotubes [60]. The stiffness of a single segment can be written as

$$K = \left[\frac{1}{k_0} + \frac{n}{k} \right]^{-1} \tag{8.21}$$

where k_0 is the stiffness of the aromatic group and k is the stiffness of a single acetylene link. If we presume that $k_0 \gg k$ (e.g., deformation occurs predominately in the link, supported by previous MD results [61, 62], then $k_0^{-1} \to 0$, such that

$$K \cong \left[\frac{n}{k} \right]^{-1} \cong \frac{k}{n} \qquad (8.22)$$

We associate this linear spring stiffness to an effective modulus, where $K = EA_0/L$ (for the 2D perspective, EA_0 can be replaced with $E_{2D}b$, where b is the planar width of the system). The force required for unit strain can be written as

$$KL = EA_0 \qquad (8.23)$$

We assume that strain predominately occurs along the acetylenic links (due to $k_0 \gg k$; see Figure 8.3(a), such that $L = L_n = nL_0$, where L_0 is the summed distance of a carbon–carbon single and carbon–carbon triple bond. Thus,

$$KL = KL_n = KnL_0 = kL_0 = \text{constant} \qquad (8.24)$$

Thus, similar to adding equivalent springs in series, the material behavior is deemed constant, regardless of the number of acetylene links. The effective area per spring, however, changes as a function of n. The associated cross-sectional area per link can be represented by $A_n^{\text{cross}} = l_n t$, where t is the effective thickness of the graphyne plane (taken as the van der Waals spacing, $t = 3.2$ Å) and l_n the equivalent length of the weighted area, and $l_n \propto a$, the lattice spacing (see Figure 8.10(a) and (b)). From Eqs (8.23) and (8.24), it follows that $Y_n A_n^{\text{cross}} = \text{constant}$, and thus

$$E_n = E_1 \left(\frac{a_1}{a_n} \right) \qquad (8.25)$$

In general, $E_n \propto Ca_n^{-1}$, where the constant here is fitted by the parameters of graphyne ($n = 1$), for example, $C = Y_1 a_1$. The simple spring formulation can be used to predict the elastic modulus of arbitrary graph-n-yne and shows that the stiffness decreases as the bond density (through n) decreases.

8.4.5 Bending Rigidity

Theoretical studies and synthesis [63, 64] have suggested that bending stiffness of monolayer graphene is critical in attaining the structural stability and morphology of graphene sheets, which in turn could have important impacts on their electronic properties. This would similarly hold for other 2D crystal structures such as silicene. Due to the relative flexibility and single atom thickness of 2D monolayer materials, a mechanical bending test is difficult to implement. It is very difficult to apply a bending moment directly to such structures as local bending on a membrane structure would induce local curvature only. To circumvent this issue, the isotropic bending modulus is determined by 1D pure bending experiments using molecular statics in which curvature is induced (and fixed) prior to energy minimization, similar to previous coarse-grain [65] and full atomistic [7, 57, 66] investigations. To calculate the bending modulus of 2D monolayer materials, a rectangular sheet is bent into a section of a cylinder with constant radius of curvature throughout the basal plane (Figure 8.11(b)). The neutral

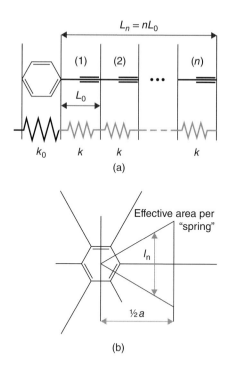

Figure 8.10 Simple spring–network model representation for scaling law. (a) Serial spring representation for the acetylene links. (b) The linear spring stiffness can be associated to an effective modulus via an effective cross-sectional area

plane for pure bending is parallel to the layer and passes through the centroid of the bending cross section.

The edges of the bent sheet are kept fixed and the bulk of the sheet is allowed to relax subject via low-temperature equilibrium and energy minimization. To avoid boundary effects at the fixed edges, the elastic energy is only considered for the interior portion of the sheet. At finite temperatures, graphene, graphyne, and silicene exhibit ripples and undulations out-of-plane (Figure 8.11(a)). As such, the material structure deviates from the ideal curvature initially imposed and the derived stiffness can be considered the effective bending modulus. However, both the low-temperature equilibrium and minimization process limit the observed ripples, and the system curvature is maintained.

The bending modulus is obtained by curve fitting with energy versus curvature data points, using the following expression for elastic bending energy

$$U = \frac{1}{2}D\kappa^2 \tag{8.26}$$

where U is the system strain energy per unit basal plane area (eV/nm^2), D is the bending modulus per unit width (eV), and κ is the prescribed beam curvature. A range of curvature is imposed on the system and the minimized energies are plotted versus κ (Figure 8.11(c)).

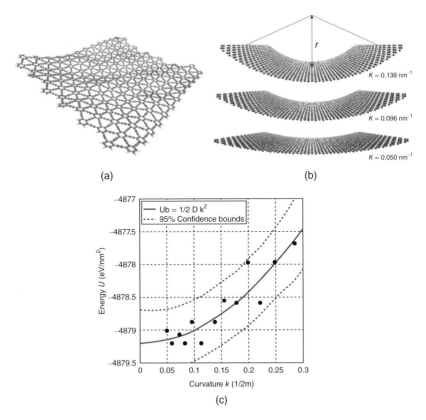

Figure 8.11 (a) Stable graphyne structure after minimization and equilibration of 0.5 ns at a temperature of 300 K. (b) Schematic illustration of a 2D silicene sheet bended with an imposed radius of curvature $\kappa = 1/r$. (c) Energy versus curvature for a silicene sheet with 95% confidence bounds

We note that the bending rigidity, D, is equivalent in function to the bending rigidity of a classical beam, EI, where E is the Young's modulus of the material and I is the second area moment of inertia. The inertia is involved due to the assumption that (1) a neutral axis exists in the cross section of the beam and (2) the section is in both tension and compression, changing once crossing the neutral axis. This core assumption, however, is meaningless for monoatomistically thin structures – is a single carbon atom, once bent in a graphene sheet, in compression or tension? The question is moot. The resistance to bending does not come from inertia effects, but rather due to electron orbital deformation and overlap. This, in turn, causes an energy increase and bending "rigidity." However, for multilayer systems (bilayer and greater), it has been shown that the bending modulus is proportional to the cube of the height of the cross section ($D \propto Eh^3$), in agreement with continuum interpretations [7, 65]. Indeed, once the system has "thickness," a neutral axis can be assumed and atoms can be in compression or tension accordingly. Note that this interpretation does not change the definition or utilization of D in the above-mentioned equation.

8.4.6 Adhesion

Due to the relatively high ratio of surface area to mass (e.g., specific surface area) and the propensity for surface interaction, the resulting adhesion plays an important role in many important technological applications of 2D materials. Graphene and other 2D materials, structures, and devices are increasingly influenced by surface forces due to the weak vdW interactions, especially as their size moves into the nanometer range. This occurs because the materials are often separated by small (interlayer) distances and are sensitive to the operant range of surface forces, and the structural stiffness decreases as its size decreases. For example, adhesive interactions are critical to nanomechanical devices [67] such as graphene switches – actuated electrostatically to bring them into, or near, contact with an electrode, while van der Waals forces can permit the release of the switch [68, 69]. Whether 2D materials are a promising nanoelectronics material depends predominantly on the nanostructure's mechanical integrity and ability to integrate or to adhere to electronic substrates.

Moreover, one of the most utilized fabrication methods of 2D materials involves CVD growth on an appropriate substrate followed by batch transferring it from the host substrate to a functional target material for device applications [70]. Another widely used technique for exfoliation of few- and mono-layered graphene is based on tearing-off graphene layers using adhesive tape [1, 71]. In both cases, the understanding of adhesion to both substrate and target material plays a crucial role in this process. The engineering of the peeling, stamping, and other fabrication processes also depends on the adhesion of the 2D layers. There is, therefore, an urgent need for experimental methods to characterize the mechanical properties and adhesion behavior.

It is well demonstrated that adhesion energies for different crystallographic stacking configurations for 2D materials such as graphene and BN show that the interlayer bonding and adhesion are due to the long-range van der Waals (vdW) forces [72]. In MD simulation, the vdW interaction is typically represented by a Lennard–Jones function – a mathematically simple model that approximates the interaction between a pair of neutral atoms and molecules. A Lennard–Jones 12:6 function is given as

$$\phi_{LJ} = 4\varepsilon_{LJ} \left[\left(\frac{\sigma_{LJ}}{r} \right)^{12} - \left(\frac{\sigma_{LJ}}{r} \right)^{6} \right] \tag{8.27}$$

where ε_{LJ} is the energetic depth of the potential well, σ_{LJ} is the finite distance at which the interparticle potential is zero, and r is the distance between the particles. The Lennard–Jones potential is a relatively good approximation to model dispersion and overlap interactions in molecular models. Applied to a system of 2D materials (e.g., a bilayer of two sheets), the interaction between the sheets at equilibrium is directly proportional to both the number of atoms, n, and the energy per atom, ε_{LJ}. It follows that the adhesion energy will scale with the area A of the sheets (assuming that the areal atomistic density, $\rho_A = n/A$, is constant). Thus, one can conclude that the adhesion energy, $U_{adhesion}$, is proportional to both ε_{LJ} and A. Note that $U_{adhesion} \neq n\varepsilon_{LJ}$ due to the long-range multineighbor interaction of the vdW forces, but $U_{adhesion} \propto \varepsilon_{LJ}$.

Using DFT or MD methods, the adhesion energy can be easily determined by direct measurement or by fitting the energy landscape. Direct measurement simply quantifies the energy change, ΔU, between an adhered system and a detached system of two sheets, sheet with substrate, sheet with molecule, and so on. Two simulations are required – one in which the 2D sheet is detached and beyond the interaction range of the vdW forces ($r \to \infty$), and the other in which the 2D sheet is adhered to the target at equilibrium ($r = r_0$). The energy change can be calculated by the simple difference of minimized system energies, where

$$U_{\text{adhesion}} = \Delta U = U(r \to \infty) - U(r = r_0) \tag{8.28}$$

If the area of 2D material is known, the adhesion energy can be calculated per area where

$$\gamma = \frac{U_{\text{adhesion}}}{A_0} \tag{8.29}$$

Beyond the simple strength of adhesion, the entire energy landscape during separation may be required to indicate the decay of interaction and shed light on phenomena such as electrostatic screening. Again, the simulation process is relatively trivial – instead of a two-point difference of energy, from $r = r_0$ to $r \to \infty$, the energy is tracked in discrete steps, such that $U_{\text{adhesion}} = f(r)$ can be plotted, for example, the energy landscape. Variations in separation procedure (e.g., shearing effects, introduction of nanoparticles) can explore different adhesion phenomena. Once the energy landscape is determined, statistical fitting methods can be used to describe the adhesion behavior. Energy can be scaled such that when $r \to \infty$, $U_{\text{adhesion}}(r) \to 0$, and in such a case:

$$\gamma = \frac{\min(U_{\text{adhesion}}(r))}{A_0} \tag{8.30}$$

Using simulation methods to determine adhesion strength is trivial, as the simulation interactions are formulated using potential energies and the values of any atom at any time can be extracted. This is not the case for experimental measures – there is no method to directly "measure" the potential energy of a single atom when adhered to a substrate. Rather, indirect measurements may be made exploiting the presumed elastic and adhesion behavior of a membrane-like system. Many ingenious methods have been derived by researchers in the field. Here, for brevity, we discuss two examples.

Zong et al. developed a technique to characterize adhesion of monolayered/multilayered graphene sheets on silicon wafer [73]. Nanoparticles trapped at graphene–silicon interface act as point wedges to support axisymmetric blisters (see Figure 8.12(a)–(c)). Local adhesion strength is found by measuring the particle height and blister radius using a scanning electron microscope. It was assumed that graphene behaved as a flexible membrane with negligible flexural rigidity. Using a continuum formulation, a governing equation was derived analogous to that of a thin membrane clamped at the periphery and being transversely loaded at the center [74]. At equilibrium, the blister contracts to a radius a and the adhesion energy can be calculated by

$$\gamma = Eh\left(\frac{w}{2a}\right)^4 \tag{8.31}$$

where E is the Young's modulus of the graphene, h the effective height of the multilayers (thus, "Eh" representing the 2D modulus as previously discussed), w the height of the blister, and a is the radius of the blister.

Alternatively, Lu and Dunn considered a 2D membrane that is adhered to a substrate over a cavity by van der Waals forces [75]. In the ground state (absence of an external load), the membrane adheres to the sidewalls over some distance, S, and is stretched flat between the walls over a distance L. This adhered length on the sidewall S and the tension in the membrane

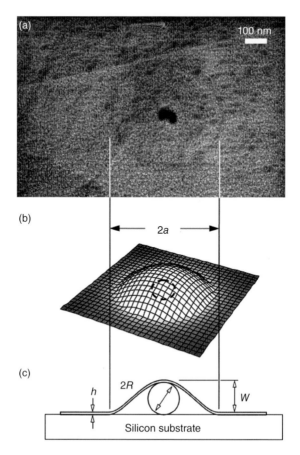

Figure 8.12 Nanoparticles trapped in a circular blister at graphene–silicon interface. (a) SEM image, where the trapped particles appear dark. (b) 3D representation. (c) Blister model sketch. [73]

are determined by the balance of surface energy due to the van der Waals forces on the sidewalls and elastic energy associated with membrane stretching. The potential energy of the system can be formulated [75], and minimizing the potential energy yields the equilibrium length:

$$S = \gamma \frac{L(1 - v^2)}{2Eh} \tag{8.32}$$

again, Eh appearing as the 2D modulus, L is the width of the cavity, and v is the Poisson's ratio. If the equilibrium length can be measured, the adhesion energy can be calculated. The critical assumptions in indirect experimental measurement of adhesion strength typically lie in the selection of a continuum approximation (e.g., membrane-like behavior for both the described approaches). Typically, such models are further validated against molecular simulation (if possible) to indicate any deviations due to atomistic phenomenal.

Adhesion is critical to the multilayered 2D structures (e.g., graphene–graphene), heterogeneous stacked layers (e.g., graphene–BN [76]), 2D materials on substrates (e.g., graphdiyne on copper), self-folded geometries (see Section 8.4.7), among others. Combined with high flexibility, 2D materials behave as ultimate thin membranes, and with adhesion conform more closely to a surface than any other solid. This provides new opportunities to study solid–solid surface interactions including the effects of even the smoothest surface topographies and potentially the nature of van der Waals and Casimir forces. The interplay of elastic and adhesion energies is shown to lead to stacking disorder and moiré structures. The dispersion, absolute band gaps, and low-energy electronic states are also dependent on the resulting stacked/adhered structures [76], once again opening yet another potential design variable.

8.4.7 Self-Adhesion and Folding

One of the consequences of a propensity of 2D layers to adhere in a multilayer manner is that, upon folding, they may also self-adhere. Although planar sheets are the most common form of such materials, scrolled and folded morphologies have attracted great interest because of their potential properties [77]. Like a sticky piece of paper folded on itself, the relative flexibility of 2D materials opens up a new field of topological possibilities and the so-called nano- or meso-origami. Indeed, the folding of paper and fabrics has been used for millennia to achieve enhanced articulation, curvature, and visual appeal for intrinsically flat, 2D materials [78]. For 2D materials, folding may transform it to complex shapes with new and distinct properties, opening up an entirely new paradigm of material's design.

From a mechanical perspective, recent studies show that suspended graphene sheets can fold and form folded edges due to van der Waals (vdW) interaction [65, 79–81]. The folded edges of graphene sheets show racket shapes with structures similar to carbon nanotube walls (Figure 8.13(a)–(d)), which can have strong influence on the electronic and magnetic properties of graphene [82–84]. The folded racket-shape graphene consists of a curved region of length $2L$ and a flat region of length L_0. The total length of the folded system is thus $L_{total} = 2L + 2L_0$. Note the adhered layers still maintain an equilibrium interlayer distance, congruent with the associated vdW adhesion. Clearly, the configuration of the folded graphene sheet results from the competition between adhesion energy $U_{adhesion}$ in the flat region and bending energy $U_{bending}$ in the curved region. This is a key behavior that also dictates the minimal crease size in multiple folded [79, 85] or crumpled 2D materials [86–88]. If the flat graphene is considered as the ground state, the energy of the folded graphene sheet is simply:

$$U_{total} = U_{bending} + U_{adhesion} \qquad (8.33)$$

The adhesion energy denotes the binding energy per unit area of graphene, and the bending energy is due to the necessary imposed curvature. The adhesion energy can be expressed as $U_{adhesion} = -\gamma L_0$ with γ denoting the binding energy per unit area of graphene (as per the previous section, $\gamma \propto \varepsilon_{LJ}$). In terms of the curved length:

$$U_{total}(L) = U_{bending} - \gamma \left(\frac{1}{2} L_{total} - L \right) \qquad (8.34)$$

Two cases arise from this simple relation:

1. If the total length of graphene is too short, the total energy $U_{total} > 0$, the resistance from the curved region can overcome the adhesion from the flat region, and therefore the folded configuration is unstable and can unfold.

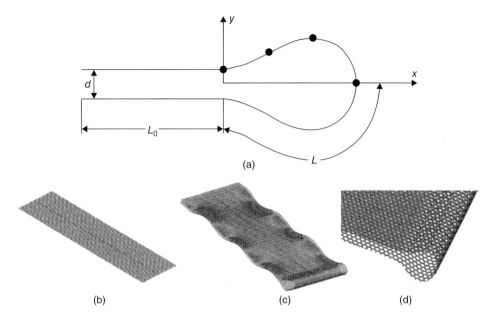

Figure 8.13 Folded configurations of graphene sheet. (a) Schematic of a folded single-layer graphene. (b) Flat state and (c) folded configuration of monolayer graphene from MD model. (d) Enlarged view of the folded graphene edge. [79, 85]

2. If the total length of graphene is long enough, the total energy $U_{total} < 0$, the adhesion energy over the flat region exceeds the resistance from the curved region, and therefore the folded graphene is energetically preferred and stable.

Obviously, there exists a critical length $L_{critical}$ such that $U_{total} = 0$, which separates the stable and the unstable folded configurations of graphene.

A recent study by Cranford et al. [65] established a small deformation mechanics model to reveal the critical folding length of multilayered graphene sheets. Modeling the folded length of the graphene sheet as two symmetric linear elastic beams, the bending energy per unit length can be defined as

$$U_{bending} \cong 2D_{multi} \int_0^L \left(\frac{d^2u}{dx^2} \right)^2 dx \tag{8.35}$$

where D_{multi} is the bending stiffness of a multilayer system (per unit width) and $u(x)$ is the presumed beam deflection. Applying the Euler–Bernoulli beam equation to derive $u(x)$ with associated small deformation boundary conditions, the total energy of the system per unit width can be expressed as

$$U_{total}(L) = \frac{2\pi^2}{L} D_{multi} - \gamma \left(\frac{1}{2} L_{total} - L \right) \tag{8.36}$$

Through minimization of the above with respect to L, we find

$$L_{critical} = \pi \sqrt{\frac{2D_{multi}}{\gamma}} \tag{8.37}$$

While approximated with small deformation assumptions, the above relation was validated with simulations of single-layer graphene and multilayer systems (up to 10 layers) with little appreciable error [65]. As predicted by Eq. (8.37), the folded length of the graphene sheet is proportional to the square root of the bending stiffness. It is also noted that the total length of the graphene, L_{total}, must be such that a folded confirmation is energetically favorable (i.e., $|U_{adhesion}| > |U_{bending}|$), but does not affect the critical folded length (however, the total length does result in a change in the absolute value of the energy minima). However, predicted shapes of the folded graphene edges are not accurately described by the above formulation, partly because small deformation models cannot accurately predict the shapes of folded graphene edges. Meng et al. [79] developed a finite deformation theoretical model to study the folding of single-layer graphene, which can accurately predict not only the critical length of single-layer graphene folding but also the shape of the folded edge. In a follow-up study, Meng et al. [80] also explored multilayered graphene-folded structures, using a similar finite deformation beam theory approach. For both single- and multilayered racket shapes, the deformed geometry of the sheets can be completely described by two quantities, κ_0 and κ_1, defined by the Cartesian coordinates of the curved region of the middle plane under curvature (e.g., the neutral axis in classical beam theory). The bending energy in the racket-shaped curved region, $U_{bending}$, can be obtained by a first-order approximation as [79, 80]:

$$U_{bending} = 2D_{multi} \int_0^\alpha \sqrt{\kappa_0^2 - (\kappa_1^2 - \kappa_0^2)\sin\theta}\, d\theta + D_{multi} \int_0^{\frac{\pi}{2}} \sqrt{\kappa_0^2 + (\kappa_1^2 - \kappa_0^2)\sin\theta}\, d\theta$$

$$(8.38)$$

where D_{multi} is the bending stiffness of the multilayer graphene, $\alpha = \sin^{-1}(\kappa_0^2/\kappa_1^2 - \kappa_0^2)$, and the governing equations for κ_0 and κ_1 can be solved numerically [79], leading to the numerical solution for $U_{bending}$. The total energy of the folded/adhered graphene sheet is still equal to Eq. (8.38), with a more complex (and accurate) term substituted for $U_{bending}$. Minimization of the total energy can then give a precise solution to multilayer folding via numerical methods. For brevity, complete solution details can be found in the original works by Meng et al. [79, 80].

Beyond racket-type folding, other self-adhering and folding phenomena has been investigated in a similar mechanistic manner, including self-scrolling graphene [89, 90], nanoscrolls with carbon nanotubes [91, 92], graphene peeling [7, 93], surface crumpling [87], surface folding [94], and self-opening capsules [95, 96], to name a few. We note that the energy balance approach described here does not consider an external energy source, which can be used as control mechanisms. Indeed, mechanical stimulation can help initiate self-folding and overcome energy barriers [97] or, alternatively, trigger unfolding [77] depending on system stability. Also complicating matters depending on the direction or location of initial folding line formations, certain directions of folding, may be more energetically favorable than others [97, 98]. Exploiting the potential of folding 2D materials is still in its infancy. In principle, single or periodic hems, pleats, creases, ripples, and ruffles can be tailored from such planar sheets. Critical to such applications is the mechanistic understanding of folding, stability, and limit states, as outlined earlier.

8.5   Failure

In addition to stiffness and strength, of utmost importance for 2D materials is how they fail. Characterized by repeating crystalline structures, such materials are typically prone to lattice defects of various types (e.g., vacancies, Stone–Wales defects) and the associated stress concentrations when subjected to load. Thus, failure is typically characterized by local fracturing and dependent on the concentration of defects – a probabilistic metric. As such, we apply the mechanical tools of fracture analysis and failure statistics to 2D materials.

Here, in this section, the failure of 2D sheets, multilayers, or related composites with 2D crystallinity in the extreme condition is addressed based on the recent theoretical deterministic or statistical approaches of (1) quantized fracture mechanics (QFM) and (2) nanoscale Weibull statistics (NWS) proposed by Pugno and collaborators [99–101] and already extensively compared with experiments and atomistic simulations (also at the nanoscale). The role of thermodynamically unavoidable atomistic defects with different size and shape is thus quantified on brittle fracture and fatigue and even elasticity.

8.5.1 Quantized Fracture Mechanics

QFM is a reformulation of classical linear fracture mechanics with an explicit treatment of the discrete nature of atomistic bonds and crystallinity, and thus a "quantized" limit of crack growth and propagation. The reader is referred to the original works of Pugno and Ruoff for the complete formulations [101]. By considering QFM, the failure stress σ_N for a 2D material having "atomic size" q (the so-called "fracture quantum") and containing an elliptical hole of half-axes a perpendicular to the applied load and b parallel can be determined including in the asymptotic solution [101] the contribution of the far-field stress. We accordingly derived

$$\frac{\sigma_N(a,b)}{\sigma_N^{(\text{theo})}} = \sqrt{\frac{1 + 2a/q(1 + 2a/b)^{-2}}{1 + 2a/q}} \tag{8.39a}$$

where

$$\sigma_N^{(\text{theo})} = \frac{K_{\text{IC}}}{\sqrt{q\pi/2}} \tag{8.39b}$$

where $\sigma_N^{(\text{theo})}$ is the theoretical (defect-free) material strength (e.g., $\sim$100 GPa for carbon nanotubes or graphene) and K_{IC} is the material fracture toughness. The self-interaction between the defect tips has been neglected here (i.e., $a \ll W$, with W the effective layer width of the 2D sheet) and would further reduce the failure stress. For atomistic defects (having characteristic length of few angstrom) in nanostructures (having characteristic size of several nanometers), this hypothesis is fully verified. However, QFM can also easily treat the self-tip interaction starting from the corresponding value of the stress-intensity factor (reported in the related Fracture Handbooks). The validity of QFM has been recently confirmed by atomistic simulations [99–102], and also at larger size scales [99, 100, 103], and for fatigue crack growth [100, 104, 105].

Regarding the defect shape, for a sharp crack perpendicular to the applied load $a/q =$ const & $b/q \to 0$, then $\sigma_N \approx \sigma_N^{(\text{theo})}/\sqrt{1 + 2a/q}$ and for $a/q >> 1$, that is, large cracks, $\sigma_N \approx K_{IC}/\sqrt{\pi a}$ in agreement with linear elastic fracture mechanics (LEFM). We note that LEFM can (1) only treat sharp cracks, and (2) unreasonably predict an infinite defect-free strength. On the other hand, for a crack parallel to the applied load $b/q =$ const and $a/q \to 0$, and thus $\sigma_N = \sigma_N^{(\text{theo})}$, as it must be. In addition, regarding the defect size, for self-similar and small holes $a/b = $ const and $a/q \to 0$ and coherently $\sigma_N = \sigma_N^{(\text{theo})}$; furthermore, for self-similar and large holes $a/b = $ const and $a/q \to \infty$, and we deduce $\sigma_N \approx \sigma_N^{(\text{theo})}/(1 + 2a/b)$ in agreement with the stress concentration posed by elasticity; but elasticity (coupled with a maximum stress criterion) unreasonably predicts (3) a strength independent of the hole size and (4) tending to zero for cracks. Note the extreme consistency of Eq. (8.39) – that removing all the limitations of LEFM – represents the first law capable of describing in a unified manner all the size and shape effects for the elliptical holes, including cracks as limit case. In other words, Eq. (8.39) shows that the two classical strength predictions based on stress intensifications (LEFM) or stress concentrations (elasticity) are only reasonable for "large" defects; Eq. (8.39) unifies their results and extends its validity to "small" defects ("large" and "small" are here with respect to the fracture quantum). Finally, Eq. (8.39) shows that even a small defect can dramatically reduce the mechanical strength.

For multilayered composites, imposing the critical force equilibrium (mean-field approach), having different layers in numerical fractions f_{abN} containing holes of half-axes a and b for each material type N, we find the bundle/composite strength σ_C (ideal if $\sigma_C^{(\text{theo})}$) in the following form:

$$\frac{\sigma_C}{\sigma_C^{(\text{theo})}} = \sum_{a,b,N} f_{abN} \frac{\sigma_N(a,b)}{\sigma_N^{(\text{theo})}} \tag{8.40}$$

The summation is extended to all the different holes and material types; the numerical fraction f_{00N} is that of defect-free material N and $\sum_{a,b,N} f_{abN} = 1$. If all the defective layers are of the same material and contain identical holes $f_{abN} = f = 1 - f_{00N}$, and the following simple relation between the strength reductions holds:

$$1 - \sigma_C/\sigma_C^{(\text{theo})} = f\left(1 - \sigma_N/\sigma_N^{(\text{theo})}\right) \tag{8.41}$$

The previous equations are based on linear elasticity, that is, on a linear relationship $\sigma \propto \varepsilon$ between stress σ and strain ε. In contrast, let us assume $\sigma \propto \varepsilon^\kappa$, where $\kappa > 1$ denotes hyperelasticity, as well as $\kappa < 1$ elastic–plasticity. The power of the stress-singularity will accordingly be modified [106] from the classical value $1/2$ to $\alpha = \kappa/(\kappa + 1)$. Thus, the problem is mathematically equivalent to that of a re-entrant corner [107], and consequently we predict

$$\frac{\sigma_N(a,b,\alpha)}{\sigma_N^{(\text{theo})}} = \left(\frac{\sigma_N(a,b)}{\sigma_N^{(\text{theo})}}\right)^{2\alpha} \tag{8.42}$$

A crack with a self-similar roughness, mathematically described by a fractal with noninteger dimension $1 < D < 2$, would similarly modify the stress-singularity, according to [108] $\alpha = (2 - D)/2$; thus, with Eq. (8.42), we can also estimate the role of the crack roughness. Both plasticity and roughness reduce the severity of the defect, whereas hyperelasticity enlarges its effect. For example, for a crack composed by n adjacent vacancies, we find $\sigma_N/\sigma_N^{(\text{theo})} \approx (1 + n)^{-\alpha}$.

According to LEFM and assuming the classical hypothesis of self-similarity ($a_{\max} \propto L$), that is, the largest crack size is proportional to the characteristic structural size L, we expect a size effect on the strength in the form of the power law $\sigma_C \propto L^{-\alpha}$. For linear elastic materials $\alpha = 1/2$ as classically considered, but for elastic–plastic materials or fractal cracks $0 \leq \alpha \leq$

1/2, whereas for hyperelastic materials $1/2 \leq \alpha \leq 1$, suggesting an unusual and superstrong size effect.

Equations (8.39)–(8.42) do not consider the defect–boundary interaction. The finite width $2W$ can be treated by applying QFM starting from the related expression of the stress-intensity factor (reported in relevant Handbooks). However, to have an idea of the defect–boundary interaction, we applied an approximated method, deriving the following correction $\sigma_N(a, b, W) \approx C(W)\sigma_N(a, b)$, $C(W) \approx (1 - a/W)/(\sigma_N(a, b)|_{q \to W-a}/\sigma_N^{(\text{theo})})$ (note that such a correction is valid also for $W \approx a$, whereas for $W \gg a$ it becomes $C(W \gg a) \approx 1 - a/W$). Similarly, the role of the defect orientation β could be treated by QFM considering the related stress-intensity factor; roughly, one could use the self-consistent approximation $\sigma_N(a, b, \beta) \approx \sigma_N(a, b)\cos^2\beta + \sigma_N(b, a)\sin^2\beta$.

By integrating the quantized Paris' law, that is, an extension of the classical Paris' law recently proposed especially for nanostructure or nanomaterial applications [100, 104, 105], we derive the following number of cycles to failure (or life time):

$$\frac{C_N(a)}{C_N^{(\text{theo})}} = \frac{(1 + q/W)^{1-m/2} - (a/W + q/W)^{1-m/2}}{(1 + q/W)^{1-m/2} - (q/W)^{1-m/2}}, m \neq 2 \qquad (8.43a)$$

and

$$\frac{C_N(a)}{C_N^{(\text{theo})}} = \frac{\ln\{(1 + q/W)/(a/W + q/W)\}}{\ln\{(1 + q/W)/(q/W)\}}, m = 2 \qquad (8.43b)$$

where $m > 0$ is the material Paris' exponent. Note that according to Wöhler $C_N^{(\text{theo})} = K\Delta\sigma^{-k}$, where K and k are material constants and $\Delta\sigma$ is the amplitude of the stress range during the oscillations.

Even if fatigue experiments in 2D materials are still to be performed, their behavior is expected to be intermediate between those of Wöhler and Paris, as displayed by all the known materials, and the quantized Paris' law basically represents their asymptotic matching (as QFM basically represents the asymptotic matching between the strength and toughness approaches). Only defects remaining self-similar during fatigue growth have to be considered, thus only a crack (of half-length a) is of interest in this context. By means of Eq. (8.43) the time to failure reduction can be estimated, similar to the brittle fracture treated by Eq. (8.39).

For a bundle or multilayered composite, considering a mean-field approach (similar to Eq. 8.40) yields

$$\frac{C_C}{C_C^{(\text{theo})}} = \sum_{a,N} f_{aN} \frac{C_N(a)}{C_N^{(\text{theo})}} \qquad (8.44)$$

Better predictions could be derived integrating the quantized Paris' law for a finite width strip. However, we note that the role of the finite width is already included in Eq. (8.44), even if these are rigorously valid in the limit of W tending to infinity.

Regarding elasticity, interpreting the incremental compliance, due to the presence of the crack, as a Young's modulus (here denoted by E) degradation, we find $\frac{E(a)}{E^{(\text{theo})}} = 1 - 2\pi\frac{a^2}{A}$ [109]. Thus, recursively, considering Q cracks having sizes a_i or, equivalently, M different cracks with multiplicity Q_i ($Q = \sum_{i=1}^{M} Q_i$), noting that $n_i = \frac{2a_i}{q}$ represents the number of adjacent vacancies in a crack of half-length a_i, with q atomic size, and $v_i = \frac{Q_i n_i}{A/q^2}$ its related numerical (or volumetric) vacancy fraction, we find [109]:

$$\frac{E}{E^{(\text{theo})}} = \prod_{i=1}^{Q} \frac{E(a_i)}{E^{(\text{theo})}} \approx 1 - \xi \sum_{i=1}^{M} v_i n_i \qquad (8.45)$$

with $\xi \geq \pi/2$, where the equality holds for isolated cracks. Equation (8.45) can be applied to 2D materials and the related bundles or composites containing defects in volumetric percentage v_i.

Forcing the interpretation of our formalism, we note that $n_i = 1$ would describe a single vacancy, that is, a small hole. Thus, as a first approximation, different defect geometries from cracks to circular holes, for example, elliptical holes, could in principle be treated by Eq. (8.45); we have to interpret n_i as the ratio between the transversal and longitudinal (parallel to the load) defect sizes ($n_i = a_i/b_i$). Introducing the ith defect eccentricity e_i as the ratio between the lengths of the longer and shorter axes, as a first approximation $n_i(\beta_i) \approx e_i \cos^2 \beta_i + 1/e_i \sin^2 \beta_i$, where β_i is the defect orientation. For a single-defect typology $\frac{E}{E^{(\text{theo})}} \approx 1 - \xi v n$, in contrast to the common assumption $\frac{E}{E^{(\text{theo})}} \approx 1 - v$, rigorously valid only for the density, for which $\frac{\rho_C}{\rho_C^{(\text{theo})}} \equiv 1 - v$. Note that the failure strain for a defective 2D materials or related bundle can also be predicted by $\varepsilon_{N,C}/\varepsilon_{N,C}^{(\text{theo})} = (\sigma_{N,C}/\sigma_{N,C}^{(\text{theo})})/(E/E^{(\text{theo})})$. In contrast to what happens for the strength, large defectiveness is required to have a considerable elastic degradation, even if we have shown that sharp transversal defects could have a role.

8.5.2   Nanoscale Weibull Statistics

The discussed tremendous defect sensitivity, described by Eq. (8.39), is confirmed by a statistical analysis based on NWS [110], applied to the nanotensile tests. According to this treatment, the probability of failure P for a nearly defect-free structure under a tensile stress σ_N is independent of its volume (or surface), in contrast to classical Weibull statistics, namely:

$$P = 1 - \exp{-N_N \left(\frac{\sigma_N}{\sigma_0}\right)^w} \qquad (8.46)$$

where w is the nanoscale Weibull modulus, σ_0 is the nominal failure stress (i.e., corresponding to a probability of failure of 63%), and $N_N \equiv 1$. In classical Weibull statistics $N_N = V/V_0$ for volume-dominating defects (or $N_N = A/A_0$ for surface-dominating defects), that is, N_N is the ratio between the volume (or surface) of the structure and a reference volume (or surface).

Moreover, defects are thermodynamically unavoidable, especially at the large size scale. At the thermal equilibrium, the vacancy fraction $f = n/N << 1$ (n is the number of vacancies and N is the total number of atoms) can be estimated as

$$f \approx e^{-E_1/k_B T_a} \qquad (8.47)$$

where E_1 is on the order of a few eV, the energy required to remove one atom of the system to create the vacancy and T_a is the absolute temperature at which the material is assembled (e.g., for carbon $E_1 \approx 7\text{eV}$ and T_a typically in the range between 2000 and 4000 K, thus $f \approx 2.4 \times 10^{-18} - 1.6 \times 10^{-9}$). The strength of the structure will be dictated by the largest transversal crack on it, according to the weakest link concept. The probability of finding a nanocrack of size m in a bundle with vacancy fraction f is $P(m) = (1 - f)f^m$, and thus the number M of such nanocracks in a bundle or composite composed by N atoms is $M(m) = P(m)N$. The size of the largest nanocrack, which typically occurs once, is found from the solution to the equation $M(m) \approx 1$, which implies [111]:

$$m \approx -\ln[(1 - f)N]/\ln f \approx -\ln N/\ln f \tag{8.48}$$

Inserting Eqs (8.47) and (8.48) into Eq. (8.39) evaluated for a transversal crack ($b \approx 0$ and $2a/q \approx m$), we deduce the statistical counterpart of Eq. (8.39) and thus the following thermo-dynamical maximum achievable strength:

$$\frac{\sigma_N(N)}{\sigma_N^{(\text{theo})}} \leq \frac{\sigma_N^{(\text{max})}(N)}{\sigma_N^{(\text{theo})}} = \frac{1}{\sqrt{1 + \frac{k_B T_a}{E_1}\ln N}} \tag{8.49}$$

The fracture mechanics approach could be of interest to evaluate the strength or assuming a different failure mechanism, not intrinsic fracture as previously treated but rather as a sliding failure mode [112]. Thus, we assume the interactions between adjacent layers as the weakest links, that is, the fracture of the bundle or composites is caused by layers sliding rather than by their intrinsic fracture.

Accordingly, the energy balance during a longitudinal delamination (here "delamination" has the meaning of Mode II crack propagation at the interface between adjacent nanotubes) dz under the applied force F is

$$d\Phi - Fdu - 2\gamma(P_C + P_{vdW})dz = 0 \tag{8.50}$$

where $d\Phi$ and du are the strain energy and elastic displacement variation due to the infinitesimal increment in the compliance caused by the delamination dz; P_{vdw} describes the still existing van der Waals attraction (e.g., attractive part of the Lennard–Jones potential) for vanishing nominal contact perimeter (e.g., the shear force between two graphite single layers becomes zero for nominally negative contact area); γ is the surface energy of the layer–layer or layer–matrix interactions in bundles or composites. Elasticity poses $\frac{d\Phi}{dz} = -\frac{F^2}{2ES}$, where S is the cross-sectional surface area of the layers, whereas according to Clapeyron's theorem $Fdu = 2d\Phi$. Thus, the following simple expression for the bundle or composite strength ($\sigma_C = F_C/S$, effective stress and cross-sectional surface area are considered here; F_C is the force at fracture) is predicted:

$$\sigma_C^{(\text{theo})} = 2\sqrt{E\gamma\frac{P}{S}} \tag{8.51}$$

in which it appears the ratio between the effective perimeter P in contact and the cross-sectional surface area of the layers. Equation (8.51) can also be considered valid for the entire bundle, since we are assuming here the same value P/S for all the layers in the bundle. Note that Eq. (8.51) is basically the asymptotic limit for sufficiently long overlapping length, that is, the length along with two adjacent layers is nominally in contact; for overlapping length smaller than a critical value the strength increases by increasing the overlapping length; for a single atomic layer this overlapping length is of the order of 10 nm, [113] it is expected to be larger for layers in bundles or composites, for example, of the order of several millimeters, as confirmed experimentally. This critical length is $\ell_C \approx 6\sqrt{\frac{hES}{PG}}$ where h and G are the thickness and shear modulus of the interface. It suggests that increasing the size-scale $L \propto \sqrt{S} \propto P \propto h$ this critical length increases too, namely $\ell \propto L$, thus the strength increases by increasing the overlapping length in a wider range; however, note that the achievable strength is reduced since $\sigma_C^{(theo)} \propto \sqrt{h}\ell^{-1} \propto \sqrt{P/S} \propto L^{-1/2}$, if $L \propto \ell \propto h$: increasing the overlapping length *ad infinitum* is not a way to indefinitely increase the strength. The real strength could be significantly smaller, not only because $\ell < \ell_C$ but also as a consequence of the misalignment of the layers with respect to the load axis. Assuming a nonperfect alignment of the layers in the bundle, described by a nonzero angle β, the longitudinal force carried by the layers will be $F/\cos\beta$; thus, the equivalent Young' modulus of the bundle or composites will be $E\cos^2\beta$, as can be evinced by the corresponding modification of the energy balance during delamination; accordingly

$$\sigma_C = 2\cos\beta\sqrt{E\gamma\frac{P}{S}} \tag{8.52}$$

For example, for graphene, the surface area for load transfer is doubled with respect to the case of the single carbon nanotube (the inner surface area does not contribute), the previous equation predicts for bundles or composites:

$$\sigma_C^{(max,G)} = \sqrt{2}\sigma_C^{(max,CNT)} \tag{8.53}$$

For a comparison of QFM or NWS with experiments or simulations, see also [114–116].

8.6 Multilayers and Composites

Two-dimensional nanostructures such as graphene, hBN, and molybdenum disulfide (MoS_2) have attracted considerable attention because of their novel properties and versatile potential applications. These novel nanostructures have complementary physical properties. Therefore, 2D layers can be integrated into a multilayer stack to create 2D heterostructures that mitigate the negative properties of each individual constituent [117]. Layered crystals are characterized by strong intralayer covalent bonding and relatively weak interlayer van der Waals bonding. Since various methods have been proposed to make atomic layered 2D materials, such as graphene, hBN, TMDs, and oxides, it becomes possible to stack monolayers into variable configurations to build unique structures with desired functionalities [41].

Because of their equivalent structural parameters and distinct electronic properties, graphene and BN have been used in several studies to explore the possibility of making a graphene/BN composite. Several configurations of layered graphene/BN had been studied, such as graphene/BN to form a bilayer system [118], two or more BN layers [119], and 3D superlattice with alternate stacking of graphene and BN monolayers [120]. Other heterostructures such as monolayer [121] and bilayer [122] graphene sandwiched between two BN layers or monolayer BN sandwiched between two graphene layers [123] have also been constructed. Similarly, Jiang and Park [117] performed MD simulations to investigate the mechanical properties of single-layer MoS_2 and a graphene/MoS_2/graphene heterostructure under uniaxial tension. The results show that the Young's modulus of the heterostructure is about three times that of MoS_2. Although the stiffness is enhanced, the yield strain of the heterostructure is considerably smaller than the MoS_2 due to lateral buckling of the outer graphene layers owing to the applied mechanical tension. Creating artificial heterostructures via stacking different 2D nanostructures on top of each other establishes a whole family of amazing materials with unusual characteristics and exciting possibilities for novel 2D devices [41].

Composite materials have advantages over traditional materials, where nanosized reinforcements are arranged in the matrix, presenting superior properties and added functionalities [2]. The excellent mechanical properties of 2D nanostructures, such as graphene, graphdiyne, and MoS_2, are promising reinforcements in high performance composites. The potential benefits of high strength and stiffness of 2D nanomaterial sheets in nanocomposites, combined with its low density compared with the density of metallic substrates, are easily recognized and are powerful incentives for the use of such materials in structural applications.

Metal–graphene nanolayered composites have been studied recently due to its promising applications (i.e., nuclear reactor structural material) and the effectiveness of graphene of significantly strengthening of metals. A reported study has demonstrated that the 2D geometry and low atomistic thickness of graphene can effectively constrain dislocation motion in polycrystalline metals, resulting in an engineered strengthening mechanism [124]. In a similar graphene/polymer composite (carbon monolayer + thin substrate), Gong et al. [125] demonstrated unambiguously that stress transfer takes place from the polymer matrix to monolayer graphene, showing that the graphene acts as a reinforcing phase. Graphene-like metal nanocomposites represent one of the most technologically promising developments of high-performance material systems.

Roman and Cranford [126] evaluated the mechanical properties of graphdiyne–copper nanocomposites, combining varied numbers of layers of graphdiyne sheets on copper substrate, as well as sandwich-structured copper/graphdiyne layers (see Figure 8.14(a)–(d)).

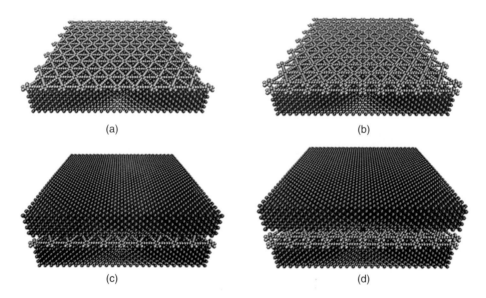

(a)   (b)

(c)   (d)

Figure 8.14 Nanocomposite configurations: (a) single graphdiyne with copper substrate, (b) bilayer graphdiyne with copper substrate, (c) single graphdiyne with copper sandwich, (d) bilayer graphdiyne with copper sandwich. [126]

Using full atomistic MD, the elastic stiffness, toughness, and limit states of these nanocomposite materials were explored considering an atomistically large model at finite temperature through direct tensile loads, using defect-free and defective copper substrates. The nature of the bonding interaction between the copper and graphdiyne is noncovalent; therefore, there is no energetic cost to bond the composite. The results show that a relatively slight amount of graphdiyne can dramatically increase the tensile strength and toughness of copper, although it is only a single atomic layer in thickness, representing nominal increases in both system size and weight. Also, it is demonstrated that combining copper with layers of graphdiyne can increase substantially the failure strain by confining the copper crystal during deformation. More interestingly is the effect on the copper component in the composite – the graphdiyne allows the copper to carry stress as if the imperfections were not present. This strengthening mechanism (e.g., constraining dislocations) acts in combination with classical rule-of-mixture enhancements (e.g., proportional stiffness) such that the contribution of copper is near-optimal in the composite system – graphdiyne improves the response of the composite material [126].

Proving the applicability of continuum mechanics, including micromechanics, to composites at the atomic level, many recent works have reported that continuum mechanics can be employed to analyze the behavior of nanocomposites. The well-known rule of mixture (RoM) for composites can be utilized to predict the tensile modulus of the nanocomposites. The modulus of the composite, $Y_{composites}$, can be calculated from the equation:

$$Y_{composite} = \frac{1}{V}\sum_i Y_i V_i = Y_{2D}n_{2D} + Y_{matrix}n_{matrix} \tag{8.54}$$

where Y_{2D} is the tensile modulus of the 2D nanostructure sheet, Y_{matrix} is the tensile modulus of the matrix, n_{2D} and n_{matrix} are the volume fractions ($n_i = V_i/V$) of the 2D nanostructure sheet and matrix, respectively. Equation (8.54) can be extended to any two-phase composites, regardless the shape of the reinforcement. Note that in these formulas, only three parameters are involved, that is, modulus of the 2D nanomaterial layer and the matrix, and the volume fraction [2, 126]. Similarly, Gong et al. [125] have modeled the behavior of graphene monolayer nanocomposites using shear-lag theory, where it is assumed that the graphene is a mechanical continuum and surrounded by a layer of elastic polymer resin, and that there is an elastic stress transfer from the matrix to the graphene layer through a shear stress at the graphene–matrix interface. For a given level of matrix strain, e_m, it is predicted that the variation of strain in the graphene monolayer, e_f, with position, x, across the monolayer will be of the form:

$$e_f = e_m\left[1 - \frac{\cos h\left(ns\frac{x}{l}\right)}{\cos h\left(\frac{ns}{2}\right)}\right] \tag{8.55}$$

where

$$n = \sqrt{\frac{2G_m}{E_g}\left(\frac{t}{T}\right)} \tag{8.56}$$

and G_m is the matrix shear modulus, E_f is the Young's modulus of the graphene flake, l is the length of the graphene flake in the x direction, t is the thickness of the graphene, T is the total resin thickness, and s is the aspect ratio of the graphene (l/t) in the x-direction. The parameter n is widely accepted in composite micromechanics to be an effective measure of the interfacial stress transfer efficiency, so ns depends on both the morphology of the graphene monolayer and the degree of interaction it has with the matrix. The variation of shear stress, τ_i, at the polymer graphene interface is given by

$$\tau_i = nE_g e_m \left[\frac{\sin h \left(ns\frac{x}{l} \right)}{\cos h \left(\frac{ns}{2} \right)} \right] \tag{8.57}$$

Reviewing these equations, it can be noted that the graphene monolayer is subjected to the highest level of stress, when the value of the product n_s is high. This implies that for good reinforcement a high aspect ratio, s, is desirable along with a high value of n. This analysis relies on the assumption that both the graphene monolayer and polymer behave as linear elastic continua [2, 125, 127].

8.7 Conclusion

Herein, we have discussed the emergence of 2D materials and the basic mechanical characterization thereof. Clearly then, the importance of graphene is not only that it has unique properties but also that it has paved the way for, and promoted interest in, the isolation and synthesis of many other 2D materials. We can now talk about a whole new class of materials, 2D atomic crystals, and already have examples with a large variety of properties (from large band-gap insulators to the very best conductors, the extremely mechanically strong to the soft and fragile, and the chemically active to the very inert). Furthermore, many of the properties of these 2D materials are very different from those of their 3D counterparts, leading to interesting discoveries in previously assumed 3D crystals (e.g., silicon to silicene). Because of their unique properties, 2D materials have attracted significant interest for both fundamental research and practical applications. In addition to the research on individual graphene-like 2D materials, a lot of effort has also been focused on the creation of multilayered heterostructures and composites as well. With most of these extraordinary materials characterized, experimentally and theoretically, we are getting to the point of convergence where the focus will be to turn these wonder properties in useful applications. The mechanical behavior of such materials is intriguing, acting as both crystals in membranes, requiring prudence in assigning traditional properties such as a Young's modulus (e.g., in MPa or N-m). Indeed, many of the low hanging fruits of these materials – characterizing and investigating the fundamental mechanical properties as discussed – have been harvested. At the same time, in terms of their material potential in future technological applications, we have merely scratched the 2D surface.

Acknowledgment

N.M.P. is supported by the European Research Council (ERC StG Ideas 2011 BIHSNAM No. 279985 on Bio-Inspired hierarchical supernanomaterials, ERC PoC 2013-1 REPLICA2 No. 619448 on Largearea replication of biological antiadhesive nanosurfaces, ERC PoC 2013-2 KNOTOUGH No. 632277 on Supertough knotted fibers), by the European Commission under the Graphene Flagship (WP10 "Nanocomposites," No. 604391) and by the Provincia Autonoma di Trento ("Graphene Nanocomposites," No. S116/2012-242637 and delib. reg. No. 2266).

References

[1] Novoselov, K.S. et al. (2004) Electric field effect in atomically thin carbon films. *Science*, **306**, 666–669.

[2] Hu, H., Onyebueke, L. and Abatan, A. (2010) Characterizing and modeling mechanical properties of nanocomposites: Review and evaluation. *Journal of Minerals and Materials Characterization and Engineering*, **9** (4), 275–319.

[3] Miro, P., Audiffred, M. and Heine, T. (2014) An atlas of two-dimensional materials. *Chemical Society Reviews*, **43** (18), 6537–6554.

[4] Tang, Q., Zhou, Z. and Chen, Z. (2013) Graphene-related nanomaterials: Tuning properties by functionalization. *Nanoscale*, **5** (11), 4541–4583.

[5] Lee, C., Wei, X., Kysar, J.W. and Hone, J. (2008) Measurement of the elastic properties and intrinsic strength of monolayer graphene. *Science*, **321**, 385–388.

[6] Koenig, S.P., Boddeti, N.G., Dunn, M.L. and Bunch, J.S. (2011) Ultrastrong adhesion of graphene membranes. *Nature Nanotechnology*, **6**, 543–546.

[7] Sen, D. et al. (2010) Tearing graphene sheets from adhesive substrates produces tapered nanoribbons. *Small*, **6** (10), 1108–1116.

[8] Haley, M.M., Brand, S.C. and Pak, J.J. (1997) Carbon networks based on dehydrobenzoannulenes: Synthesis of graphdiyne substructures. *Angewandte Chemie International Edition in English*, **36** (8), 836–838.

[9] Li, G. et al. (2010) Architecture of graphdiyne nanoscale films. *Chemical Communications (Cambridge, England)*, **46** (19), 3256–3258.

[10] Ivanovskii, A.L. (2013) Graphynes and graphdiynes. *Progress in Solid State Chemistry*, **41** (1-2), 1–19.

[11] Cranford, S.W., Brommer, D.B. and Buehler, M.J. (2012) Extended graphynes: Simple scaling laws for stiffness, strength and fracture. *Nanoscale*, **4** (24), 7797–7809.

[12] Enyashin, A.N. and Ivanovskii, A.L. (2011) Graphene allotropes. *Physica Status Solidi*, **248** (8), 1879–1883.

[13] Takeda, K. and Shiraishi, K. (1994) Theoretical possibility of stage corrugation in Si and Ge analogs of graphite. *Physical Review B*, **50** (20), 14916–14922.

[14] Guzman-Verri, G.G. and Voon, L.C.L.Y. (2007) Electronic structure of silicon-based nanostructures. *Physical Review B*, **76** (7), 075131.

[15] Tsai, W.-F. et al. (2013) Gated silicene as a tunable source of nearly 100% spin-polarized electrons. *Nature Communications*, **4**, 1500.

[16] Roman, R.E. and Cranford, S.W. (2014) Mechanical properties of silicene. *Computational Materials Science*, **82**, 50–55.

[17] Le, M.Q. and Nguyen, D.T. (2014) The role of defects in the tensile properties of silicene. *Applied Physics A*, **118**, 1437–1445.

[18] Song, L. et al. (2010) Large scale growth and characterization of atomic hexagonal boron nitride layers. *Nano Letters*, **10** (8), 3209–3215.

[19] Tang, Q. and Zhou, Z. (2013) Graphene-analogous low-dimensional materials. *Progress in Materials Science*, **58**, 1244–1315.

[20] Kostoglou, N., Polychronopoulou, K. and Rebholz, C. (2015) Thermal and chemical stability of hexagonal boron nitride (h-BN) nanoplatelets. *Vacuum*, **112**, 42–45.

[21] Peng, Q. et al. (2012) Elastic properties of hybrid graphene/boron nitride monolayer. *Acta Mechanica*, **223** (12), 2591–2596.

[22] Yue, Q. et al. (2012) Mechanical and electronic properties of monolayer MoS2 under elastic strain. *Physics Letters A*, **376** (12–13), 1166–1170.

[23] Dang, K.Q., Simpson, J.P. and Spearot, D.E. (2014) Phase transformation in monolayer molybdenum disulphide (MoS2) under tension predicted by molecular dynamics simulations. *Scripta Materialia*, **76**, 41–44.

[24] Peng, Q. and De, S. (2013) Outstanding mechanical properties of monolayer MoS2 and its application in elastic energy storage. *Physical Chemistry Chemical Physics*, **15** (44), 19427–19437.

[25] Gan, Y. and Zhao, H. (2014) Chirality effect of mechanical and electronic properties of monolayer MoS2 with vacancies. *Physics Letters A*, **378** (38–39), 2910–2914.

[26] Balendhran, S. et al. (2014) Elemental analogues of graphene: Silicene, germanene, stanene, and phosphorene. *Small*, **11**, 640–652.

[27] Cahangirov, S. et al. (2009) Two- and one-dimensional honeycomb structures of silicon and germanium. *Physical Review Letters*, **102** (23), 236804.

[28] Bianco, E. et al. (2013) Stability and exfoliation of germanane: A germanium graphane analogue. *ACS Nano*, **7** (5), 4414–4421.

[29] Li, L. et al. (2014) Buckled Germanene Formation on Pt(111). *Advanced Materials*, **26** (28), 4820–4824.

[30] van den Broek, B. et al. (2014) Two-dimensional hexagonal tin: ab initio geometry, stability, electronic structure and functionalization. *2D Materials*, **1** (2), 021004.

[31] Manjanath, A., Kumar, V. and Singh, A.K. (2014) Mechanical and electronic properties of pristine and Ni-doped Si, Ge, and Sn sheets. *Physical Chemistry Chemical Physics*, **16** (4), 1667–1671.

[32] Brent, J.R. et al. (2014) Production of few-layer phosphorene by liquid exfoliation of black phosphorus. *Chemical Communications*, **50** (87), 13338–13341.

[33] Lange, S., Schmidt, P. and Nilges, T. (2007) Au3SnP7@black phosphorus: An easy access to black phosphorus. *Inorganic Chemistry*, **46** (10), 4028–4035.

[34] Koski, K.J. and Cui, Y. (2013) The new skinny in two-dimensional nanomaterials. *ACS Nano*, **7** (5), 3739–3743.

[35] Li, Y. et al. (2014) Mechanical properties of hydrogen functionalized graphene allotropes. *Computational Materials Science*, **83**, 212–216.

[36] Zhao, H. (2012) Strain and chirality effects on the mechanical and electronic properties of silicene and silicane under uniaxial tension. *Physics Letters A*, **376** (46), 3546–3550.

[37] Leggett, G.J. (2009) Scanning probe microscopy, in *Surface Analysis – The Principal Techniques*, John Wiley & Sons, pp. 479–562.

[38] Cao, G. (2004) *Nanostructures and Nanomaterials – Synthesis, Properties and Applications*, World Scientific.

[39] Frank, I.W. et al. (2007) Mechanical properties of suspended graphene sheets. *Journal of Vacuum Science & Technology B*, **25** (6), 2558–2561.

[40] Ferralis, N. (2010) Probing mechanical properties of graphene with Raman spectroscopy. *Journal of Materials Science*, **45** (19), 5135–5149.

[41] Wang, H., Liu, F., Fu, W. et al. (2014) Two-dimensional heterostructures: Fabrication, characterization, and application. *Nanoscale*, **6**, 12250–12272.

[42] Bertolazzi, S., Brivio, J. and Kis, A. (2011) Stretching and breaking of ultrathin MoS2. *ACS Nano*, **5** (12), 9703–9709.

[43] Poot, M. and van der Zant, H.S.J. (2008) Nanomechanical properties of few-layer graphene membranes. *Applied Physics Letters*, **92** (6), 063111.

[44] Vogt, P. et al. (2012) Silicene: Compelling experimental evidence for graphenelike two-dimensional silicon. *Physical Review Letters*, **108** (15), 155501.

[45] Gouadec, G. and Colomban, P. (2007) Raman Spectroscopy of nanomaterials: How spectra relate to disorder, particle size and mechanical properties. *Progress in Crystal Growth and Characterization of Materials*, **53** (1), 1–56.

[46] Dresselhaus, M.S. et al. (2010) Perspectives on carbon nanotubes and graphene raman spectroscopy. *Nano Letters*, **10** (3), 751–758.

[47] Butler, S.Z. et al. (2013) Progress, challenges, and opportunities in two-dimensional materials beyond graphene. *ACS Nano*, **7** (4), 2898–2926.

[48] Nguyen, B.-S., Lin, J.-F. and Perng, D.-C. (2014) Microstructural, electrical, and mechanical properties of graphene films on flexible substrate determined by cyclic bending test. *ACS Applied Materials & Interfaces*, **6** (22), 19566–19573.

[49] Frenkel, D. and Smit, B. (2002) Molecular dynamics simulations, in *Understanding Molecular Simulation*, Second edn (eds D. Frenkel and B. Smit), Academic Press, San Diego, pp. 63–107.

[50] Cranford, S.W. and Buehler, M.J. (2012) *Biomateriomics*, Springer.

[51] Tsai, D.H. (1979) The virial theorem and stress calculation in molecular dynamics. *The Journal of Chemical Physics*, **70** (3), 1375–1382.

[52] Zimmerman, J.A. et al. (2004) Calculation of stress in atomistic simulation. *Modelling and Simulation in Materials Science and Engineering*, **12** (4), S319.

[53] Lu, Q., Gao, W. and Huang, R. (2011) Atomistic simulation and continuum modeling of graphene nanoribbons under uniaxial tension. *Modelling and Simulation in Materials Science and Engineering*, **19** (5), 054006.

[54] Huang, Y., Wu, J. and Hwang, K.C. (2006) Thickness of graphene and single-wall carbon nanotubes. *Physical Review B*, **74** (24), 245413.

[55] Baughman, R.H., Eckhardt, H. and Kertesz, M. (1987) Structure-property predictions for new planar forms of carbon: Layered phases containing sp2 and sp atoms. *The Journal of Chemical Physics*, **87** (11), 6687.

[56] Wei, X. et al. (2009) Nonlinear elastic behavior of graphene: *Ab initio* calculations to continuum description. *Physical Review B*, **80** (20), 205407.

[57] Lu, Q. and Huang, R. (2009) Nonlinear mechanical properties of graphene nanoribbons. *International Journal of Applied Mechanics*, **1** (3), 443–467.

[58] Zhou, J. and Huang, R. (2008) Internal lattice relaxation of single-layer graphene under in-plane deformation. *Journal of the Mechanics and Physics of Solids*, **56** (4), 1609–1623.

[59] Scarpa, F., Adhikari, S. and Phani, A.S. (2009) Effective elastic mechanical properties of single layer graphene sheets. *Nanotechnology*, **20** (6), 065709.

[60] Nasdala, L. and Ernst, G. (2005) Development of a 4-node finite element for the computation of nano-structured materials. *Computational Materials Science*, **33** (4), 443–458.

[61] Cranford, S.W. and Buehler, M.J. (2011) Mechanical properties of graphyne. *Carbon*, **49** (13), 4111–4121.

[62] Yang, Y. and Xu, X. (2012) Mechanical properties of graphyne and its family – A molecular dynamics investigation. *Computational Materials Science*, **61**, 83–88.

[63] Fasolino, A., Los, J.H. and Katsnelson, M.I. (2007) Intrinsic ripples in graphene. *Nature Materials*, **6** (11), 858–861.

[64] Kim, E.-A. and Neto, A.H.C. (2008) Graphene as an electronic membrane. *EPL*, **84** (5), 57007.

[65] Cranford, S., Sen, D. and Buehler, M.J. (2009) Meso-origami: Folding multilayer graphene sheets. *Applied Physics Letters*, **95** (12), 123121.

[66] Lu, Q., Arroyo, M. and Huang, R. (2009) Elastic bending modulus of monolayer graphene. *Journal of Physics D: Applied Physics*, **42** (10), 102002.

[67] Bunch, J.S. et al. (2007) Electromechanical resonators from graphene sheets. *Science*, **315** (5811), 490–493.

[68] Wang, X.L. et al. (2009) First-principles study on the enhancement of lithium storage capacity in boron doped graphene. *Applied Physics Letters*, **95** (18), 183103.

[69] Shi, Z.W. et al. (2012) Studies of graphene-based nanoelectromechanical switches. *Nano Research*, **5** (2), 82–87.

[70] Li, X.S. et al. (2009) Large-area synthesis of high-quality and uniform graphene films on copper foils. *Science*, **324** (5932), 1312–1314.

[71] Novoselov, K.S. et al. (2005) Two-dimensional atomic crystals. *Proceedings of the National Academy of Sciences of the United States of America*, **102** (30), 10451–10453.

[72] Hod, O. (2012) Graphite and hexagonal boron-nitride have the same interlayer distance. Why? *Journal of Chemical Theory and Computation*, **8** (4), 1360–1369.

[73] Zong, Z. et al. (2010) Direct measurement of graphene adhesion on silicon surface by intercalation of nanoparticles. *Journal of Applied Physics*, **107** (2), 026104.

[74] Wan, K.T. and Mai, Y.W. (1996) Fracture mechanics of a shaft-loaded blister of thin flexible membrane on rigid substrate. *International Journal of Fracture*, **74** (2), 181–197.

[75] Lu, Z.X. and Dunn, M.L. (2010) van der Waals adhesion of graphene membranes. *Journal of Applied Physics*, **107** (4), 044301.

[76] Sachs, B. et al. (2011) Adhesion and electronic structure of graphene on hexagonal boron nitride substrates. *Physical Review B*, **84** (19), 195414.

[77] Yi, L.J. et al. (2014) Temperature-induced unfolding of scrolled graphene and folded graphene. *Journal of Applied Physics*, **115** (20), 204307.

[78] Kim, K. et al. (2011) Multiply folded graphene. *Physical Review B*, **83** (24), 245433.

[79] Meng, X.H. et al. (2013) Mechanics of self-folding of single-layer graphene. *Journal of Physics D: Applied Physics*, **46** (5), 055308.

[80] Meng, X.H. et al. (2014) Folding of multi-layer graphene sheets induced by van der Waals interaction. *Acta Mechanica Sinica*, **30** (3), 410–417.

[81] Le, N.B. and Woods, L.M. (2012) Folded graphene nanoribbons with single and double closed edges. *Physical Review B*, **85** (3), 035403.

[82] Son, Y.W., Cohen, M.L. and Louie, S.G. (2006) Energy gaps in graphene nanoribbons. *Physical Review Letters*, **97** (21), 216803.

[83] Nakada, K. et al. (1996) Edge state in graphene ribbons: Nanometer size effect and edge shape dependence. *Physical Review B*, **54** (24), 17954–17961.

[84] Ritter, K.A. and Lyding, J.W. (2009) The influence of edge structure on the electronic properties of graphene quantum dots and nanoribbons. *Nature Materials*, **8** (3), 235–242.

[85] Zheng, Y.P. et al. (2011) Mechanical properties of grafold: A demonstration of strengthened graphene. *Nanotechnology*, **22** (47), 405701.

[86] Cranford, S.W. and Buehler, M.J. (2011) Packing efficiency and accessible surface area of crumpled graphene. *Physical Review B*, **84** (20), 205451.

[87] Zang, J.F. et al. (2013) Multifunctionality and control of the crumpling and unfolding of large-area graphene. *Nature Materials*, **12** (4), 321–325.

[88] Becton, M., Zhang, L.Y. and Wang, X.Q. (2014) Mechanics of graphyne crumpling. *Physical Chemistry Chemical Physics*, **16** (34), 18233–18240.

[89] Chen, Y., Lu, J. and Gao, Z.X. (2007) Structural and electronic study of nanoscrolls rolled up by a single graphene sheet. *Journal of Physical Chemistry C*, **111** (4), 1625–1630.

[90] Wang, Y. et al. (2015) Formation of carbon nanoscrolls from graphene nanoribbons: A molecular dynamics study. *Computational Materials Science*, **96**, 300–305.

[91] Xia, D. et al. (2010) Fabrication of carbon nanoscrolls from monolayer graphene. *Small*, **6** (18), 2010–2019.

[92] Xie, X. et al. (2009) Controlled fabrication of high-quality carbon nanoscrolls from monolayer graphene. *Nano Letters*, **9** (7), 2565–2570.

[93] Shi, X.H., Yin, Q.F. and Wei, Y.J. (2012) A theoretical analysis of the surface dependent binding, peeling and folding of graphene on single crystal copper. *Carbon*, **50** (8), 3055–3063.

[94] Chen, X.M. et al. (2014) Graphene folding on flat substrates. *Journal of Applied Physics*, **116** (16), 164301.

[95] Kwan, K. and Cranford, S.W. (2014) Scaling of the critical free length for progressive unfolding of self-bonded graphene. *Applied Physics Letters*, **104** (20), 203101.

[96] Zhang, L.Y., Zeng, X.W. and Wang, X.Q. (2013) Programmable hydrogenation of graphene for novel nanocages. *Scientific Reports*, **3**, 3162.

[97] Zhang, J. et al. (2010) Free folding of suspended graphene sheets by random mechanical stimulation. *Physical Review Letters*, **104** (16), 166805.

[98] Hiura, H. et al. (1994) Role of Sp(3) defect structures in graphite and carbon nanotubes. *Nature*, **367** (6459), 148–151.

[99] Pugno, N.M. (2006) Dynamic quantized fracture mechanics. *International Journal of Fracture*, **140** (1-4), 159–168.

[100] Pugno, N. (2006) New quantized failure criteria: Application to nanotubes and nanowires. *International Journal of Fracture*, **141** (1–2), 313–323.

[101] Pugno, N.M. and Ruoff, R.S. (2004) Quantized fracture mechanics. *Philosophical Magazine*, **84** (27), 2829–2845.

[102] Ippolito, M. et al. (2006) Role of lattice discreteness on brittle fracture: Atomistic simulations versus analytical models. *Physical Review B*, **73** (10), 104111.

[103] Taylor, D., Cornetti, P. and Pugno, N. (2005) The fracture mechanics of finite crack extension. *Engineering Fracture Mechanics*, **72** (7), 1021–1038.

[104] Pugno, N. et al. (2006) A generalized Paris' law for fatigue crack growth. *Journal of the Mechanics and Physics of Solids*, **54** (7), 1333–1349.

[105] Pugno, N., Cornetti, P. and Carpinteri, A. (2007) New unified laws in fatigue: From the Wöhler's to the Paris' regime. *Engineering Fracture Mechanics*, **74** (4), 595–601.

[106] Rice, J.R. and Rosengren, G.F. (1968) Plane strain deformation near a crack tip in a power-law hardening material. *Journal of the Mechanics and Physics of Solids*, **16** (1), 1–12.

[107] Carpinteri, A. and Pugno, N. (2005) Fracture instability and limit strength condition in structures with re-entrant corners. *Engineering Fracture Mechanics*, **72** (8), 1254–1267.

[108] Carpinteri, A. and Chiaia, B. (1986) Crack-resistance behavior as a consequence of self-similar fracture topologies. *International Journal of Fracture*, **76** (4), 327–340.

[109] Pugno, N.M. (2007) Young's modulus reduction of defective nanotubes. *Applied Physics Letters*, **90** (4), 043106.

[110] Pugno, N.M. and Ruoff, R.S. (2006) Nanoscale Weibull statistics. *Journal of Applied Physics*, **99** (2), 024301.

[111] Beale, P.D. and Srolovitz, D.J. (1988) Elastic fracture in random materials. *Physical Review B*, **37** (10), 5500–5507.

[112] Pugno, N.M. (2010) The design of self-collapsed super-strong nanotube bundles. *Journal of the Mechanics and Physics of Solids*, **58** (9), 1397–1410.

[113] Pugno, N. M., Yin, Q., Shi, X., Capozza, R., (2013) *A generalization of the Coulomb's friction law: from graphene to macroscale. Meccanica*, **48**, 1845–1851.

[114] Pugno, N.M. (2007) The role of defects in the design of space elevator cable: From nanotube to megatube. *Acta Materialia*, **55** (15), 5269–5279.

[115] Pugno, N.M. (2007) Space elevator: out of order? *Nano Today*, **2** (6), 44–47.

[116] Pugno, N.M. (2006) On the strength of the carbon nanotube-based space elevator cable: From nanomechanics to megamechanics. *Journal of Physics: Condensed Matter*, **18** (33), S1971.

[117] Jiang, J.W. and Park, H.S. (2014) Mechanical properties of MoS$_2$/graphene heterostructures. *Applied Physics Letters*, **105** (3), 033108.

[118] Giovannetti, G. et al. (2007) Substrate-induced band gap in graphene on hexagonal boron nitride: *Ab initio* density functional calculations. *Physical Review B*, **76** (7), 073103.

[119] Kaloni, T.P., Cheng, Y.C. and Schwingenschlogl, U. (2012) Electronic structure of superlattices of graphene and hexagonal boron nitride. *Journal of Materials Chemistry*, **22** (3), 919–922.

[120] Sakai, Y., Koretsune, T. and Saito, S. (2011) Electronic structure and stability of layered superlattice composed of graphene and boron nitride monolayer. *Physical Review B*, **83** (20), 205434.

[121] Zhong, X. et al. (2012) Electronic structure and quantum transport properties of trilayers formed from graphene and boron nitride. *Nanoscale*, **4** (17), 5490–5498.

[122] Ramasubramaniam, A., Naveh, D. and Towe, E. (2011) Tunable band gaps in bilayer graphene−BN heterostructures. *Nano Letters*, **11** (3), 1070–1075.

[123] Li, Y.-J. et al. (2012) Hexagonal boron nitride intercalated multi-layer graphene: A possible ultimate solution to ultra-scaled interconnect technology. *AIP Advances*, **2** (1), 012191.

[124] Kim, Y. et al. (2013) Strengthening effect of single-atomic-layer graphene in metal-graphene nanolayered composites. *Nature Communications*, **4**, 2114.

[125] Gong, L., Kinloch, I.A., Young, R.J. et al. (2010) Interfacial stress transfer in a graphene monolayer nanocomposite. *Advanced Materials*, **22**, 2694–2697.

[126] Roman, R.E. and Cranford, S.W. (2014) Strength and toughness of graphdiyne/copper nanocomposites. *Advanced Engineering Materials*, **16** (7), 862–871.

[127] Young, R.J., Kinloch, I.A., Gong, L. and Novoselov, K.S. (2012) The mechanics of graphene nanocomposites: A review. *Composites Science and Technology*, **72**, 1459–1476.

9

The Effect of Chirality on the Mechanical Properties of Defective Carbon Nanotubes

Keka Talukdar

Department of Physics, Nadiha High School, Durgapur, West Bengal, India

9.1   Introduction

Carbon nanotubes (CNTs) have been identified as promising materials for reinforcing composite materials. The remarkable mechanical properties of carbon nanotubes have raised the expectation of today's scientists and industrialists to build superstrong composite materials for various applications. A lot of experimental as well as theoretical studies have been carried out until now, which point towards many fascinating mechanical characteristics of the CNTs. Their tensile strength, Young's modulus, stiffness, toughness and ability to withstand large angle bending and twisting have caught the attention of the new-generation scientists to expect much improvement in the elastic modulus and strength of the future composites. But CNTs can only be properly utilised as reinforcing agents by controlling their chirality as this structural speciality of the CNTs govern the mechanical, electrical and optical properties of individual nanotubes. So, in this work, the effect of chirality on the mechanical properties of the CNTs has been investigated thoroughly by molecular dynamics (MD) simulation. Single-walled CNTs (SWCNTs), multi-walled CNTs (MWCNTs) and SWCNT bundles are taken as sample and their mechanical properties are analysed with pristine as well as defective tubes. Their breaking and buckling behaviour are also modelled.

   After the discovery of the CNTs by Iijima [1], a lot of experimental and theoretical investigations have been reported so far to find out their exact strength and stiffness. Few studies reveal the effect of chirality on the mechanical properties of the CNTs. The effect of chirality on the buckling behaviour of SWCNTs is investigated by Zhang and Wang [2], and the investigators have come to the conclusion that chirality plays an important role in the

Advanced Computational Nanomechanics, First Edition. Edited by Nuno Silvestre.
© 2016 John Wiley & Sons, Ltd. Published 2016 by John Wiley & Sons, Ltd.

buckling property of SWCNTs with small chiral angles and the effect can be neglected for large chiral angle. Dependence of Poisson's ratio on the chirality of SWCNTs is observed by Natsuki et al. [3]. No influence of chirality on the intertube distances and binding strengths between the tubes of a bundle is observed by Dumlich et al. [4]. However, strong dependence of the rippled structure of the collapsed CNTs on chirality is revealed by the molecular dynamics approach of Baimova et al. [5]. Also, in another study, the clear dependence of critical buckling loads on the chirality of zigzag double-walled CNTs is reported [6]. By finite element modelling, Zuberi and Esat [7] investigated the size effect and chirality on the mechanical properties of the SWCNTs. *Ab initio* calculations [8] for the effect of chirality on the phonon spectra, mechanical and thermal properties of the narrow SWCNTs found the dependence of Young's modulus and compressive modulus on chirality. Molecular mechanics (MM) simulations with Compass force field [9] showed that for large deformations, accompanied by compression and bending, the buckling of the SWCNTs largely depends on tube chirality. Not only that, the stability of the SWCNTs under ion irradiation depends on the tube diameter and chirality [10]. In this work, prediction of the greater stability of zigzag tube than that of an armchair tube is made for two same diameter tubes.

The investigations on the role of chirality and defects on the mechanical and electrical properties of CNTs also tell us many unknown facts [11–19]. Defects and chirality in turn bring many changes to the properties of CNT composite materials [20–23]. Experimental progress regarding measurement or assigning chirality of the CNT samples is remarkable. The measurement of chiral angle by Scanning Tunnelling Microscope [24], experimental observation to assign the chirality of the CNTs [25], the observation of the change of chirality of CNTs through near-field Raman [26] and so on should be mentioned here. Determination of chiral structure of SWCNTs is successfully done by Chen et al. [27]. To get control over chirality is a promising challenge to the experimentalists during the last decade. For selective growing of SWCNTs with specific chirality, considerable and successful attempts are taken up so far [28–31]. As nanotube prefers to grow minimally chiral (near-armchair) due to some specific reason [32] and since for proper exploitation of the CNTs, we need to have control over their chirality, experimentalists are giving more emphasis on such types of experiments.

The properties of the CNTs are mainly governed by the CNT topology, more precisely speaking, their chirality. There are many evidences of the change of electrical properties on mechanical deformation [33, 34]. In applications related to nanoelectromechanical systems such as actuators, motors can be envisaged with the CNTs where both the mechanical and electrical properties are important. Zhang et al. [22] in their excellent work demonstrated photoactuators, oscillators and motors based on polymer–carbon nanotube bilayer where the selected CNTs are of different chirality distributions.

However, in spite of many above-mentioned studies, the experimental values of the mechanical properties of the CNTs vary in magnitude up to one order as the actual internal structure of the CNTs are still not completely known and the properties vary largely with their chirality and size [35]. So, for better understanding of the internal molecular details, computer modelling and simulation may serve an important tool. Moreover, MD simulation can tell us the internal dynamical change that the system is running through in course of time under various external forces. Hence, a comprehensive investigation is required to find out the influence of chirality on the defective CNTs as in reality, CNTs get associated with defects in the process of manufacturing or purification. They often form bundles and multi-walled tubes that are used in building composites. There is some available literature [2–10] where the influence of diameter and chirality on the mechanical properties of SWCNTs is addressed. Nowhere, nanotube bundle or MWCNTs are taken for investigation. So, here we have taken seven SWCNTs of diameter 4-10 Å of different chirality, seven set of SWCNT bundles made with same diameter SWCNTs, and seven three-walled MWCNTs to investigate the effect of chirality for all types of CNTs in detail.

9.2   Carbon Nanotubes, Their Molecular Structure and Bonding

9.2.1   Diameter and Chiral Angle

The microscopic structure of carbon nanotubes is closely related to graphene. So the structure of a single-walled carbon nanotube can be expressed in terms of graphene lattice vectors. One-dimensional unit cell, defined by the vector

$$c_h = na_1 + ma_2 \qquad (9.1)$$

known as a chiral vector, where a_1 and a_2 are two unit vectors at 60° angles to each other as shown in Figure 9.1. The basis vectors have the same length, $|a_1| = |a_2| = a_0 = 2.461$ Å. A CNT with (n, m) indexes can be visualised by drawing na_1 and ma_2 vectors at 60° angle to each other and then by adding the two vectors. The line representing the sum of the vectors defines the circumference of the CNT along the plane perpendicular to its axis. The diameter of the CNT with (n,m) indices is

$$d = \frac{|c_h|}{\pi} = \frac{a_0}{\pi}\sqrt{(n^2 + nm + m^2)} \qquad (9.2)$$

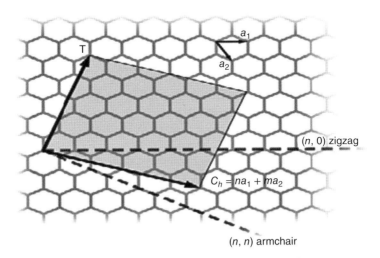

Figure 9.1   Schematic diagram showing how a hexagonal sheet of graphite is "rolled up" to form a CNT

Table 9.1 Structural parameters of armchair (A), zigzag (Z) and chiral (C) SWCNTs [36]

Tube with indices	$N = n^2 + nm + m^2$	Diameter d	Chiral angle θ	Value of a
$A(n,n)$	$3n^2$	$\sqrt{3}.na_0/\pi$	$30°$	a_0
$Z(n,0)$	n^2	$n\,a_0/\pi$	$0°$	$\sqrt{3}.a_0$
$C(n,m)$	$n^2 + nm + m^2$	$\sqrt{N}.a_0/\pi$	arc cos $(n+m/2)/\sqrt{N}$	$\sqrt{(3N)}a_0/(nR)$

Here a_0 is the magnitude of any one of the two equal basis vectors $\boldsymbol{a}_1$ and $\boldsymbol{a}_2$. The chiral angle θ is defined as the angle between $\boldsymbol{a}_1$ and $\boldsymbol{c}_h$ and can be expressed as

$$\cos \theta = \frac{(a_1.c_h)}{|a_1|.|c_h|} = \frac{\left(m + \frac{n}{2}\right)}{\sqrt{(n^2 + nm + m^2)}} \tag{9.3}$$

Helix of the graphene lattice points around the tube axis changes when θ changes from $0°$ and $30°$ to $30°$ and $60°$. Due to the sixfold rotational symmetry of graphene, to any chiral vector an equivalent one exists with $\theta \leq 60°$.

The smallest graphene lattice vector $\boldsymbol{a}$ perpendicular to $\boldsymbol{c}$ defines the translational period a along the tube axis. In terms of chiral indices (n,m), translational vector may be defined by

$$a = -\left[\frac{(2m + n)}{nR}\right]a_1 + \left[\frac{(2n + m)}{nR}\right]a_2 \tag{9.4}$$

where $R = 3$ if $(m - n)/3k$ is integer and $R = 1$ otherwise. So the nanotube unit cell is formed by a cylindrical surface with height a and diameter d. The chiral vectors and translational vectors for chiral and achiral tubes can be simplified as

$$a_z = \sqrt{3}.a_0|c_z| = na_0 \text{ (zigzag)} \tag{9.5}$$

$$a_A = a_0|c_A| = \sqrt{3}.na_0 \text{ (armchair)} \tag{9.6}$$

Table 9.1 shows the basic structural parameters of different types of CNTs.

9.2.2 Bonding Speciality in CNTs

CNTs are tubular structure with hexagonal pattern repeating itself in space. In carbon nanotubes, each atom is bonded with other three carbon atoms by sp^2 hybridisation. One s and two p orbitals combine to form three sp^2 orbitals at $120°$ to each other within a plane. The covalent σ bond is a strong chemical bond. In contrary, there exists a weak π bond out of plane, which acts in between different shells of an MWCNT or in between different SWCNTs in a SWCNT bundle. The bonding in a CNT is shown in Figure 9.2.

Due to the surface curvature of the nanotubes, a rehybridisation process, including a certain amount of σ character in a π-type orbital, changes both their chemical and physical properties.

9.2.3 Defects in CNT Structure

Normally, CNTs contain defects in their structure. Defects may be of various types, namely, Pentagons, heptagons, vacancies, lattice-trapped states, ad-dimers, Stone–Wales (SW) defects

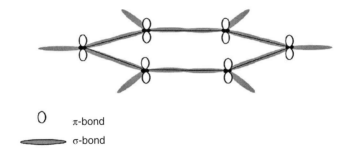

O   π-bond

〜〜〜   σ-bond

Figure 9.2   Bonding in carbon nanotubes

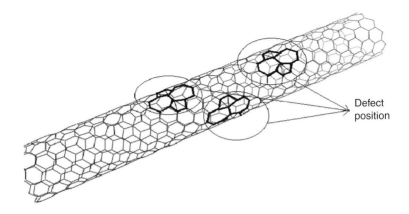

Defect position

Figure 9.3   A (15,0) SWCNT with three SW defects

and so on. Here, we have only studied the effect of SW defects. The SW defects and vacancies have been found to degrade the mechanical properties of CNTs in much respect, so these two defects are the major source of changing nanotube properties to a large extent. An SW defect is formed when a C–C bond in the hexagonal honeycomb lattice structure of the CNT is rotated by 90°. The bond rotation changes the hexagonal pattern into two pentagons and two heptagons. Such defect is also called 5-7-7-5 defect. The formation of the defect is depicted in Figure 9.3.

9.3   Methods and Modelling

9.3.1   Simulation Method

All calculations are done by MD simulation done on a 64-bit Windows machine. Nanotube coordinates are generated by online TubGen software [37], Fortran codes for MD simulation are run in GFortran for windows7, MINGW-w64. For visualisation of results, education version of PyMol [38] is used. After generating CNT coordinates, codes are written where one end of the tube is kept fixed and at the other end load is applied. The strain increment is a step-by-step process. Time step of 0.5 fs is chosen for the molecular dynamics simulation presented here. It is tested that slight changes in the time step affect the simulation result negligibly. The evolution of the system energy with respect to time is noticed, and it is observed that 30,000–50,000 time steps are required to achieve convergence in energy minimisation and hence to get equilibrium condition. To maintain an average temperature, Berendsen thermostat is applied. Coordinates $r_i(t)$ and velocities $v_i(t)$ are upgraded by velocity verlet algorithm. The algorithm is given as

Velocity verlet algorithm

$$r_i(t + \Delta t) = r_i(t) + v_i(t)\Delta t + \frac{1}{2}a_i(t)\Delta t^2 \tag{9.7}$$

$$v_i(t + \Delta t) = v_i(t) + \frac{1}{2}\left(a_i(t) + a_i(t + \Delta t)\right)\Delta t \tag{9.8}$$

$a_i(t)$ is the acceleration and Δt is the small increment of time.

9.3.2 Berendsen Thermostat

To modify the equations of motion to obtain a specific thermodynamical ensemble, the velocities of the atoms are rescaled by a scaling factor. Maxwell–Boltzmann distribution gives the distribution of velocities as

$$P(v_i) = \left[\frac{m}{2\pi k_B T}\right]^{\frac{1}{2}} \exp\left[-\left(\frac{mv_i^2}{2k_B T}\right)\right] \tag{9.9}$$

The temperature of a system is related to the velocity of atoms by the relation

$$\frac{1}{2k_B T} = \left\langle \frac{1}{2}\sum m_i v_i^2 \right\rangle \tag{9.10}$$

The kinetic energy corresponds to the average kinetic energy per degree of freedom, which is here the ensemble average for all atoms. So for a finite size system, the instantaneous temperature $T(t)$ may be defined by

$$k_B T = \frac{1}{N_f}\sum_i^N m_i v_i^2 \tag{9.11}$$

where N_f is the number of degrees of freedom.

For a dynamical system, velocity can be continuously upgraded such that total kinetic energy is constant. Velocity is updated each time and the temperature of the system is measured. In this thermostat, velocity is rescaled by a factor

$$\lambda = \sqrt{\frac{T_0}{T(t)}} \tag{9.12}$$

New velocity vectors are given by

$$\lambda_{\text{final}} = \lambda_{\text{old}} \tag{9.13}$$

So the associated change in temperature can be calculated as

Or, $$\Delta T = (\lambda^2 - 1)T(t) \tag{9.14}$$

In Berendsen thermostat [39], the temperature of the system is controlled by the weak coupling of the system to an external heat bath with fixed temperature T_0. In each step, velocity is rescaled so that the rate of increase in temperature is proportional to the difference in temperature.

$$\frac{dT(t)}{dt} = \frac{1}{\tau}(T_0 - T(t)) \tag{9.15}$$

So the scale factor is given by

$$\lambda = \left[1 + \frac{\Delta T}{\tau}\left(\frac{T}{T_0} - 1\right)\right]^{\frac{1}{2}} \tag{9.16}$$

when Δt is the time step and τ is a coupling constant. It is called the rise time, which gives the strength of the coupling of the system with the hypothetical heat bath. Larger τ means weaker coupling, that is larger τ means, the system will take a long time to achieve the final temperature. T_0 is the set point temperature. If $\Delta t \ll \tau$, then the Eq. (9.16) implies that no rescaling takes place and we recover microcanonical ensemble. The temperature fluctuations will grow until they reach the appropriate value for a microcanonical ensemble. But appropriate value for a canonical ensemble is not reached. Again when τ is very small, very low temperature fluctuation occurs. Finally, when $\tau = \Delta t$, it gives simple velocity scaling. For most temperature fluctuations, τ is taken as 0.4 ps. Berendsen thermostat does not strictly fix the temperature to a certain value, rather it exponentially decays and finally reach a target value.

9.3.3   Second-Generation REBO Potential

In second-generation REBO potential [40], total energy E of the system is given as

$$E = \sum E_i = \frac{1}{2}\sum_i \sum_{i \neq j} V(r_{ij}) \tag{9.17}$$

$$V(r_{ij}) = f_c(r_{ij})\left[V^R(r_{ij}) + \overline{b}_{ij}V^A(r_{ij})\right] \tag{9.18}$$

E is decomposed into a site energy E_i which is a function of bond energy $V(r_{ij})$, where $f_c(r_{ij})$ is a cut-off function that reduces to zero interaction beyond 2.0 Å. $V^R(r_{ij})$ is a pair-wise term that models the core–core and electron–electron repulsive interactions and $V^A(r_{ij})$ is a pair-wise term that models core–electron attractive interactions, where r_{ij} is the distance between nearest neighbour atoms i and j, and $\overline{b}_{ij}$ is a many-body, bond-order term that depends on the number and types of neighbours and the bond angles. $V^R(r_{ij})$ is the sum of attractive potential $V^A(r_{ij})$ and repulsive potential $V^R(r_{ij})$

$$V^R(r_{ij}) = f_c(r_{ij})\left(1 + \frac{Q_{ij}}{r_{ij}}\right)A_{ij}e^{-\alpha_{ij}r_{ij}} \tag{9.19}$$

$$V^A(r_{ij}) = -f_c(r_{ij})\sum_{(n=1,3)} B_n e^{-\beta_{ijn}r_{ij}} \tag{9.20}$$

$$\overline{b}_{ij} = \frac{\left(p_{ij}^{\sigma\pi} + p_{ji}^{\sigma\pi}\right)}{2} - 2 + p_{ij}^{\pi} \tag{9.21}$$

$$p_{ij}^{\pi} = \pi_{ij}^{rc} + \pi_{ij}^{dh} \tag{9.22}$$

$$p_{ji}^{\sigma\pi} = \left[1 + G_{ij} + P_{ij}\left(N_i^{(H)}, N_i^{(C)}\right)\right]^{-\frac{1}{2}} \tag{9.23}$$

$$G_{ij} = \sum_{k \neq i,j} f_c(r_{ik}) G_i\left(\cos\left(\theta_{ijk}\right)\right) e^{\lambda_{ijk}(r_{ij}-r_{ik})} \tag{9.24}$$

$$\pi_{ij}^{rc} = F_{ij}\left(N_i^{(t)}, N_j^{(t)}, N_{ij}^{conj}\right) \tag{9.25}$$

The angular function

$$g_c = G_c(\cos\theta) + Q(N_i^t).[\gamma_c(\cos\theta) - G_c(\cos\theta)] \tag{9.26}$$

$\gamma_c(\cos\theta)$ is a second spline function. $Q(N_i^t)$ is defined as

$$Q\left(N_i^t\right) = \begin{bmatrix} 1 & N_i^t \leq 3.2 \\ \frac{1}{2} + \frac{1}{2}\cos\pi\frac{(N_i^t-3.2)}{(3.7-3.2)} & 3.2 < N_i^t < 3.7 \\ 0 & N_i^t \geq 3.7 \end{bmatrix} \tag{9.27}$$

$$N_{ij}^{conj} = 1 + \left[\sum_{k \neq i,j}^{\text{carbon atoms}} f_c\left(r_{ik}\right) F(\chi_{ik})\right]^2 + \left[\sum_{l \neq i,j}^{\text{carbon atoms}} f_c\left(r_{jl}\right) F(\chi_{jl})\right]^2 \tag{9.28}$$

$$\pi_{ij}^{hd} = T_{ij}\left(N_i^{(t)}, N_j^{(t)}, N_{ij}^{conj}\right)\left[\sum_{k \neq i,j}\sum_{l \neq i,j}\left(1 - \cos^2\omega_{ijkl}\right) f_c(r_{ik}) f_c(r_{jl})\right] \tag{9.29}$$

$$\omega_{ijkl} = e_{ijk}.e_{ijl} \tag{9.30}$$

The function $f_c(r)$ includes the interactions of nearest neighbour only. P represents bicubic spline. N_{ij}^{conj} is one when all of the carbon atoms that are bonded to a pair of carbon atoms i and j have four or more neighbours. As the coordination number of the neighbouring atoms decreases, N_{ij}^{conj} becomes greater than 1. Here, the term N_{ij}^{conj} differentiates between different configurations. Also the term accounts for bond formation and breaking. The function $T_{ij}(N_i^{(t)}, N_j^{(t)}, N_{ij}^{conj})$ is a tricubic spline and e_{ijk} and e_{ijl} are the unit vectors in the direction of $R_{ji} \times R_{ik}$ and $R_{ij} \times R_{jl}$, respectively, where the R vectors connect the subscripted atoms, and π_{ij}^{dh} depends on the dihedral angle for the C–C double bonds. $G_c(\cos\theta_{ijk})$ modulates the contribution that each nearest-neighbour makes to b_i. The modified potential includes both modified analytical functions for the intramolecular interactions and an expanded fitting database. This gives a better description of bond energies, lengths, and especially force constants for carbon–carbon bonds. It has produced an improved fit to the elastic properties of diamond and graphite. Better prediction can be made for the energies of several surface reconstructions and interstitial defects. The rotation about dihedral angles for carbon–carbon double bonds and the angular interactions associated with hydrogen centres have been modelled in the second-generation REBO potential.

In the present simulation, we have used smoothing as described in Brenner's code of the cut-off in the rotational potential in the second-generation REBO potential to avoid energy non-conservation.

9.3.4 C–C Non-bonding Potential

In some simulation studies, the non-bonding interactions between carbon atoms are required. Various types of potentials are there to model the non-bonding potential. In the present study, Lennard–Jones potential [41] is adopted to describe the van der Waals inter-molecular interactions between different CNTs in a multi-walled carbon nanotube or in between the tubes of a single-walled carbon nanotube bundle. The form of the potential was first proposed by John Lennard–Jones and is given by

$$V_{LJ} = 4\varepsilon_{ij} \left[\left(\frac{\sigma_{ij}}{r_{ij}} \right)^{12} - \left(\frac{\sigma_{ij}}{r_{ij}} \right)^{6} \right] \tag{9.31}$$

ε is the depth of the potential well, σ is the distance at which the inter-particle potential is zero and r is the distance between the particles. i, j stand for ith and jth atoms. The term r^{-12} describes Pauli repulsion at short ranges due to the overlapping of electronic orbitals and r^{-6} describes the long-range attraction. According to Pauli exclusion principle, the energy of the system increases due to the overlapping of the electronic wave functions surrounding the atoms.

9.3.5 Method of Calculation

Simulations are carried out for different CNTs under consideration. Equilibrium energy is noted for each step. To calculate stress from the energy–strain curve, we have used a linear relationship for the elastic region and appropriate non-linear equations for the segments of high-strain deformation regions. Young's modulus was found from the slope of the linear portion of the stress–strain curve.

Stress was calculated from the energy–strain curve as

$$\sigma = \frac{1}{V} \left(\frac{dE}{d\varepsilon} \right) \tag{9.32}$$

where σ is the longitudinal stress, V the volume of the tube, ε the strain and E the strain energy of the tube. Volume of the tube is found as $V = 2\pi r \delta r l$ where r is the inner radius of the tube, δr its wall thickness and l the length of the tube. We have taken δr as 0.34 nm, which has been the standard value, used by most of the authors.

9.4 Results and Discussions

9.4.1 Results for SWCNTs

To analyse the result systematically, the SWCNTs are first simulated with and without defects. (15,0), (10,1), (5,1), (9,3), (6,3), (5,4) and (9,9) SWCNTs are taken for test samples. The length of the tubes are taken in such a way that aspect ratio is always greater than 10. The CNTs are subjected to tensile loading and then compressive loading separately. Their breaking and buckling behaviour are modelled. The stress–strain curves show differences with the chiral angle. It is observed that almost same diameter tubes show difference in their maximum tensile stress,

Young's modulus, compressive modulus and critical buckling stress. To compare the results for different chiralities but of same diameter, Table 9.2 is used. In Table 9.2, the variation of different mechanical properties of the tubes of nearly equal diameter, but of different chiralities is shown for the defective SWCNTs. It is noteworthy that three SW defects in each tube are introduced with the same orientation. The orientation of the defects are set by changing the θ value in (r, θ, z) positions. Figure 9.4 shows the defect positions in each tube. It is clear from the figure that the line joining the two pentagons in each defect is almost parallel. Hence, their orientations are same.

(5,4) and (6,3) SWCNTs of radii 3.11 and 3.21 Å have Young's moduli of 0.813 and 0.815 TPa and their compressive moduli are 0.798 and 0.815 TPa, respectively. Much difference is observed in their maximum tensile strengths and critical buckling stresses. Differences are also observed in the values for the (9,3) and (10,1) SWCNTs. The stress–strain curves for defect-free and defective SWCNTs are depicted in Figures 9.5 and 9.6, respectively. The armchair (9,9) SWCNT shows maximum breaking stress with and without defect. Tube with small chiral angle, that is the (9,3) tube shows higher Young's modulus. The chiral tubes, from the beginning, are in a strained condition. Evidence for which is the formation of minimally chiral SWCNTs in the experiments [32]. Looking at the (5,4) SWCNT with one defect, the

Table 9.2 Variation of mechanical properties of the defective SWCNTs with nearly equal radius but of different chiralities. [Y.M. – Young's modulus, F.S. – failure stress, F.Sr. – failure strain, C.M. – compressive modulus, C.B.S. – critical buckling stress, C.B.Sr. – critical buckling strain]

SWCNT	Radius of SWCNT (Å)	Chiral angle (°)	Y.M. (TPa)	F.S. (GPa)	F.Sr	C.M. (TPa)	C.B.S. (GPa)	C.B.Sr
(5,4)	3.11	26.3295	0.813	97.05	0.22	0.798	38.76	0.07
(6,3)	3.21	19.1066	0.815	89.90	0.20	0.815	50.49	0.07
(9,3)	4.3891	13.8922	0.829	80.54	0.14	0.829	45.76	0.06
(10,1)	4.173	4.1750	0.911	81.61	0.16	0.969	30.57	0.06

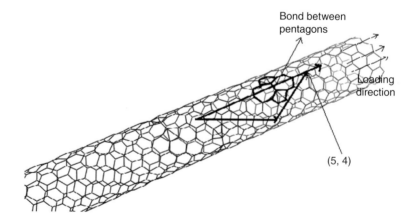

Figure 9.4 Chiral vector, orientation of SW defect and loading direction

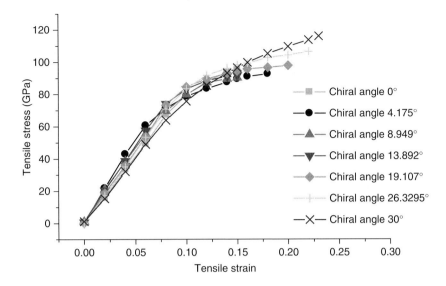

Figure 9.5   Stress–strain curves for defect-free SWCNTs under tensile loading

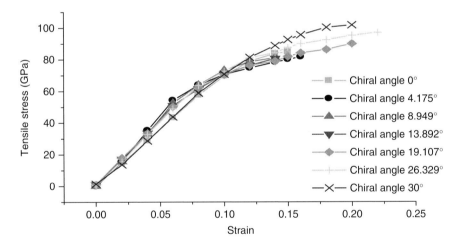

Figure 9.6   Stress–strain curves for defective SWCNTs under tensile loading

bond between the pentagons is in perpendicular direction to the external strain and also almost perpendicular to the chiral vector (neglecting small distortion due to two-dimensional picture of the three-dimensional object). So while straining the tube, say from the right, the weakest bond, that is the bond between the pentagons, does not break so fast as other tubes, resulting the elongated portion of its stress–strain curve (Figure 9.6) unlike the other defective tubes. The SWCNTs with large chiral angle, for the same reason, show higher maximum tensile stress than the other tubes with smaller chiral angles. Figure 9.7 gives the profile of pair energy of a defective (5,4) SWCNT.

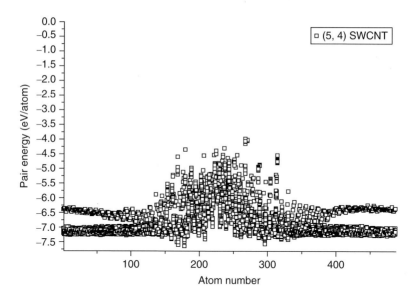

Figure 9.7   Pair-energy of a (5,4) SWCNT with defects

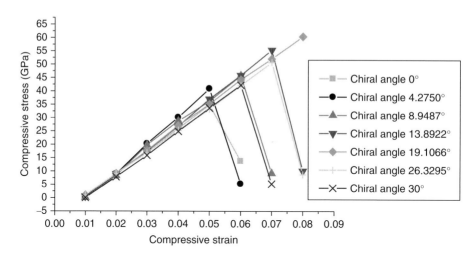

Figure 9.8   Stress–strain response for the defect-free SWCNTs under compressive loading

Under compressive stress, the stress–strain response shows linear character (Figure 9.8). Higher critical buckling stress is observed for chiral angles 13.8922° and 19.1066°. Sharp fall of the compressive stress indicates the collapsing of the tubes. Figure 9.8 gives the compressive stress versus strain curve for the defect-free SWCNTs. The same behaviour is exhibited by the defective SWCNTs.

Striking breaking and buckling patterns are observed for the defect-free and defective SWCNTs, which strongly suggest their dependence on chirality. Sharp breaking modes are observed for chiral tubes. Defective chiral tubes break more suddenly in a brittle manner

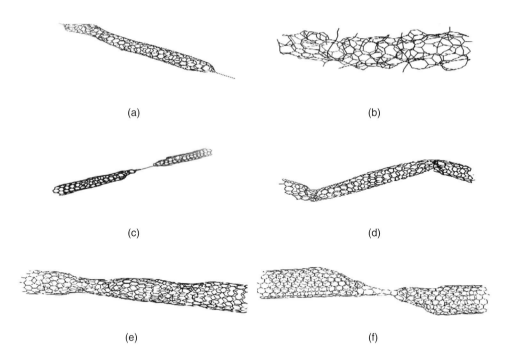

(a)   (b)

(c)   (d)

(e)   (f)

Figure 9.9   (a) Breaking of a (5,1) SWCNT without defect; (b) collapsing of a (6,3) SWCNT without defect: (c) sharp breaking of a (6,3) SWCNT with three SW defects; (d) kink formation in a (5,4) SWCNT with three defects; (e) necking of a defective (15,0) tube at 12% strain; (f) the defective (9,9) SWCNT just before failure

(Figure 9.9(c)). The narrow and pristine (5,1) SWCNT of chiral angle 8.9487° breaks from the edge (Figure 9.9(a)). Under compressive force, the chiral tubes collapse at a certain compression. Any other buckling modes are not so prominent before failure. Figure 9.9(b) is the picture of a collapsing pristine (6,3) tube. With defects, kinks are formed (Figure 9.9(d)) in all tubes, except for the tubes with large chiral angle. Only the zigzag (15,0) tube shows necking before failure under tensile loading (Figure 9.9(e)). The breaking of a defective (9,9) SWCNT is given in Figure 9.9(f).

With an increase in chiral angle, the plastic flow region before failure is more elongated than that of the SWCNTs of small chiral angle. CNTs with small chiral angle show necking before failure, and their necking lasts for a span of 3–7% strain. For the SWCNTs with large chiral angle, the tube breaks suddenly without necking, or even if necking is observed, it fails within 1–2% more strain after formation of neck. On the other hand, the defective (9,9) SWCNT breaks at 20% strain and necking starts at the strain of 19%. Without defect, the same tube breaks at 23% strain with a neck forming at a strain of 22%.

9.4.2   Results for SWCNT Bundle and MWCNTs

Nanotubes form bundles as they interact with each other by van der Waals interaction. Isolation of single tube is an important issue for the experimentalists to exploit the extraordinary property of a single tube. Due to bundling the mechanical as well as electrical properties change remarkably. It is hence an important matter to study the effect of chirality on the bundling of the tubes. In the present work, bundle of three tubes is taken in each time with three defects in each tube. So, as a whole, 7 bundles of defective tubes are taken and simulated to find the change in their property from the isolated defective tubes. The bundle of three (6,3) SWCNTs with defects is shown in Figure 9.10. The tubes are separated by 3.4 Å from each other. They are staying together due the presence of van der Waals force in between the tubes.

MWCNTs are also modelled and defects are introduced that are energetically favourable. In Figures 9.11 and 9.12, stress–strain curves for the SWCNT bundles and MWCNTs with defects are described. Figure 9.13 shows the compressive stress versus strain response of the bundle of tubes. Unlike the defective SWCNTs, here for the bundles, the failure stresses are lowered by a considerable amount. But the Young's modulus of the bundles has higher values than that of

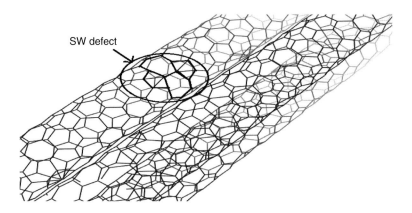

Figure 9.10   A three-membered (6,3) SWCNT bundle with defects

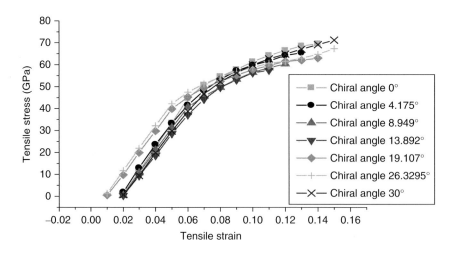

Figure 9.11   Stress–strain curves for SWCNT bundles with defects under tensile loading

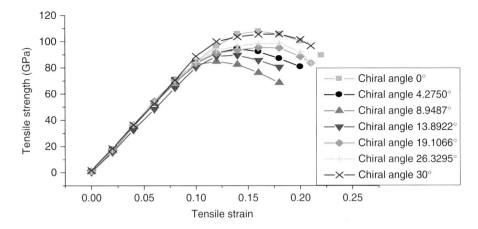

Figure 9.12   Stress–strain curves for MWCNTs with defects under tensile loading

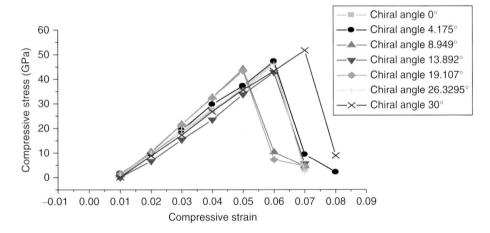

Figure 9.13   Stress–strain curves for SWCNT bundles with defects under compressive loading

the SWCNTs. Due to bundling, the properties of the single tube is changed [42, 43]. The failure stresses vary with the chiral angle. The failure stresses are dropped to much lower value than that of the isolated tubes. The lowering of failure stresses is not so prominent for the MWCNTs. But remarkable extension of the plastic flow region is reported for the defective MWCNTs. For the zigzag MWCNT, the plastic flow region is mostly extended. For it, necking lasts for 4–6% strain. The same thing happens for tubes with higher chiral angle; that is the inclusion of defects and intertube interaction on the MWCNTs has made the tubes more ductile. MWCNTs with intermediate values of chiral angle exhibit less ductility. The pair energy of the defective bundle of (5,4) SWCNTs and defective (5,4) MWCNTs are shown in Figures 9.14 and 9.15, respectively. The Young's modulus and tensile strength of SWCNT bundles can be compared to the experimental values obtained by Yu et al. [44]. However, for single tubes, the value of tensile strengths is higher than that of the bundled tubes due to absence of intertube interaction.

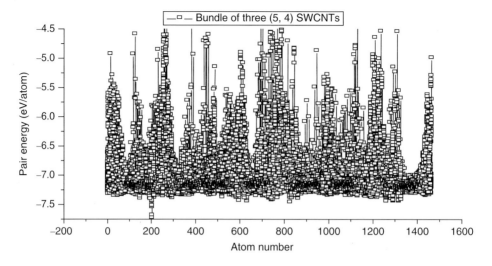

Figure 9.14   Pair energy of a (5,4) SWCNT bundle with defects

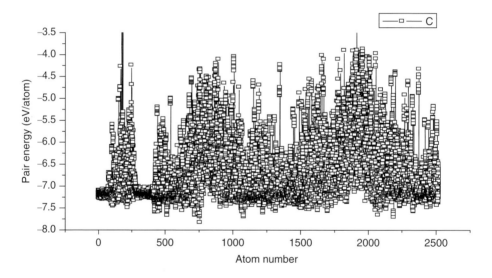

Figure 9.15   Pair energy of a (5,4) MWCNT with defects

The drop of failure stress for SWCNT bundles and increase in elastic modulus may be explained by overlapping of electronic bands and hence the breaking of symmetry due to the presence of other tubes in proximity [42, 43]. Due to the overlapping of electronic bands, attraction or repulsion arises inside a bundle due to the relative distance between molecules. All intermolecular or van der Waals forces are anisotropic, which means that they depend on the relative orientation of the molecules. The induction and dispersion interactions are always attractive, irrespective of orientation, but the electrostatic interaction changes sign upon rotation of the molecules. That is the electrostatic force can be attractive or repulsive,

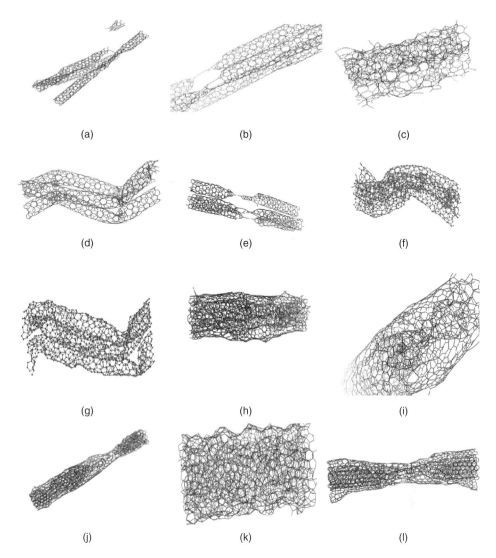

(a) (b) (c)

(d) (e) (f)

(g) (h) (i)

(j) (k) (l)

Figure 9.16 (a) Breaking of (5,1) bundle with defects; (b) breaking of (6,3) bundle with defects; (c) collapsing of (5,4) MWCNT under pressure; (d) buckling of (5,4) bundle with defects; (e) breaking of (5,4) bundle with defects; (f) large angle buckling of (6,3) bundle without defect; (g) (6,3) buckling with defects; (h) buckling of defective (9,3) MWCNT; (i) inner-wall breaking of a (15,0) MWCNT without defect; (j) necking of (10,1) MWCNT; (k) collapsing of the (9,9) MWCNT; (l) necking before failure of defective (10,1) MWCNT

depending on the mutual orientation of the molecules giving rise to different interaction between the tubes of a bundle. Inclusion of defects changes the interaction in a complex way giving rise to such changed behaviour.

Various breaking and buckling features (Figure 9.16(a)–(l)) of the defective bundles and multi-walled tubes are observed in this study. Buckling and ultimate collapse of the

MWCNTs with large chiral angle (Figure 9.16(c) and (k)) show the same pattern. Ripples are produced under pressure. Breaking modes of the (5,1) bundle (Figure 9.16(a)) and (5,4) bundle (Figure 9.16(e)) with defects are totally different though they have slightly different radii (2.2516 and 3.11 Å, respectively). But their chiral angles are 8.9487° and 26.3295°, that is they differ a lot. Inner-wall breaking on theapplication of tensile load is observed for defect-free MWCNTs (Figure 9.16(i)). Only for defective (10,1) MWCNT (Figure 9.16(l)), vigorous bond breaking is observed in the outer layer too along with the inner wall breaking.

9.4.3   Chirality Dependence

Variations of different mechanical properties of the defective SWCNT bundles and MWCNTs for nearly equal radii are demonstrated in Figures 9.17–9.20. Failure stress values (Figure 9.19) for all cases show the same trend of gradual decrease at first and then gradual increase with chirality. Young's modulus of the bundles and MWCNTs are of more or less same character but differ with the single tube (Figure 9.17). Variation of compressive modulus of SWCNTs with and without defects with the chiral angle is of same nature. However, change of compressive modulus for the bundles and the MWCNTs with chiral angle are showing somewhat different behaviour (Figure 9.18). Resemblance of the critical compressive stress values for the bundle of the tubes and MWCNTs are also obtained, which can be observed in Figure 9.20. Table 9.3 is a comparison of the various mechanical properties of defective SWCNT bundles and MWCNTs. First two rows and last two rows of each category represent tubes of nearly equal radii.

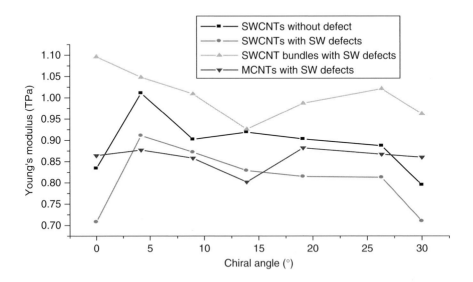

Figure 9.17   Variation of Young's modulus of the CNTs with chiral angle

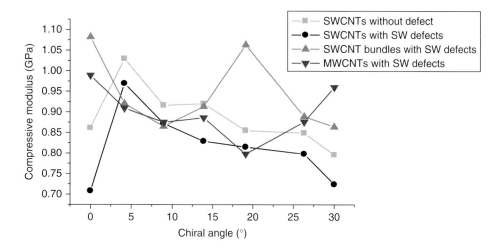

Figure 9.18   Variation of critical buckling stress of the CNTs with chiral angle

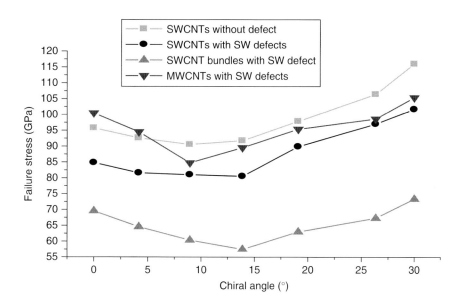

Figure 9.19   Variation of maximum tensile stress with chiral angle for different types of CNTs

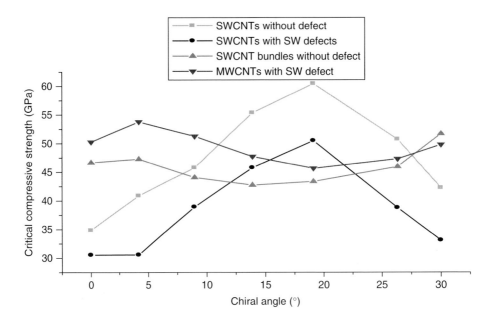

Figure 9.20   Variation of critical compressive stress with chiral angle for different types of CNTs

Table 9.3   Variation of mechanical properties of the defective SWCNT bundles and defective MWCNTs with chirality. [Four rows of each category represent (5,4), (6,3), (9,3) and (10,1) CNTs, respectively]

	Chiral angle (°)	Y.M. (TPa)	F.S. (GPa)	F.Sr.	C.M. (TPa)	C.B.S. (GPa)	C.B.Sr.
SWCNT bundle	26.3295	1.021	67.34	0.15	0.889	45.94	0.06
with defects	19.1066	0.987	62.99	0.14	1.063	43.35	0.05
	13.8922	0.926	57.47	0.11	0.913	42.73	0.06
	4.1750	1.048	64.56	0.13	0.920	47.27	0.06
MWCNT with	26.3295	0.867	98.45	0.21	0.856	47.79	0.08
defects	19.1066	0.882	95.39	0.21	0.798	45.64	0.07
	13.8922	0.802	89.56	0.18	0.886	47.72	0.07
	4.1750	0.877	94.52	0.20	0.909	53.78	0.06

9.5   Conclusions

An elaborative investigation of the influence of the mechanical properties of the CNTs on chirality is presented in this chapter. SWCNTs, either with defects or without defect, show differences in their mechanical properties with chirality. Maximum difference is observed in critical buckling strength. The value of critical buckling stress differs by 11.73 GPa for the defective (5,4) and defective (6,3) SWCNTs where their radii differ only by 0.10 Å. The drop of this value for the (9,3) and (10,1) defective SWCNTs is 15.19 GPa, and their radii differ by 0.21 Å. The former two tubes also show difference in their failure stresses.

For SWCNT bundles and MWCNTs, the variation of the tensile stress or critical buckling stress is not so high. But their Young's modulus and compressive modulus vary sufficiently. The effect of interlayer interaction in their mechanical characteristics is proved by the change in the values of elastic modulus, tensile and compressive strength for the bundle of SWCNTs and MWCNTs compared to the values obtained for the SWCNTs. The nature of tube–tube interaction in SWCNT bundles and MWCNTs is different, which can be inferred from the difference in their failure stress and failure strain values. Combined effect of chirality and orientation of defects is observed in (5,4) SWCNT, which shows little more elongation in the stress–strain curve. SWCNT bundles with large chiral angle have higher value of Young's modulus. Breaking mechanism and buckling stages of the CNTs also point towards their strong chirality-dependent nature.

References

[1] Iijima, S. (1991) Helical microtubules of graphitic carbon. *Nature*, **354**, 56–58.

[2] Zhang, Y.Y. and Wang, C.M. (2006) Effect of chirality on buckling behaviour of single-walled carbon nanotubes. *Journal of Applied Physics*, **100**, 074304.

[3] Natsuki, T., Tantrakaran, K. and Endo, M. (2004) Effects of carbon nanotube structures on mechanical properties. *Applied Physics A: Materials Science & Processing*, **79**, 117–124.

[4] Dumlich, H., Gegg, M., Hennrich, F. and Reich, S. (2011) Bundle and chirality influences on properties of carbon nanotubes studied with van der Waals density functional theory. *Physica Status Solidi B*, **248**, 2589–2592.

[5] Baimova, J.A., Fan, Q. and Zeng, L. (2015) Atomic structure and energy distribution of collapsed carbon nanotubes of different chiralities. *Journal of Nanomaterials*, **2015**, 1–5.

[6] Benguediab, S., Tounsi, A., Zidour, M. and Semmah, A. (2014) Chirality and scale effects on mechanical buckling properties of zigzag double-walled carbon nanotubes. *Composites Part B: Engineering*, **57**, 21–24.

[7] Zuberi, M.J.S., Esat, V. (2014) Estimating the effect of chirality and size on the mechanical properties of carbon nanotubes through finite element modelling. ASME 2014. 12th Biennial conference on engineering systems design and analysis, Copenhagen, Denmark, July 25-27, pp V001T04A002

[8] Tashakori, H., Khoshnevisan, B. and Kanjuri, F. (2014) Ab initio systematic study of chirality effect on phonon spectra, mechanical and thermal properties of narrow single-walled carbon nanotubes. *Computational Materials Science*, **83**, 16–21.

[9] Cao, G. and Chen, X. (2006) The effects of chirality and boundary conditions on the mechanical properties of Single-walled carbon nanotubes. *International Journal of Solids and Structures*, **44**, 5447–5465.

[10] Xu, Z., Zhang, W., Zhu, Z. et al. (2009) Effects of tube diameter ad chirality on the stability of single-walled carbon nanotubes under ion irradiation. *Journal of Applied Physics*, **106**, 043501.

[11] Gayathri, V. and Geetha, R. (2006) Carbon nanotube as NEMS sensor-effect of chirality and Stone-Wales defect intend. *Journal of Physics: Conference Series*, **34**, 824.

[12] Yang, M., Koutsos, V. and Zaiser, M. (2007) Size effect in the tensile fracture of single-walled carbon nanotubes with defects. *Nanotechnology*, **18**, 155708.

[13] Talukdar, K. and Mitra, A.K. (2010) Comparative MD simulation study on the mechanical properties of a zigzag single-walled carbon nanotube in the presence of Stone-Thrower-Wales defects. *Composite Structures*, **92**, 1701–1705.

[14] Vafek, O. and Yang, K. (2010) Many-body instability of Coulomb interacting bilayer graphene: Renormalization group approach. *Physical Review B*, **81**, 041401.

[15] Tserpes, K.I. and Papanikos, P. (2007) The effect of Stone–Wales defect on the tensile behavior and fracture of single-walled carbon nanotubes. *Composite Structures*, **79**, 581–589.

[16] Lu, Q. and Bhattacharya, B. (2005) Effect of randomly occurring Stone–Wales defects on mechanical properties of carbon nanotubes using atomistic simulation. *Nanotechnology*, **16**, 555.

[17] Chandra, N., Namilae, S. and Shet, C. (2004) Local elastic properties of carbon nanotubes in the presence of Stone-Wales defects. *Physical Review B*, **69**, 094101.

[18] Aghaei, A. and Dayal, K. (2012) Tension and twist of chiral nanotubes: torsionalbuckling, mechanical response and indicators of failure. *Modelling and Simulation in Materials Science and Engineering*, **20**, 085001.

[19] Ranjbartoreh, A.R. and Wang, G. (2011) Effect of topological defects on buckling behaviour of single-walled carbon nanotube. *Nanoscale Research Letters*, **6**, 1–6.

[20] Chen, X.Q., Liao, X.Y., Yu, J.G. et al. (2013) Chiral carbon nanotubes and carbon nanotube chiral composites. *Nano*, **08**, 1330002.

[21] Haik, M.A., Hussaini, M.Y. and Garmestani, H. (2005) Adhesion energy in carbon nanotube-polyethylene composite: Effect of chirality. *Journal of Applied Physics*, **97**, 074306.

[22] Zhang, X., Yu, Z. and Wang, C. (2014) Photoactuators and motors based on carbon nanotube with selective chirality distributions. *Nature Communications*, **5**, 2984.

[23] Joshi, U.A., Sharma, S.C. and Harsha, S.P. (2012) A multiscale approach for estimating the chirality effects in carbon nanotube reinforced composites. *Physica E: Low-dimensional Systems and Nanostructures*, **45**, 28–35.

[24] Kim, P., Odom, T.W., Huang, J. and Lieber, C.M. (2000) STM study of single-walled carbon nanotubes. *Carbon*, **38**, 1741–1744.

[25] Thomsen, C., Telg, H., Maultzsch, J. and Reich, S. (2005) Chirality assignments in carbon nanotubes based on resonant Raman scattering. *Physica Status Solidi B*, **242**, 1802–1806.

[26] Anderson, N., Hartschuh, A. and Novotny, L. (2007) Chirality changes in carbon nanotubes studied with near-field Raman spectroscopy. *Nano Letters*, **7**, 577–582.

[27] Chen, Y., Hu, Y. and Liu, M. (2013) Chiral structure determination of aligned single-walled carbon nanotubes on graphitic surface. *Nano Letters*, **13**, 5666–5671.

[28] Li, X., Tu, X., Zaric, S. et al. (2007) Selective synthesis combined with chemical separation of single-walled carbon nanotubes for chirality selection. *Journal of the American Chemical Society*, **129**, 15770–15771.

[29] Chen, Y., Shen, Z. and Xu, Z. (2013) Helicity-dependent single-walled carbon nanotube alignment on graphite for helical angle and handedness recognition. *Nature Communications*, **4**, 2205.

[30] Liu, J., Wang, C., Tu, X. et al. (2012) Chirality-controlled synthesis of single-wall carbon nanotubes using vapour phase epitaxy. *Nature Communications*, **3**, 1199.

[31] Orbaek, A.W., Owens, A.C., Crouse, C.C. et al. (2013) Single-walled carbon nanotube growth and chirality dependence on catalyst composition. *Nanoscale*, **5**, 9848–9859.

[32] Artyukhov, V.I., Penev, E.S. and Yakobson, B.I. (2014) Why nanotubes grow chiral. *Nature Communications*, **5**, 4892.

[33] Bernholc, J., Brenner, D., Nardelli, M.B. et al. (2002) Mechanical and electrical properties of nanotubes. *Annual Review of Materials Research*, **32**, 347–375.

[34] Liu, B., Jiang, H., Johnson, H.T. and Huang, Y. (2004) The influence of mechanical deformation on the electrical properties of single wall carbon nanotubes. *Journal of the Mechanics and Physics of Solids*, **52**, 1–26.

[35] Tanaka, K. and Iijima, S. (2014) *Carbon nanotube and graphene*, 2 edn, Elsevier.

[36] Buehler, M.J. (2008) *Atomistic modelling of materials failure*, Springer, New York.

[37] TubeGen-Online version 3.4, link: http://turin.nss.udel.edu/research/tubegenonline.html

[38] The PyMOL Molecular Graphics System, Version 1.7.4 Schrödinger, LLC

[39] Berendsen, H.J.C., Postma, J.P.M., van Gunsteren, W.F. et al. (1984) Molecular dynamics with coupling to an external bath. *The Journal of Chemical Physics*, **81**, 3684.

[40] Brenner, D.W., Shenderova, O.A., Harrison, J.A. et al. (2002) A second generation reactive empirical bond order (REBO) potential energy expression for hydrocarbons. *Journal of Physics: Condensed Matter*, **14**, 783–802.

[41] Jones, J.E. (1924) On the determination of molecular fields-II, from the equation of state of a gas. *Proceedings of the Royal Society of London, Series A*, **106**, 463–477.

[42] Reich, S., Thomsen, C. and Ordejon, P. (2002) Electronic band structure of isolated and bundled carbon nanotubes. *Physical Review B*, **65**, 155411.

[43] Wildöer, J.W.G., Venema, L.C., Rinzler, A.G. et al. (1998) Electronic structure of atomically resolved carbon nanotubes. *Nature*, **391**, 59–62.

[44] Yu, M.F., Files, B.S. and Arepalli, S. (2000) Tensile loading of ropes of single wall carbon nanotubes and their mechanical properties. *Physical Review Letters*, **84**, 5552–5555.

10

Mechanics of Thermal Transport in Mass-Disordered Nanostructures

Ganesh Balasubramanian
Department of Mechanical Engineering, Iowa State University, Ames, IA, USA

10.1   Introduction

Material structure drives material properties. Over the last decade, there has been an enormous interest from the scientific and the industrial communities to understand the thermal properties of materials and subsequently control the material structure to tune the properties for targeted applications. Such efforts have found stronger encouragement with the initiation of the Materials Genome Initiative by the US government to accelerate the pace of discovery by use of materials design principles. Computations play a major role in design since they are useful to explore a vast parametric space generally not feasible with experimental approaches. However, one should note that device dimensions and characteristic material dimensions in a system might not always be the same. The emergence of nanomaterials has paved the way for miniaturization of devices constituted with materials of molecular dimensions. Such advances have created opportunities to tweak properties of materials by atomic scale manipulation of their structure [1–3].

An important issue to consider is the role of defects in materials: isotopes, vacancies, dopants, and so on. Presence of defects is detrimental for heat transfer. Lattice vibrations and phonons are the dominant energy carriers in materials devoid of free electrons. Atomic impurities create local point defects that scatter the phonons, resulting in energy losses and lower thermal transport. Hence, material with more defects typically has lower thermal conductivity (TC). While on the surface it seems that the research community should focus on fabricating material structures with low-defect fractions, an alternate school of scientists have been attempting at high-resolution engineering of these defects to minimize the phonon TC of electrically conductive materials. The significance of lowering thermal transport, especially in thermoelectric devices, is paramount; the ratio of electrical properties to TC offers a

Advanced Computational Nanomechanics, First Edition. Edited by Nuno Silvestre.
© 2016 John Wiley & Sons, Ltd. Published 2016 by John Wiley & Sons, Ltd.

metric for thermoelectric efficiency that is higher for thermally insulating and electrically conductive materials. Thus, random or ordered arrangement of defects in a material structure can influence the fundamental transport mechanics in nanomaterials, resulting in thermal properties tailored for specific applications.

The scope of this chapter is to provide a flavor of a set of computational approaches that are employed to investigate heat transfer mechanisms in mass-disordered nanostructures, in particular those containing isotope impurities. Carbon nanomaterials have been attractive case studies for their remarkable properties. Although the examples below demonstrate effects of isotopes mostly in carbon nanotubes (CNTs) and graphene, the approaches are generic and applicable to a wide array of nanomaterials. This chapter begins from the fundamentals of thermal transport in nanomaterials, moves to engineering material properties and their computation, and finally concludes with data-driven methods for designing defect-engineered nanostructures.

10.2   Equilibrium Molecular Dynamics to Understand Vibrational Spectra

Equilibrium molecular dynamics (MD) simulations are based on classical mechanics principles but can be effectively employed to deterministically predict the behavior of phonons in a material. Several different programming options currently exist, one of the most popular ones being LAMMPS [4], which being freely available and open source provides the freedom to update and tune the code according to the requirements of the problem being simulated. Typically, a nanostructure is constructed using geometric parameters of the material of interest followed by energy minimization, equilibration, and subsequently nonequilibrium simulations as necessary. Appropriate boundary conditions need to be applied that describe the interactions of the material at the system peripheries. The goodness of the results is directly coupled to the use of accurate potential functions that model the interatomic interactions. While often overlooked, it is important to remember that potential functions are specific to both element and the structure and are intrinsic to the phenomena being investigated. Several empirical potentials have been proposed for, say, carbon nanomaterials, which show differences in the structure and property predictions [5, 6]. The investigations of mass-disordered structures with isotope impurities generally consider same potential function for both the isotopes since electronic structure is invariant for the different masses of the same element.

The results of vibrational density of states (DOS) obtained for a 40 nm CNT with random distribution of ^{14}C atoms in the structure is shown in Figure 10.1(a). The Adaptive Intermolecular Reactive Empirical Bond Order (AIREBO) potential model is used to describe the atomistic bonded and nonbonded interactions. After a multistage equilibration process, simulations are performed at 300 K to derive the classical vibrational DOS from the Fourier transformation of the mass-weighted velocity autocorrelation of the ensemble. The details of the procedure are described in the literature [7]. The figure shows that the modes of the nanotubes containing heavier carbon atoms are translated along the horizontal axis of wave numbers, and hence characteristic frequencies. The broad profiles below 700 cm^{-1} occur due to the radial vibrational modes while those around 1762 cm^{-1} are attributed to the short-ranged stretching modes of carbon atoms. A magnified representation of the high-frequency domain shown in Figure 10.1(b) reveals that increasing concentration of ^{14}C atoms in the nanotube structure

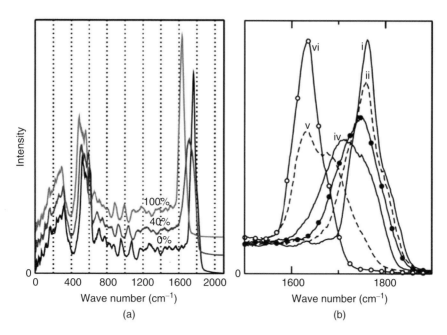

Figure 10.1 (a) Density of vibrational states (DOS) of CNTs containing 0%, 40%, and 100% of the ^{14}C isotopes in the material structure. (b) The vibrational spectra are shifted in the vertical direction to facilitate comparison. Reproduced by permission from Royal Society of Chemistry [7]

shifts the frequencies to lower values, with the peak wave number for a nanotube composed 100% with the heavier isotopes occurring around $1632\,cm^{-1}$. Simultaneously, the intensity of the phonon modes decreases with the broadening of the spectra between the extremities, the minimum observed for a mass fraction of 50%. Shift to lower wave numbers or frequencies implies reduced energy-carrying capacity, and hence a lower TC. However, it is important to realize whether it is only the mass disorder that contributes to this TC reduction, or if there are other factors that need to be accounted for to understand the mechanics of thermal transport in such defect-engineered systems.

Let us consider M_x to be the average mass of an atom in a material containing x molar fraction of heavier isotopes M_A and $(1-x)$ fraction of lighter isotopes M_B. Then,

$$M_x = xM_A + (1 - x)M_B \tag{10.1}$$

Under the harmonic oscillator approximation, the characteristic wave number or frequency of such a binary mixture depends on the spring constant and the mass of the atom. Since mass disorder is induced with isotopes, the differences in the electronic structure of the lighter and heavier atoms, and hence the interatomic interactions or simply stating the differences in spring constants, can be neglected. Thus, wave number is inversely proportional to the square root of the average mass of the material. In other words, ratio of the wave numbers for the pure material (f_0) composed of lighter isotopes and that for the mass-disordered system (f_x) can be obtained as

$$f_0/f_x = \left\{xM_A/M_B + (1 - x)\right\}^{0.5} \tag{10.2}$$

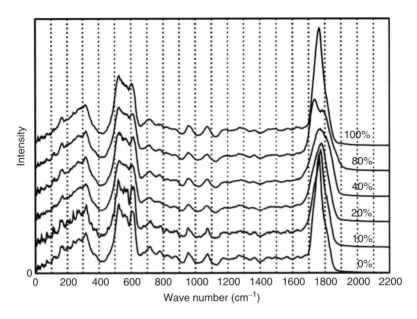

Figure 10.2 Density of vibrational states (DOS) of CNTs containing 0%, 10%, 20%, 40%, 80%, and 100% of the ^{14}C isotopes in the material structure. The wave numbers are rescaled according to Eq. 10.2. The intensity in the plot is reported in arbitrary units. Reproduced by permission from Royal Society of Chemistry [7]

Rescaling the DOS with the above-mentioned model results in normalized DOS as shown in Figure 10.2. The vibrational spectra for the different nanotubes appear to lie on top of each other, further demonstrating the shifting of frequencies is direct consequence of the mass disorder. Thus, phonon modes observed in the DOS spectra of mass-disordered CNTs could be thought of as weakly coupled harmonic oscillators. However, the monotonic variation in DOS spectra shifting is not directly reproduced for TC predictions, as discussed later in this chapter. It is important to remember that while broadening of modes is illustrated for high-frequency domain, it is the delocalized phonons in the low-frequency region that exert major contribution to heat conduction. A small fraction of the mass disorder can strongly perturb the mechanics of lattice vibrations in materials impeding thermal transport.

10.3 Nonequilibrium Molecular Dynamics for Property Prediction

Prediction of thermal conductivities of nanomaterials is typically achieved by employing MD simulations over sufficiently long computational times [8]. Although other tools such as Boltzmann transport equation exist for prediction of thermal properties in the mesoscale regime, MD holds the advantage of being deterministic and hence able to incorporate lattice heterogeneities in the predictions. Again, three different approaches have been adopted in MD for TC computation [9]. These include the use of (1) equilibrium simulations over extremely long times to determine thermal properties from heat flux autocorrelation function using the Green–Kubo approach and the linear response theory, (2) nonequilibrium MD simulations

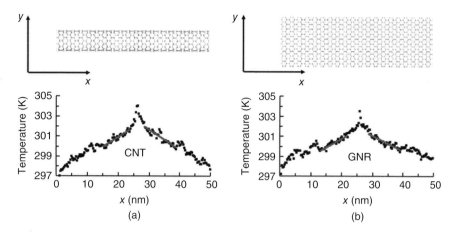

Figure 10.3 The steady-state temperature distribution along the direction of heat transfer obtained from the MD simulations is presented for (a) CNT and (b) GNR, each of length $L = 50$ nm. Reproduced by permission from Elsevier [12]

where a direct method of heating and cooling boundaries of the nanomaterial by thermostats or heat sources/sinks and consequently measuring the TC with Fourier's law using the established temperature gradient, and (3) a reverse nonequilibrium MD (RNEMD) technique where exchange of kinetic energies across the material domain establishes a thermal gradient that results in heat conduction through the material [10]. While each method has its own pros and cons, the RNEMD has found widespread interest over the recent years due to its simplicity in the application and accuracy in the predictions with reasonable demands on computing power.

In RNEMD, the total energy and momentum of the system are held constant. The kinetic energies of atoms across specified subdomains of the material are periodically exchanged to establish a heat flux across the simulated domain. The material tries to overcome this imposed energy flux by conducting heat from the hotter to the cooler regions, and a steady-state temperature gradient is realized when the rate of kinetic energy exchange is equal to the rate of heat transfer. The energy exchange is implemented along that direction where TC is to be determined. The crucial aspect of this method is that it relies on the use of Fourier's law for the calculation, and hence generating a linear profile in the temperature gradient is vital. Care should be taken to not exchange energies too frequently, which may cause deviations from the linear temperature profile, especially away from the heated and cooled subdomains. Another important concern is the definition of area in the TC computation. Most studies in the literature consider cross section of the material along the direction of heat transfer for choosing the appropriate area [7, 11]. Schematic representations of the technique for carbon nanomaterials, nanotube, and graphene nanoribbon are shown in Figure 10.3(a) and (b) [12].

Mass-disordered nanostructures have lower TC than their pure forms. Atomic impurities, for example isotopes, create scattering sites for the phonons and for the formation of long-range modes that have the dominant role in heat conduction. The greater the scattering, the lower is the TC. Thus, for a binary system, TC is minimum for a molar fraction of 0.5 for each of the species in the material composition. However, reduction of TC by isotopic impurity is

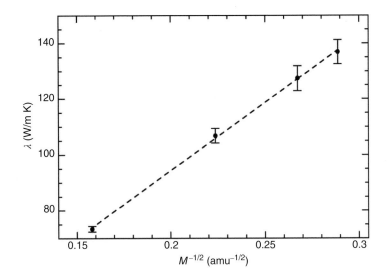

Figure 10.4   Thermal conductivity (in W/m K) of the pure CNTs with respect to the inverse of the square root of the mass of the atom (atomic mass unit) constituting the structure. The dotted line shows a fit to the data that result in a linear profile. Reproduced by permission from American Institute of Physics [13]

apparently invariant to the dimensionality of the nanostructure, as seen from a recent study comparing transport mechanisms in nanotube and graphene [12].

TC predictions of single isotope systems of the same structure reveal another interesting feature. Leroy et al. [13] showed that for CNTs composed solely of carbon atoms with atomic masses of 12, 14, 20, and 40 (the latter two being pseudoatoms chosen as case studies), the TC decreases with increasing mass. As presented in Figure 10.4, a linear correlation between TC and inverse of the square root of atomic mass supports the conclusion that magnitude of mass directly influences the characteristic phonon frequencies, which are representative of the energy-carrying capacity of materials where free electrons are nonexistent.

10.4   Quantum Mechanical Calculations for Phonon Dispersion Features

Lattice vibrations or phonons have both a particle and a wave representation. While classical simulations and velocities of the atoms provide description of the characteristic modes contributing to thermal transport, wave-vector-dependent vibrational features offer an improved understanding of the TC variations in mass-disordered nanomaterials. The phonon dispersion curves that correlate the phonon frequencies or wave numbers with the momentum vector (q) provide the bandwidths that demonstrate relative contributions of different modes [14, 15]. One has to resort to quantum calculations to accurately determine the phonon dispersion curves. Typically, density functional theory (DFT) or local density approximation (LDA) methods are considered suitable for such computations. Leroy et al. applied the LDA technique for explaining the phonon dispersion behavior of nanotubes composed of different atomic masses [13]. A few important items should be borne in mind when employing quantum mechanical calculations. Use of an appropriate exchange correlation potential is key to generating quantitatively correct results. Also, the basis set to describe the role of valence electrons and potential energy surface should be large enough to incorporate effects of polarization, especially in systems with partial atomic charges. While these computations are extremely resource demanding, using small repetitive periodic simulation cells is able to generate predictions for reasonably sized material structures.

Quantitatively, the TC in CNTs can be predicted from

$$\lambda = \sum_i \sum_q C_i(q) v_i^2(q) \tau_i(q) \tag{10.3}$$

where $C_i(q)$ denotes the specific heat per unit volume, $v_i(q)$ is the group velocity, and $\tau_i(q)$ the life time of the ith phonon with momentum vector q, respectively [6]. The summation in Eq. 10.3 runs over each of the phonon mode i and for each mode over the momentum vector $|q|$ from 0 to π/a with a denoting the magnitude of the translational vector. In the classical limit of our MD simulations, each phonon specific heat capacity is k_B/V, with k_B the Boltzmann constant and V the volume of the system. Moreover, each phonon has a mean free path $\Lambda_{i,}$, which can be obtained from $\Lambda_i = v_i \tau_i$. Since for pure systems devoid of mass defects and of sub-100 nm dimensions, thermal transport occurs ballistically and there is negligible diffusive effects [16, 17], in which phonon mean free paths can be assumed to be independent of the atomic masses. Thus, variation in TC can be attributed solely to the phonon group velocities.

The LDA-based prediction of phonon DOS of ^{12}C nanotube presented in Figure 10.5(a) shows that the transverse (TA) and longitudinal (LA) acoustic branches exist with group

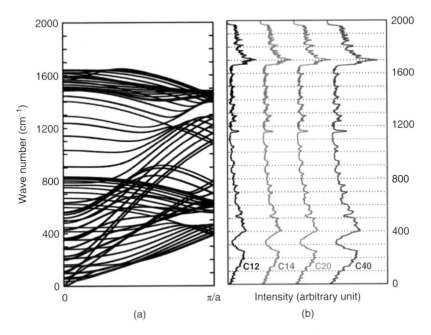

(a)   (b)

Figure 10.5 (a) The one-dimensional phonon dispersion behavior for a ^{12}C nanotube as determined by LDA calculations. (b) The classical DOS of pure ^{X}C nanotube, where $X = 12, 14, 20,$ and 40, with wave numbers rescaled by $(12/X)^{0.5}$. Reproduced by permission from American Institute of Physics [13]

velocities of 9.4 and 26.6 km/s, respectively. Group velocities are determined by finite difference analysis of the phonon dispersion curves at the gamma point. As mentioned earlier, this spectrum approximates the phonon energy when averaged over the momentum space. The DOS maxima beyond $1600 \, cm^{-1}$ correspond to the optical modes that appear with weak dispersion between 1200 and $1700 \, cm^{-1}$. In the low wave number region of the DOS, the intensity minima around $300 \, cm^{-1}$ can be found in the region around $500 \, cm^{-1}$ in the phonon-band structure.

A combined MD–band structure approach study of CNTs [18] has highlighted the restrictions imposed by potential functions on the reproduction of high-energy phonon modes. It is not wrong to think that different force fields can result in strikingly different vibrational spectra and characteristic frequencies. Nevertheless, low-frequency delocalized acoustic modes with high group velocities are an important aspect of structures with notable thermal conductivities. Figure 10.5(b) shows the DOS of the ^{14}C, ^{20}C, and ^{40}C nanotubes with wave number axis normalized with $(12/X)^{0.5}$. The vibrational spectra illustrate similar trends, as do the phonon dispersion curves. Thus, one can safely summarize that group velocities of phonons in mass-disordered nanomaterials are proportional to $(12/X)^{0.5}$. Extending the scaling feature to Eq. 10.3 results in

$$\frac{\lambda_{12C}}{\lambda_{XC}} = \sqrt{12/X} \qquad (10.4)$$

that correlates the TC of a carbon nanomaterial composed of atoms with mass X a.m.u. to that of a corresponding nanostructure made of pure ^{12}C atoms. Thus, interpretation of TC predictions from MD with observations from dispersion curves provides a consistent and exhaustive analysis of the mechanics of heat transfer in mass-disordered materials.

10.5   Mean-Field Approximation Model for Binary Mixtures

Effect of mass disorder on thermal transport of nanomaterials reflects the features of the Klemens' theory proposed nearly half a century ago [19]. Both experiments and simulated results suggest substantial reduction in heat transfer due to isotope substitution with the greatest decrease occurring for small fractions of mass disorder. Additionally, the material composed of the lighter atoms conducts energy better than that made of heavier isotopes. The lowest TC is observed for binary mixtures containing equimolar amounts of the different masses. As observed earlier in Sections 1.2 and 1.4, scaling the vibrational states by the average mass illustrates that only the difference in mass and their relative percentages are important to understand variation in thermal transport. Hence, in alignment with Klemens' theory, a mean-field approximation-based model for mass distribution should replicate the effects of isotopic disorder in nanomaterials. It is important to note that such an analysis is for systems where lattice vibrations dominate the mechanics of heat transfer and contribution of free electrons is negligible.

The heat rate q_i through a material containing a binary mixture of two different bonded atomic masses M_a and M_b if assumed to be proportional to the characteristic vibrational frequency w_i of a mass(M_a)–spring–mass(M_b) oscillator, then the TC of the mass-disordered structure relative to the pure material can be defined as

$$k_r(x) = \frac{q_x}{q_0} = \frac{w_x}{w_0} \qquad (10.5)$$

where $i = x, 0$, respectively, denote the cases with the x fraction of heavier isotopes A and the case where the pure material is considered. Thus, $M_a = xM_A$ and $M_b = (1-x)M_B$. A detailed mathematical treatment as presented in [7] shows that

$$k_r(x) = \text{fn.}(x, M_A, M_B) = \frac{\sqrt{n(1+n)\{(1-x+n)M_A + (x+n)M_B\}}}{\sqrt{(x+n)(1-x+n))\{(1+n)M_A + nM_B\}}} \tag{10.6}$$

The constraints imposed on k_r include it being 1 for the pure material, while at 50% isotope impurity one can assume k_r to be minimum under a homogeneous mass distribution. The results of such a predictive model are compared against earlier simulation results [20–24] in Figure 10.6 and show remarkable agreement validating the applicability of Klemens' theory for mass-disordered structures. The differences in the quantitative predictions of the simulations can be attributed to the lack of sampling structures in the calculations. The similarity in the nonmonotonic behavior of the model and simulations enables one to assert the significant role of long-range modes in comparison to the localized phonons for TC.

A closer examination of Figure 10.6 shows that for carbon nanomaterials, addition of heavier isotopes successively reduces the TC until about mass fraction $x = 0.5$ when the heat transfer rate is minimum. Further inclusion of isotopes essentially reduces the mass disorder and improves thermal transport, and it continues to increase the TC till the material is completely made of the heavier mass. Important observation to note is that the overall heat transfer rate is more for the system containing lighter isotopes than that composed solely of the heavier ones. This can be thought in terms of the vibrational frequencies; the lighter atoms have higher characteristic frequencies than the heavier ones, and hence better energy transfer capacity. The mean-field approximation approach is not limited to carbon nanomaterials. As shown in

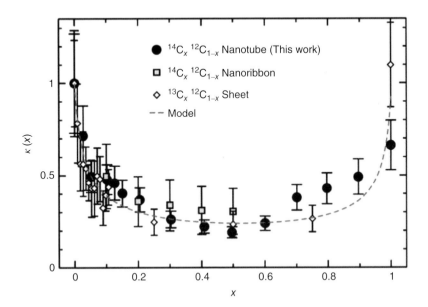

Figure 10.6 Comparison between RNEMD calculations and other simulations of carbon nanostructures (shown as data points), and the mean-field approximation model represented as a dashed curve shows good agreement. Reproduced by permission from Royal Society of Chemistry [7]

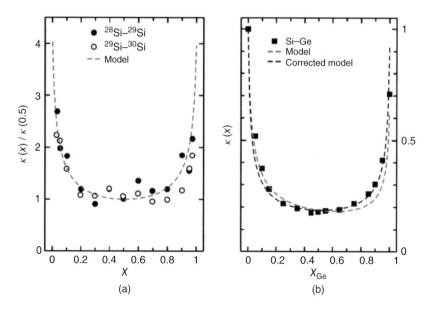

Figure 10.7 (a) Comparison between the model and the MD simulations of silicon crystals with ^{28}Si and ^{29}Si substituted with ^{29}Si and ^{30}Si atoms, respectively, shows good agreement. x denotes the fraction of the lightest isotope, that is, ^{28}Si or ^{29}Si. The ratio $k_r(x)/k_r(0.5)$ is presented instead of $k_r(x)$ because data for the pure components are unavailable. The dashed line corresponds to mean-field approximation model described in Eq. 10.6. (b) Comparison between MD simulations in silicon–germanium nanowires and the prediction of Eq. 10.6 shows differences that are eliminated by correcting effects of multielemental composition. Reproduced by permission from Royal Society of Chemistry [7]

Figure 10.7(a) and (b), the model is able to replicate the findings from molecular simulations of TC of silicon containing various isotopes [25].

The lack of experiments, especially for isolated freely suspended nanomaterials, containing isotopic defects prevents a direct comparison among simulations, model, and measurements. The mean-field approximation model parameterization has an inherent assumption in its derivation that by simply modifying the stretching modes of the bonded masses one can evaluate the contribution of mass disorder. This does not hold for structures with large radius of curvature, such as armchair nanotubes where the chiral indices are, for example (5,5), where coupling of the radial breathing and bending modes needs to be accounted for in the description of heat transfer. The model can, however, be extended to systems containing masses of two different elements, as shown in Figure 10.7, for instance nanowires containing both silicon and germanium atoms. While some differences exist in the electronic structure of the atoms, those selected from the same group of the periodic table can be approximated under the mean-field theory. It must be noted that unlike isotopes of the same species, binary mixture of dissimilar elements strongly perturb the potential energy landscape of the material, which is also not accounted for in such a simplified model. Differences between the model and predictions exist because the constraint at $x = 0.5$ of minimum TC would be violated in case of such bi-elemental mixtures. Accurate determination of mass fraction x corresponding to the minimum TC yields modified model, which is able to reproduce results of simulations.

Does this mean that we no longer require detailed experimental measurements or computationally intensive simulations? Can such a predictive model solve challenges in mechanics of heat transfer of impure material systems? Well, the answer is no, and there are several reasons for that. In the first place, the model is insensitive to dimensional effects; it has been shown repeatedly that the TC varies differently in the ballistic (sub-mean free path system dimension) and diffusive regimes (system sizes longer than mean free path of energy carriers). The mean-field model fails to capture those differences since the variations are not always linear especially for complex geometries. Second, long-range order in the distribution of mass defects is not accounted for in the model. If several different structures are considered, each with same fraction of mass disorder but different arrangement of atoms (both ordered and random), the model predictions will be similar because it depends only on the fraction and not distribution. This is a strong limitation, especially considering the design of nanoscale materials. Third, the model has a simple representation for binary mixtures; it is not difficult to anticipate that for materials with multielemental composition, the model variation will be more complicated and may even depend on relative material phases of the different constituents. Nonetheless, the model is a quick and easy prediction tool for TC of mass-disordered structures that is qualitatively able to reproduce the role of different masses or atoms on the overall mechanics of thermal transport.

10.6   Materials Informatics for Design of Mass-Disordered Structures

Eigenvalue decomposition methods have been useful in the analyses of power spectra to interlink electronic states to the atomic structure [26–30]. One of the most popular tools employed in data interpretation and pattern extraction is principal component analysis (PCA), which can correlate characteristics of the DOS spectra, vibrational properties to isotope percentages, and thermal properties to derive information through machine learning [27]. For instance, PCA is applied to the DOS profile of isotope-substituted graphene sheets where the spectrum for each of the different mass-disorder cases is decomposed as 104 discrete points. An overall of 105 data values are included in the data analyses. Different cases of carbon isotope substitution are considered as a sample condition, while each characteristic wave number is deemed as a unique descriptor for the mass disorder. The intensity of the discretized DOS is the response to the specific descriptor. Using an eigenvalue decomposition methodology, the DOS spectra are broken down into (1) primary spectral patterns consisting of the DOS and ranked according to their relative importance in contributing to changes in the DOS with mass disorder, and (2) weight functions for each of the patterns that enable reproduction of the original DOS. Subsequently, a predictive model correlating the weights and TC is constructed efficiently through decomposition of the DOS into a parameter matrix. The predictions of TC from the informatics when compared against simulation results and experimental measurements show reasonable agreement. With data science techniques, hidden patterns in DOS profiles are revealed and mathematical relationship between structure and property is developed. The outcomes of the PCA and high-dimensional regression used to minimize the root-mean-square error on the computed data are shown in Figure 10.8.

The agreement in the property predictions facilitates decoupling the roles of short- and long-ranged modes on the TC. The eigenvalue spectrum in Figure 10.9(a)–(c) illustrates the contributions of localized and delocalized phonons to the original DOS. The shape of peak

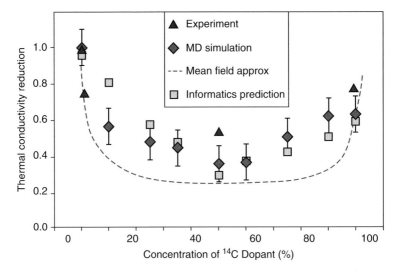

Figure 10.8   There is a good agreement between informatics predictions with MD simulation, and both capture the same minima in conductivity with 50% isotope concentration. The predictions also capture the same trends as previous computational and experimental results, with differences existing due to different atomic masses of the carbon isotopes. Using informatics, we are able to predict the thermal conductivity of isotope-substituted graphene of all possible binary compositions without requiring additional MD calculations or measurements. Reproduced by permission from American Institute of Physics [27]

intensities, such as those between the 1200 and 2000 cm^{-1} range, facilitates the derivation of mathematical relations between vibrational features and nanomaterial properties. TC is estimated as

$$TC = -64.10\,x\,PC1 + 374.28\,x\,PC2 + 0.55 \qquad (10.7)$$

Delocalized modes are known to exert the dominant effect on heat transfer through nanomaterials and, likewise, in graphene. Information science, however, reveals that phonons with characteristic frequencies above 1200 cm^{-1} are the ones that impart the stronger contribution when isotope or mass-disorder effects are considered. PC2 loading spectra with the larger coefficient, presented in Eq. 10.7, are vital for the TC of mass-disordered graphene sheets. As shown in Figure 10.9, the highlighted sections reflect the significant features of the principal components and hence the thermal transport. These frequencies lie between 1200 and 1800 cm^{-1}, which implies TC is influenced by localized modes, contrary to the long-ranged modes being important for heat transfer in nanomaterials.

To put the results of PCA into a broader perspective, energy transport in 2D carbon nanomaterial, that is, graphene, can be realized as a combined effect of both high-frequency in-plane stretching modes that are localized and the long-range flexural out-of-plane phonons that although have a low frequency can conduct heat to larger distances without scattering. The role of mass disorder in the form of ^{14}C isotopes, which have similar electronic structure to their naturally abundant ^{12}C counterpart but heavier mass, can be understood in the stronger phonon–phonon and phonon–impurity scattering of the localized modes. This leads to a notable translation of the characteristic vibrational frequencies to lower values. This shift results in an effective impedance to the low-frequency delocalized modes that ultimately reduce TC.

Informatics creates a new avenue for engineering transport and mechanics through nanostructures by consciously designing the mass disorder. On the one hand, such a data-driven method offers reasonably accurate predictions for thermal properties without resorting to time-intensive computations and fills the structure–property landscape with results for cases that do not have prior experimental or simulation-based analyses. Experiments on graphene have shown a 35% reduction in TC for 50% ^{13}C isotope-substituted graphene, while

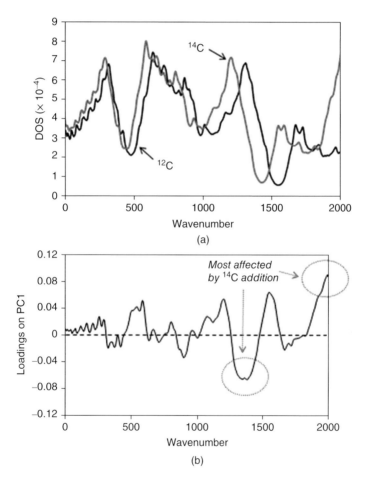

Figure 10.9 The modes of the vibrational spectra most affected by changing isotope percentages and resulting in thermal conductivity variation. In isotope-substituted materials, high-frequency modes are strongly influenced to induce a reduction in thermal conductivity. (a) The localized modes are defined as those with wavenumber greater than $800\,cm^{-1}$, while those below $800\,cm^{-1}$ are delocalized. In the loadings spectra, the circled peaks (highest magnitude) are those that are (shown in (b)) most impacted by changing composition and (shown in (c)) most significant in determining thermal conductivity. Without employing informatics, we could not have otherwise picked up on the relationship between localized modes and thermal conductivity. Reproduced by permission from American Institute of Physics [27]

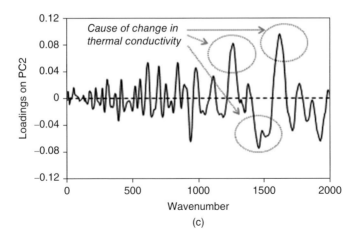

Figure 10.9 (*Continued*)

computations have predicted 33% decrease in heat-conducting capacity for 1% substitution of graphene atoms by ^{14}C. Machine learning predictions, unaware of the physics of heat transfer, are able to provide similar estimates. On the other hand, the relative importance of short-ranged phonons with respect to the delocalized modes in mass-disordered nanostructures is revealed through PCA. Thus, information science not only expedites the pace of property prediction but also provides physical interpretation of the mechanics of thermal transport.

These efforts are timely and useful to explore a large search space of different materials and construct elemental designer combinations to discover materials with targeted properties. For instance, given a targeted value of material property, say TC, the above-described statistical learning tool is able to recommend optimal concentration of ^{14}C isotopes to facilitate innovation of hypothesis-driven materials.

10.7   Future Directions in Mass-Disordered Nanomaterials

Effects of defects in materials are profound not only for mechanics of thermal transport but also for other structural and functional properties. At the nanoscale, the interplay of physics and chemistry of materials driven by the fundamental energy carriers gives rise to novel phenomena, and interesting features that if thoroughly understood and suitably engineered could lead to ground-breaking effects in materials design and discovery. The mass-disordered structures can be assimilated as the building blocks of the next-generation material systems and can create devices that are tailor-made to achieve targeted performance metrics.

Computations play a significant role in understanding thermal transport phenomena in nanomaterials, and more so with the innumerable possibilities that can exist for structures containing mass impurities. Numerical experiments provide a robust platform to probe potential structures that have not yet been synthesized in the laboratory, or investigate processes at the scales where experimental efforts are limited due to miniaturization. While computing costs are noteworthy, they are well short of the expenditures associated with sophisticated characterization and measurement tools, and hence provide an economic route to examine millions of possibilities to narrow down the selection process for experiments. However, as seen in the preceding sections, given the vast spatial dimension from the molecular structure to the device geometry and the related differences in materials phenomena across the nano–micro–meso–macro regime, no single computational approach is deemed suitable to tackle the wide array of challenges in transport mechanics. Several challenges can be immediately foreseen in the area of computational mechanics of transport in nanomaterials: (1) hierarchical modeling: how can information derived from quantum and molecular simulations of mass-disordered structures be transferred to device scale computations; (2) multiscale modeling: can dynamics of transport in the nanomaterial with defects be described concurrently with the macroscale changes on the material system; (3) optimization: how large a sample space is required to identify the best defect-engineered candidate materials that possess the desired material properties toward diminishing or enhancing heat transfer; (4) validation and integration: how can computational results be compared against experiments, especially since most of the simulations are at scales much smaller than the structures synthesized for measurements; how can the experimental and computational approaches be holistically integrated to provide novel paradigms for reverse engineering in materials discovery; (5) model development: can predictive frameworks be constructed for thermal conductivities of materials with varied distribution of mass impurities, and over a range of thermophysical parameters; (6) defect characteristics: can the contribution of a variety of elemental impurities such as isotopes, vacancies, dopants be described by a uniform model or if different underlying principles guide the role of different point defects, and so on. While role of impurities on heat transfer has been investigated by a range of computational methodologies over the past decade, and experimentalists are encouraged to fabricate defect-engineered structures, the knowledge base is still only at a surface level and deeper insights need to sought to completely explore and exploit the characteristics of mass-disordered nanomaterials. With the emergence of BIGDATA tools and information science based on data-driven learning, it is easy to envision that the mechanics of heat transfer will benefit from the computational advances.

References

[1] Chen, S., Wu, Q., Mishra, C. et al. (2012) Thermal conductivity of isotopically modified graphene. *Nature Materials*, **11**, 203–207.
[2] Hao, F., Fang, D. and Xu, Z. (2011) Mechanical and thermal transport properties of graphene with defects. *Applied Physics Letters*, **99**, 041901.
[3] Yang, N., Zhang, G. and Li, B. (2008) Ultra low thermal conductivity of isotope-doped silicon nanowires. *Nano Letters*, **8**, 276–280.
[4] Plimpton, S. (1995) Fast parallel algorithms for short-range molecular dynamics. *Journal of Computational Physics*, **117**, 1–19.
[5] Tersoff, J. (1989) Modeling solid-state chemistry: Interatomic potentials for multicomponent systems. *Physical Review B*, **39**, 5566–5568.
[6] Lindsay, L. and Broido, D.A. (2010) Optimized Tersoff and Brenner empirical potential parameters for lattice dynamics and phonon thermal transport in carbon nanotubes and graphene. *Physical Review B*, **81**, 205441.
[7] Balasubramanian, G., Puri, I.K., Böhm, M.C. and Leroy, F. (2011) Thermal conductivity reduction through isotope substitution in nanomaterials: predictions from an analytical classical model and nonequilibrium molecular dynamics simulations. *Nanoscale*, **3**, 3714–20.
[8] Poulikakos, D., Arcidiacono, S. and Maruyama, S. (2003) Molecular dynamics simulation in nanoscale heat transfer: A review. *Microscale Thermophysical Engineering*, **7** (3), 181–206.
[9] Schelling, P.K., Phillpot, S.R. and Keblinski, P. (2002) Comparison of atomic-level simulation methods for computing thermal conductivity. *Physical Review B*, **65** (14), 144306.

[10] Müller-Plathe, F. (1997) A simple nonequilibrium molecular dynamics method for calculating the thermal conductivity. *Journal of Chemical Physics*, **106**, 6082.

[11] Alaghemandi, M., Leroy, F., Algaer, E. et al. (2010) Thermal rectification in mass-graded nanotubes: A model approach in the framework of reverse non-equilibrium molecular dynamics simulations. *Nanotechnology*, **21**, 75704.

[12] Ray, U. and Balasubramanian, G. (2014) Reduced thermal conductivity of isotope substituted carbon nanomaterials: Nanotube versus graphene nanoribbon. *Chemical Physics Letters*, **599**, 154–158.

[13] Leroy, F., Schulte, J., Balasubramanian, G. and Böhm, M.C. (2014) Influence of longitudinal isotope substitution on the thermal conductivity of carbon nanotubes: Results of nonequilibrium molecular dynamics and local density functional calculations. *Journal of Chemical Physics*, **140**, 144704.

[14] Tan, Z.W., Wang, J.S. and Gan, C.K. (2011) First-principles study of heat transport properties of graphene nanoribbons. *Nano Letters*, **11**, 214–219.

[15] Donadio, D. and Galli, G. (2007) Thermal conductivity of isolated and interacting carbon nanotubes: Comparing results from molecular dynamics and the Boltzmann transport equation. *Physical Review Letters*, **99**, 255502.

[16] Chang, C.W., Okawa, D., Garcia, H. et al. (2008) Breakdown of Fourier's law in nanotube thermal conductors. *Physical Review Letters*, **101**, 075903.

[17] Mingo, N. and Broido, D.A. (2005) Length dependence of carbon nanotube thermal conductivity and the "Problem of Long Waves". *Nano Letters*, **5** (7), 1221–1225.

[18] Alaghemandi, M., Schulte, J., Leroy, F. and Böhm, M.C. (2011) Correlation between thermal conductivity and bond length alternation in carbon nanotubes: A combined reverse nonequilibrium molecular dynamics—Crystal orbital analysis. *Journal of Computational Chemistry*, **32**, 121–133.

[19] Klemens, P.G. (1955) The scattering of low-frequency lattice waves by static imperfections. *Proceeding of the Physical Society, London, Section A*, **68**, 1113–1128.

[20] Zhang, G. and Li, B. (2010) Impacts of doping on thermal and thermoelectric properties of nanomaterials. *Nanoscale*, **2**, 1058–1068.

[21] Hu, M., Giapis, K.P., Goicochea, J.V. et al. (2011) Significant reduction of thermal conductivity in si/ge core-shell nanowires. *Nano Letters*, **11** (2), 618–623.

[22] Maruyama, S., Igarashi, Y., Taniguchi, Y. and Shiomi, J. (2006) Anisotropic heat transfer of single-walled carbon nanotubes. *Journal of Thermal Science and Technology*, **1** (2), 138–148.

[23] Jiang, J.-W., Lan, J., Wang, J.-S. and Li, B. (2010) Isotopic effects on the thermal conductivity of graphene nanoribbons: Localization mechanism. *Journal of Applied Physics*, **107**, 054314.

[24] Zhang, H., Lee, G., Fonseca, A.F. et al. (2010) Isotope effect on the thermal conductivity of graphene. *Journal of Nanomaterials*, **2010**, 537657.

[25] Zhang, G. and Li, B. (2005) Thermal conductivity of nanotubes revisited: Effects of chirality, isotope impurity, tube length, and temperature. *Journal of Chemical Physics*, **123** (11), 114714.

[26] Broderick, S.R. and Rajan, K. (2011) Eigenvalue decomposition of spectral features in density of states curves. *Europhysics Letters*, **95**, 57005.

[27] Broderick, S., Ray, U., Srinivasan, S. et al. (2014) An informatics based analysis of the impact of isotope substitution on phonon modes in graphene. *Applied Physics Letters*, **104**, 243110.

[28] Broderick, S.R., Aourag, H. and Rajan, K. (2011) Classification of oxide compounds through data-mining density of states spectra. *Journal of American Ceramic Society*, **94**, 2974.

[29] Broderick, S.R., Suh, C., Provine, J. et al. (2012) Application of principal component analysis to a full profile correlative analysis of FTIR spectra. *Surface and Interface Analysis*, **44**, 365.

[30] Broderick, S.R., Aourag, H. and Rajan, K. (2009) Data mining density of states spectra for crystal structure classification: An inverse problem approach. *Statistical Analysis and Data Mining*, **6**, 353.

11

Thermal Boundary Resistance Effects in Carbon Nanotube Composites

Dimitrios V. Papavassiliou[1,2], Khoa Bui[1] and Huong Nguyen[1]

[1]*School of Chemical, Biological and Materials Engineering, The University of Oklahoma, Norman, OK, USA*

[2]*Division for Chemical, Bioengineering and Environmental Transport Systems, National Science Foundation, Arlington, VA, USA*

11.1 Introduction

Nanocomposites (materials with fillers with at least one dimension less than 100 nm) offer multifunctional properties, that is, electrical and thermal conductivities in addition to enhanced mechanical strength [1–6] in combination with light weight. Carbon nanotubes (CNTs) have been suggested as fillers in composite materials due to their exceptional electrical and mechanical properties. In terms of thermal properties, incorporating CNTs into a polymer matrix is expected to increase the effective thermal conductivity (K_{eff}) of the resulting composite [7, 8]. Such materials are needed, since progress in solid-state electronics demands transformative thermal-management materials (e.g., resins with high thermal conductivity that maintain electrical insulation).

Calculations based on simple-weighted average rules, such as the Maxwell/Rayleigh equation [9, 10]

$$K_{eff} = (1 - \varphi) K_m + \varphi K_{CNT} \tag{11.1}$$

lead to the expectation that, even at low %wt concentration of CNTs in a polymer composite, the effective thermal conductivity of the resulting material can be at least one order of magnitude higher than the thermal conductivity of the matrix material, since the thermal conductivity of the CNTs is three to four orders of magnitude higher than the thermal conductivity of the matrix. In Eq. (11.1), K_m and K_{CNT} are the thermal conductivities of the matrix material and the CNT, respectively, and φ is the volume fraction of the CNTs. For example,

Advanced Computational Nanomechanics, First Edition. Edited by Nuno Silvestre.
© 2016 John Wiley & Sons, Ltd. Published 2016 by John Wiley & Sons, Ltd.

the thermal conductivity of 0.1% wt of single-walled CNTs (with thermal conductivity in the range 1750–6600 W/m K) in epoxy can be calculated from Eq. (11.1) to be between 40 and 150 W/m K, while the thermal conductivity of pure epoxy is approximately 0.198 W/m K[7] Experiments, however, indicate that K_{eff} increases on the order of two to three times the value of K_m. Such findings have been reported for composite materials and suspensions of CNTs.

The reason for this observed discrepancy is partially the difficulty of dispersing CNTs in a homogeneous pattern, but more importantly the fact that there is a resistance to the transfer of heat that is quite important in the case of nanoscale inclusions in composites. This resistance appears at the interface between two materials, and it is known as thermal boundary resistance (TBR) or Kapitza resistance. It represents a barrier to the heat flow associated with the differences in the phonon spectra and the possibly imperfect contacts at the interface, and it was discovered in 1941 by Kapitza [11] to affect heat transfer at very low temperatures at the interface of liquid helium and copper. At room temperatures, the effects of the TBR on heat transfer from one material to another are not detectable, unless the area through which heat is transferred is large relative to the volume of the heated material [12]. This is exactly the case in CNT composites, where the effective thermal conductivity of the composite is dominated by the TBR between the inclusions and the matrix material, and also by the dispersion pattern and geometry of the CNTs.

Theoretical interpretation of the TBR has been obtained by using the acoustic mismatch theory and the diffusive mismatch theory [12]. Molecular dynamics studies have also confirmed the existence of TBR at the interface of CNTs in epoxy resins or in suspension and have resulted to estimations of the value of the TBR [13, 14]. An elegant picosecond transient absorption experiment conducted for CNTs suspended in water allowed Huxtable et al. [15] to monitor the flow of heat in real time from the CNTs to the surrounding surfactants. This experiment yielded an estimate for the CNT–surfactant interfacial conductance (i.e., the inverse of the TBR) of approximately 12 MW/m^2 K. This low value supported the hypothesis that Kapitza resistances are responsible for the poor thermal performance of CNT-based nanocomposites. To model the effects of the CNT length and dispersion pattern at the mesoscale, methods such as Monte Carlo (MC) simulations have been applied by our laboratory [16–18], or methods that are based on conventional simulations such as simulations with finite elements [19, 20]. Possible strategies to diminish the effects of TBR could be (1) to increase the volume fraction of the inclusions, in order to break through the percolation limit for thermal transfer, (2) to affect the interface by either functionalizing the CNTs, or by coating them in a different material and creating a gradient of stiffness between the CNT and the matrix material so that the effective TBR becomes smaller [21].

In this work, we review briefly predictive models for the thermal behavior of CNTs in composites, and simulation efforts to investigate improvements in K_{eff} with an increase in CNT volume fraction. Emphasis is placed in the investigation of the TBR in CNTs coated with silica in order to reduce the overall TBR of the composites.

11.2 Background

Composite materials consist of two or more materials combined in such a way that each constituent remains distinguishable. Most composites have two constituents: a matrix and a reinforcement. The reinforcement is usually much stronger and stiffer than the matrix, and gives the composite its improved properties. For the case of nanocomposites, the reinforcements, also called inclusions, are CNTs that are dispersed inside common polymers. In terms of thermal conductivity, each constituent has its own, but the composite can be treated as a homogeneous material with an effective thermal conductivity, K_{eff}. Estimation of the conductivity of heterogeneous solids was first contributed by Maxwell for the case of spheres [16]. For nonspherical inclusions, the composite is anisotropic and K_{eff} becomes a tensor. When square arrays of long cylinders are parallel to the z-axis (i.e., the direction of the heat flow), Rayleigh [9] showed that the zz component of the thermal conductivity tensor is

$$\frac{K_{eff,zz}}{K_m} = 1 + \frac{(K_p - K_m)\varphi}{K_m} \tag{11.2}$$

and the other two components are

$$\frac{K_{eff,xx}}{K_m} = \frac{K_{eff,yy}}{K_m} = 1 + \frac{2\varphi}{\dfrac{K_p + K_m}{K_p - K_m} - \varphi + \dfrac{K_p - K_m}{K_p + K_m}(0.30584\varphi^4 + 0.013363\varphi^8 + \cdots)} \tag{11.3}$$

where K_p is the thermal conductivity of the inclusions. For complex nonspherical inclusions, no exact treatment is available. Intended for simple unconsolidated granular beds, the following expression has proven successful [10]:

$$\frac{K_{eff}}{K_m} = \frac{(1 - \varphi) + \eta\varphi(K_p/K_m)}{(1 - \varphi) + \eta\varphi} \tag{11.4}$$

in which

$$\eta = \frac{1}{3}\sum_{k=1}^{3}\left[1 + \left(\frac{K_p}{K_m} - 1\right)g_k\right]^{-1} \tag{11.5}$$

The factor g_k is the shape factor for the granules of the medium, where $g_1 + g_2 + g_3 = 1$ must be satisfied. For spheres, $g_1 = g_2 = g_3 = 1/3$, and Eq. (11.3) reduces to Eq. (11.2). For unconsolidated solids, $g_1 = g_2 = 1/8$ and $g_3 = 3/4$. The above formulations would be sufficient to estimate K_{eff} in nanocomposites when one did not account for Kapitza's TBR at the interface between the two different materials.

The acoustic mismatch model (AMM) is a model that was proposed to predict TBR [22]. In this model, the heat flux is carried through the interface by acoustic waves propagating by the same law as the transmission of sound from one medium to another. The AMM provides a possible explanation of the principle cause of Kapitza resistance. However, for some cases, predictions are almost one order of magnitude higher than experimental values [23]. Conserving the main idea that a major part of the heat flux through the interface is carried by acoustic waves, the diffuse mismatch model (DMM) developed by Swartz et al. [12] gives another estimation of the Kapitza resistance. A crucial assumption in the AMM model is that no scattering occurs at the interface. There were, however, experimental evidences that high-frequency phonons transferring heat got scattered at the interface and opened up new channels for heat transport [24]. Hence, this assumption was replaced with the opposite extreme in the DMM model: all phonons are assumed to be diffusely scattered at the interface [12]. There have been lots of theoretical work to improve the acoustic models for the better estimation of Kapitza resistance [23–31].

In general, this resistance appears at the interface between any materials, and its value varies differently for each system [32]. Based on acoustic theory [12], the more the differences in density and sound velocity of two materials, the greater the value of Kapitza resistance will be. This is why the Debye temperature, the temperature at which the maximum frequency of phonon is activated, is often used to qualitatively compare the Kapitza resistance. For example, at room temperature, R_{bd} of Bi/diamond was measured to be 11.7×10^{-8} W^{-1}m^2 K [33, 34], which is among the highest resistances for bulk materials. Bismuth is a metal containing free electrons as primary heat carriers, while carbon atoms in diamond do not have many free electrons. As a result, the energy transport by means of electrons between these materials is suppressed. Furthermore, bismuth, which has low Debye temperature, has many phonons at low frequencies. Diamond, on the other hand, has very high Debye temperature with high-frequency phonons. The significant difference in Debye temperature means that phonons do not efficiently couple across the Bi/diamond interface. In nanocomposites, the Debye temperature of common polymers is about 400 K or lower, while it is higher than 2000 K for CNTs and graphene sheets (GSs). This also explains why the Kapitza resistance is a bottleneck for the heat transfer in nanocomposites.

Since Maxwell's and Rayleigh's models originally could not take into account the Kapitza resistance, Hasselman and Johnson [35] extended these theories and derived an effective medium approximation (EMA) for calculating K_{eff} that includes interfacial effects and particle size for simple inclusions. Nan et al. [36, 37] generalized the EMA for various particle geometries and orientations. Considering the CNTs as straight rods that are randomly dispersed in a matrix, results from this approach agree well with experiments for CNT suspension in oil [37, 38]. The calculation also shows that, when TBR $\geq 10^{-8}$ W^{-1}m^2 K, single-walled nanotubes with high thermal conductivity, that is, 6000 W/Mk [39], do not induce larger thermal conductivity enhancement than multiwalled nanotubes with slightly lower thermal conductivity, that is, 3000 W/mK [40]. Xue et al. [41] employed average polarization theory to develop analytical formulas for calculating K_{eff} of CNT-based composites. The theoretical work conducted by Xue et al. allows one to quantify the dependence of K_{eff} on TBR, length, diameter, and concentration of CNTs. However, this approach does not explicitly take into account the realistic configurations or the placement of CNTs inside the polymer matrix. Instead, the CNTs are considered as perfectly straight rods that are either parallel or perpendicular to the heat flux.

Heat transfer in nanocomposites has also been simulated numerically with conventional numerical methods [19, 20, 42–46]. A solution of the steady-state heat conduction equation with finite element methods (FEM) taking into account the Kapitza resistance was first performed by Davis et al. [43], who considered a spherical inclusion embedded inside a matrix. Results from this approach for spherical inclusions were in good agreement with experimental data and EMA. Kumar et al. employed a similar approach to estimate K_{eff} of thin-film CNT-based composites [19, 20]. Based on the Fourier conduction equation in the CNTs and the substrate, the temperature distribution was calculated by a finite volume discretization (FVD) scheme and K_{eff} was calculated as

$$K_{eff} = \frac{\sum_{tubes} K_{CNT} A \frac{dT_i}{ds}\big|_{x=0} + \sum_{subtrates} K_s \Delta y \Delta z \frac{dT_s}{ds}\big|_{x=0}}{Ht \left(\dfrac{T_{drain} - T_{source}}{L_C} \right)} \qquad (11.6)$$

The area of each CNT is given by A. This formula was derived by assuming each CNT as a straight cylinder in a thin box (2D). The first term in the numerator is the heat flow through the tubes in the lateral direction, while the second term represents the heat flow in the substrate. T_i and T_s are the temperature of tube segment and of the substrate, respectively. The computational domain was represented by height (H), width (L_C), and thickness (t). K_t and K_s are the thermal conductivities of CNTs and substrate, respectively. Nishimura et al. [44] employed boundary integral equation (BIE) method to numerically solve the heat conduction problems in the 2D finite medium. Values of K_{eff} computed by this approach at low volume fraction of CNTs were reported to match experimental data, although the two deviated at higher volume fractions.

Note that Fourier's law of heat conduction assumes diffusive transport, while heat transfer in nanocomposites takes place in diffusive-ballistic regime due to the nano-sized inclusions [47–49]. Furthermore, measuring the electrical conductivity of CNT-based nanocomposites usually reveals a significant jump due to interconnecting network of tubes at very low wt% of CNTs (0.05–4.0 wt% [50, 51]), while there is no analogous observation for the case of thermal conductivity [8, 13].

In summary, it has been more than 10 years since early experimental attempts [38, 52, 53] to produce CNT-based composites to improve the thermal conductivity, and the knowledge of heat transfer in nanocomposites has evolved significantly to explain/predict the anomalous behavior of K_{eff}. Current theoretical models in the literature allow one to quantify the dependence of K_{eff} on TBR, length, diameter, and concentration of CNTs. Nevertheless, these models are limited to simple cases and agree with experimental data at low volume fraction of CNTs only. The primary question of how to improve K_{eff} of nanocomposites requires further multiscale investigations of not only the Kapitza resistance at the interface but also the realistic configurations of nanoinclusions inside the polymer matrix. In this paper, recent advances in atomistic simulations and mesoscale simulations addressing that question are reviewed systematically.

11.3   Techniques to Enhance the Thermal Conductivity of CNT Nanocomposites

Because the Kapitza resistance at the CNT–polymer interface is the bottleneck for the heat transfer in nanocomposites, the trivial question is how much can one improve K_{eff} by suppressing TBR. It has been reported that the overlap of the thermal vibrational spectra between two materials is the key point to control the Kapitza resistance at the interface [54–57]. Varshney et al. [58] studied the effect of different organic linkers – connecting the two nanotubes – on the thermal interface conductance using nonequilibrium molecular dynamics (MD) simulations. The thermal conductance, inverse of TBR, was found to increase with the number of linking functionality but show an opposite trend with respect to the linker's length, that is, the longer the linker is, the lower the conductance. Roy et al. [59] reported that the effect of surface functionalization toward establishing covalent bonding between the nanoconstituent surface the matrix (such as polymers) is extremely important in enhancing the interface thermal conductance.

Balandin et al. [60] reported extremely high values of the thermal conductivity of single-layer graphene sheet (GS) that outperform CNTs in terms of heat conduction. This result gives rise to the expectation that GS composites might be able to fulfill the promise of thermally conductive carbon-based nanocomposites. Konatham and Striolo [55, 56] calculated TBR at the GS–octane interface and found that it is three times smaller than

SWNT–octane interface. Furthermore, when the alkane chains are covalently bonded at the edges of these GSs, the value of TBR in that case was reported to be 10 times smaller than that at a single-walled CNT–octane interface. While MD simulations have answered the question of whether GS has less Kapitza resistance than CNT or not [56], the question of how much the macroscopic thermal properties of resulting nanocomposites are affected by the Kapitza resistance was addressed by Bui et al. [61] with mesoscopic, off-lattice Monte Carlo simulations. They reported that controlling the orientation of the GS in the composite can lead to the manufacturing of composites that have quite different thermal behavior in different directions. For example, the material can be a thermal insulator in one direction and have a thermal conductivity 350 times larger than in a direction perpendicular to that [61, 62].

Another way to improve the thermal properties of heat-conducting CNT composites that one could suggest is the "densification" of the composite [63, 64] by increasing the volume fraction of the CNTs [8, 17], as an alternative to the modification of the properties of the matrix–CNT interface described earlier. However, increasing the thermal transport of nanocomposites by increasing CNT loading has been found ineffective, because of the tendency of nanotubes to form bundles, and because the CNT–CNT TBR is often higher than that between the CNT and the surrounding matrix [65–67]. This is a counter-intuitive finding, but it appears that it is true. Yang et al. [68] concluded that the CNT–CNT TBR is strongly dependent on contact area. They reported a value of CNT–CNT TBR equal to 1.2×10^{-8} Km^2/W when two CNTs are aligned and partially overlap with each other. On the other hand, this value is 10 times smaller, 1.2×10^{-9} Km^2/W, when two CNTs are crossing each other. Functional groups have been found to significantly reduce the TBR between CNTs and between graphene sheets, both in computations [55, 56, 58, 69, 70], and in experiments [71].

11.4 Dual-Walled CNTs and Composites with CNTs Encapsulated in Silica

Based on the discussion of the previous section, where evidence has been presented about the increased TBR at the CNT–CNT interface, we investigate here the value of the TBR within a multiwalled CNT, and more specifically between the walls of a dual-walled carbon nanotube (DWNT). In prior work out of our laboratory, Bui et al. [61] investigated TBR between the two concentric walls of DWNTs via nonequilibrium MD simulations. They examined the TBR as a function of the chirality of the CNT walls, the length of the DWNT, and examined the transfer of heat from the inner nanotube to the outer and vice versa. They found that the values of the Kapitza resistance obtained in all cases were always larger than those typically observed at the CNT–polymer interface (e.g., 2.61×10^{-8} $W^{-1}m^2$ K for CNT–epoxy [9], 0.95×10^{-8} $W^{-1}m^2$ K for CNT–PMMA [72], 2.21×10^{-8} $W^{-1}m^2$ K for CNT–polystyrene [17]). It was also found that high-frequency optical phonons in the range 60–65 THz govern heat transfer in the system considered, unlike soft phonons of frequency less than 5 THz that were found to govern heat transfer between CNTs and octane [14, 56, 67]. Another interesting conclusion of that work was that heat transfer occurs much more easily from a hot outer CNT to a cold inner one than vice versa, yielding thermal rectification factors in the radial direction in the range 103%–114 % for longitudinal symmetric, defect-free DWNTs of length exceeding 10 nm.. We defined the thermal rectification factor, R_{REC}, as

$$R_{REC} = \frac{K_+ - K_-}{K_-} \times 100\% \qquad (11.7)$$

In Eq. (11.7), K_+ is the thermal conductance (i.e., the inverse of the TBR) as heat transfers from the outer to the inner SWNT and K_- is thermal conductance in the opposite direction. Konatham et al. [73] employed MD simulations to study TBR between GSs either in octane or in vacuum. A higher value of TBR for GS in vacuum was found compared to that in octane because the GS–GS interface has larger Kapitza resistance than the GS–octane interface. More importantly, TBR for the GS–GS contact can be 30% lower than values reported for the CNT–CNT contact.

The TBR between the concentric nanotubes was not only larger than that between CNT and polymers but also chirality-dependent. These findings suggest that when DWNTs are used in thermally conducting nanocomposites, heat travels predominantly along the outer nanotube, while the inner nanotube does not contribute much to thermal management. By examining the relative rotation of the two nanotubes, it was found that those CNTs that seem to have frictionless movement (faster rotation) were characterized by higher TBR, suggesting that mechanically coupling the vibrations of carbon atoms in the two concentric CNTs facilitates heat transfer [61]. Herein, we seek to answer the question of whether nonbonded, amorphous SiO_2 coating can help to reduce the TBR between the CNT and the polymer matrix. Also, we will calculate the local TBR, between the edge of a DWNT and SiO_2, and compare with average TBR between the whole DWNT and SiO_2. Encapsulating the CNT in a material such as SiO_2 has been experimentally found to improve heat transfer by reducing the effective TBR between the CNT and the polymer [21].

11.4.1 Simulation Setup

The molecular dynamics (MD) simulations were performed using the package LAMMPS [74]. The Tersoff potential [75, 76] was employed to describe the interaction among carbon atoms within one nanotube as well as the Si–Si, Si–O, and O–O atomic interactions in SiO_2. Details about the parameters used have been described elsewhere [61, 77]. The nonbonded interaction between carbon atoms of different nanotubes was described using the 12-6 Lennard–Jones (LJ) potential. The van der Waals interactions between DWNT and SiO_2 was modeled by 12-6 LJ potential, as proposed by Hertel et al. [78]. The LJ parameters for the Si–C and O–C interactions are calculated using the Lorentz–Berthelot mixing rules. The cutoff distances of the LJ potential were equal to 2.5σ.

DWNTs in vacuum were generated with different lengths, ranging from 10 to 40 nm. The chirality of the two nanotubes in a DWNT was either the same [i.e., a (5,5) single-walled CNT (SWNT) inside a (10,10) SWNT, indicated as (5,5)–(10,10) DWNT] or different [i.e., (6,6)–(19,0) DWNT and (5,5)–(16,5) DWNT]. In all cases the intertube distance was 0.34 nm. The total number of carbon atoms in the DWNTs varied from 2460 atoms to 11056 atoms depending on the length and chirality.

To generate the amorphous SiO_2 coat enclosing the DWNT, we started with a rectangular slab of β-cristobalite SiO_2. Next, a cylindrical hollow was made at the center of the SiO_2 block to insert the DWNT in the center of the slab (see Figure 11.1(a) and (b)) so that the distance between the outer wall of the DWNT and the silica slab was also 0.34 nm. The size of the silica slabs were adjusted in order to keep the thickness of silica walls around DWNTs constant when DWNTs length increased. The total number of Si and O atoms in the silica slab varied from 22198 to 106785 atoms. We annealed the crystalline SiO_2 slab to 6000 K for 200 ps, and then slowly quenched the system back to 300 K to obtain the amorphous structure.

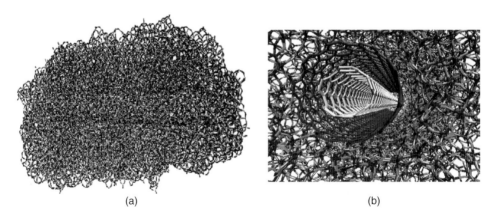

(a)   (b)

Figure 11.1 Snapshot of the DWNT coated by amorphous SiO_2. (a) View from the side and along the z-axis; (b) Close up view showing the inserted nanotube to β-cristobalite SiO_2

Both the heating and cooling rates were 10^{12} K/s, which is similar to the procedure described by Ong et al. [79]. Note that the DWNT was frozen during this process, but there existed the van der Waals interaction between the nanotube and SiO_2. By initially inserting the nanotube to β-cristobalite SiO_2 before the annealing step, we obtained better contact between the nanotube and the coat than just simply placing the DWNT inside the amorphous SiO_2 (Figure 11.1).

In order to calculate TBR simultaneously between two nanotubes as well as between the outer nanotube and the amorphous SiO_2, the transient method [14, 56, 58, 61, 62] is no longer applicable. This is because the lumped capacitance model consists of only two components: the hot (or cold) object and the environment. For this system of interest, there are three components (the amorphous SiO_2, inner, and outer nanotube) that contribute to thermal transfer. Instead, the Green–Kubo method, which is based on equilibrium MD simulations, was used. The interfacial thermal conductance across the interface, G, which is the inverse of Kapitza resistance, is defined as [79]

$$G = \frac{1}{Ak_bT^2} \int_0^\infty dt \langle Q(0)Q(t) \rangle \tag{11.8}$$

where A is area of the interface and Q is the total energy flux from a group of atoms A to another group of atoms B as

$$Q_{A \to B} = -\frac{1}{2} \sum_{i \in A} \sum_{j \in B} Q_{ij} = -\frac{1}{2} \sum_{i \in A} \sum_{j \in B} (v_i + v_j) \vec{f}_{ij} \tag{11.9}$$

where f_{ij} is the interatomic force between the ith and the jth atoms, and v_i is the velocity of the ith atom. For the system of interest, groups A and B can be either the inner and outer nanotube or the outer nanotube and amorphous SiO_2. In Eq. (11.8), $\langle Q(0)Q(t) \rangle$ is an ensemble-averaged time autocorrelation function of the energy flux, Q. The system of DWNTs and amorphous SiO_2 is fixed at 300 K (NVT ensemble) for 2 ns, and then equilibrated for 5 ns (NVE ensemble). During the equilibrium stage, 10 different configurations of the system are extracted during the interval of 50 ps. Each configuration is then simulated with NVE ensemble for 5 ns and at a time step of 0.5 fs. During that, Q is recorded and numerically integrated to obtain TBR between outer wall of the DWNT and amorphous SiO$_2$, as well as between outer wall and inner wall of the DWNT.

11.4.2   Results

The values of the TBR between the outer wall of the DWNT and amorphous SiO$_2$ with different length are plotted in Figure 11.2. When the nanotube is 10 nm long, TBR is chirality-dependent and it is almost two times larger for the case of zigzag nanotube as the outer wall. For longer nanotubes, this difference is depleted. The TBR is also observed to be independent of nanotube length at 30 nm and higher. As explained in our previous work [61], the reason for length dependence is the availability of more long-wavelength phonons for thermal transfer as the tube length increases and TBR will reach a plateau at 30 nm.

The TBR between the inner and outer nanotubes is presented in Figure 11.3. It is significantly higher than that between the outer nanotube and amorphous SiO$_2$. Note that these values of TBR in Figure 11.3 are slightly smaller than those calculated by the transient method, as they were reported in [61]. The reason of discrepancy is the temperature of the system. While the equilibrium method used in this work is controlled at 300 K for both DWNT and SiO$_2$, the transient method required a temperature difference between the hot side (i.e., from 400 to 500 K) and cold side (300 K). Since TBR is reported to increase with temperature [80], it is reasonable to obtain smaller TBR using equilibrium method in this work.

To address the question whether TBR is the same over the whole nanotube, we calculated the TBR between the edge of the nanotube (the portion at the edge is chosen to be 1 nm long) and the amorphous silica. As we can see in Figure 11.4, the TBR at the edge is smaller than that of the whole nanotube and the difference increases for longer nanotube. Similar tendency is also observed for TBR between the outer wall and inner wall of a DWNT (Figure 11.5).

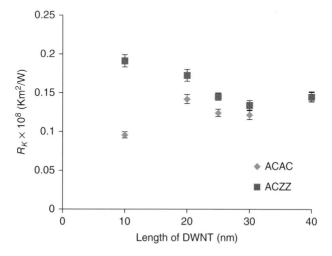

Figure 11.2   Kapitza resistance between outer wall of DWNT and amorphous SiO$_2$. "ACAC" and "ACZZ" indicate (5,5)–(10,10) and (6,6)–(19,0) DWNTs, respectively

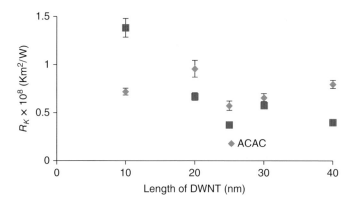

Figure 11.3 Kapitza resistance between the inner wall and outer wall of the DWNT

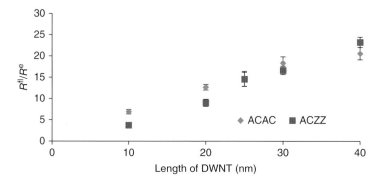

Figure 11.4 The ratio of Kapitza resistance between the full-length outer wall of DWNTs and amorphous silica (R^{fl}), and the Kapitza resistance between the edges of the outer wall and amorphous silica (R^{e})

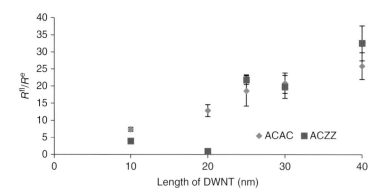

Figure 11.5 The ratio of Kapitza resistance between the full-length inner wall and outer wall of the DWNTs, and the Kapitza resistance between their edges

11.5 Discussion and Conclusions

Recent advances in simulations and theoretical work in this field have resulted in significant contributions toward the understanding of heat transfer in nanocomposites. Reviews of the computational investigations have been recently offered in the literature [81, 82]. Overcoming the Kapitza resistance is important if one were to design CNT composites that will be able to come close to the theoretical maximum of heat transfer performance. As described earlier, different approaches to this problem have been tried, but with limited results. Densification of the composite with an increase in the CNT concentration leads to overall improvements in the thermal conductivity, but the net improvement is hampered by the CNT–CNT resistance to heat transfer. Using functionalized CNTs, or other types of nanoinclusions, such as GS, can also lead to an enhancement of effective heat transfer. This is encouraging, since functionalization of CNTs, if it can be achieved so that they can even be chemically connected and can also improve heat transfer from CNT to CNT [58]. It appears that the use of multi-walled CNTs instead of SWNTs can be advantageous, because heat can be transferred mainly through the outer wall offering the advantage of a larger area of heat transfer. Other ideas, such as three-phase composites, with a combination of spherical inclusions with CNTs can lead to other possibilities [83]. However, a better approach might be to design composites where the CNT orientation, when it leads to anisotropic thermal properties, or thermal rectification could be explored. Thermal rectification appears to occur not only in simulations of DWNTs in vacuum but also in simulations of DWNTs coated with silica.

Acknowledgments

Parts of the work were supported by DOD-EPSCOR: FA9550-10-1-0031. The use of computing facilities at the University of Oklahoma Supercomputing Center for Education and Research (OSCER) and at XSEDE (under allocation CTS-090025) is also gratefully acknowledged. This work was done while Dimitrios Papavassiliou was serving at the National Science Foundation (NSF). Any opinion, findings, and conclusions or recommendations expressed in this material are those of the authors and do not necessarily reflect the views of the NSF.

References

[1] Sumita, M., Tsukishi, H. and Miyasaka, K. (1984) Dynamic mechanical properties of polypropylene composites filled with ultrafine particles. *Journal of Applied Polymer Science*, **29**, 1523.

[2] Ash, B.J., Stone, J., Rogers, D.F. et al. (2001) Investigation into the thermal and mechanical behavior of PMMA/alumina. *Materials Research Society Symposium Proceedings*, **441**, 661.

[3] Ramanathan, T., Liu, H. and Brinson, L.C. (2005) Functionalized SWNT/polymer nanocomposites for dramatic property improvement. *J. Polym. Sci., Part B*, **43**, 2269.

[4] Pan, Y., Yu, Z., Ou, Y. and Hu, G. (2000) A new process of fabricating electrically conducting nylon 6/graphite nanocomposites via intercalation polymerization. *J. Polym. Sci., Part B*, **38**, 1626.

[5] Chen, G., Weng, W., Wu, D. and Wu, C. (2003) PMMA/graphite nanosheets composite and its conducting properties. *European Polymer Journal*, **39**, 2329.

[6] Ramanathan, T., Stankovich, S., Dikin, D.A. et al. (2007) Graphitic nanofillers in PMMA nanocomposites—an investigation of particle size and dispersion and their influence on nanocomposite properties. *J. Polym. Sci., Part B*, **45**, 2097.

[7] Bryning, M.B., Milkie, D.E., Kikkawa, J.W. and Yodh, A.G. (2005) Thermal conductivity and interfacial resistance in single-wall carbon nanotube epoxy composites. *Applied Physics Letters*, **87**, 161909.

[8] Peters, J.E., Papavassiliou, D.V. and Grady, B.P. (2008) Unique thermal conductivity behavior of single-walled carbon nanotube-polystyrene composites. *Macromolecules*, **41**, 7274–7277.

[9] Rayleigh, L. (1892) On the influence of obstacles arranged in rectangular order upon the properties of a medium. *Philosophical Magazine*, **34**, 481.

[10] Bird, R.B., Stewart, W.S. and Lightfoot, E.N. (2002) *Transport Phenomena*, Wiley, New York, p. 282.

[11] Kapitza, P.L. (1941) The study of heat transfer in helium II. *Journal of Physics (Moscow)*, **4**, 181.

[12] Swartz, E.T. and Pohl, R.O. (1989) Thermal boundary resistance. *Reviews of Modern Physics*, **61**, 605.

[13] Shenogina, N., Shenogin, S., Xue, L. et al. (2005) On the lack of thermal percolation in carbon nanotube composites. *Applied Physics Letters*, **87**, 133106.

[14] Shenogin, S., Xue, L., Ozisik, R. et al. (2004) Role of thermal boundary resistance on the heat flow in carbon-nanotube composites. *Journal of Applied Physics*, **95**, 8136.

[15] Huxtable, S.T., Cahill, D.G., Shenogin, S. et al. (2003) Interfacial heat flow in carbon nanotube suspensions. *Nature Materials*, **2**, 731.

[16] Duong, H.M., Papavassiliou, D.V. and Lee, L.L. (2005) Random walks in nanotube composites: improved algorithms and the role of thermal boundary resistance. *Applied Physics Letters*, **87**, 013101.

[17] Bui, K., Grady, B.P. and Papavassiliou, D.V. (2011) Heat transfer in high volume fraction CNT nanocomposites: effects of inter-nanotube thermal resistance. *Chemical Physics Letters*, **508** (4–6), 248–251.

[18] Duong, H.M., Papavassiliou, D.V., Mullen, K.J. et al. (2008) Computational modeling of the thermal conductivity of single-walled carbon nanotube-polymer composites. *Nanotechnology*, **19** (6), 065702.

[19] Kumar, S., Alam, M.A. and Murthy, J.Y. (2007) Computational model for transport in nanotube-based composites with applications to flexible electronics. *Journal of Heat Transfer*, **129**, 500.

[20] Kumar, S., Alam, M.A. and Murthy, J.Y. (2007) Effect of percolation on thermal transport in nanotube composites. *Applied Physics Letters*, **90**, 104105.

[21] Jiaxi Guo, P.S., Liang, J., Saha, M. and Grady, B.P. (2013) Multi-walled carbon nanotubes coated by multi-layer silica for improving thermal conductivity of polymer composites. *Journal of Thermal Analysis and Calorimetry*, **113** (2), 467.

[22] Khalatnikov, I.M. (1952) Teploobmen mezhdu tverdym telom I geliem-II. *Zhurnal Eksperimentalnoi I Teoreticheskoi Fiziki*, **22**, 687.

[23] Pollack, G.L. (1969) Kapitza resistance. *Reviews of Modern Physics*, **41**, 48.

[24] Swartz, E.T. and Pohl, P.O. (1987) Thermal resistance at interfaces. *Applied Physics Letters*, **51**, 2200.

[25] Budaev, B.V. and Bogy, D.B. (2011) An extension of Khalatnikov's theory of Kapitza thermal resistance. *Annalen der Physik*, **523**, 208.

[26] Budaev, B.V. and Bogy, D.B. (2010) On interface thermal resistance associated with thermal vibrations. *Physics Letters A* , **374**, 4774.

[27] Budaev, B.V. and Bogy, D.B. (2010) A self-consistent acoustics model of interface thermal resistance. *SIAM Journal on Applied Mathematics*, **70**, 1691.

[28] Alzina, A., Toussaint, E., Béakou, A. et al. (2006) Multiscale modelling of thermal conductivity in composite materials for cryogenic structures. Skoczen. *Composite Structures*, **74**, 175.

[29] Balasubramanian, G. and Puri, I.K. (2011) Heat conduction across a solid-solid interface: understanding nanoscale interfacial effects on thermal resistance. *Applied Physics Letters*, **99**, 013116.

[30] Pettersson, S. and Mahan, G.D. (1990) Theory of the thermal boundary resistance between dissimilar lattices. *Physical Review B*, **42**, 7386.

[31] Landry, E.S. and McGaughey, J.H. (2009) Thermal boundary resistance predictions from molecular dynamics simulations and theoretical calculations. *Physical Review B*, **80**, 165304.

[32] Patel, H.A., Garde, S. and Keblinski, P. (2005) Thermal resistance of nanoscopic liquid-liquid interfaces: dependence on chemistry and molecular architecture. *Nano Letters*, **5**, 2225.

[33] Lyeo, H.-K. and Cahill, D.G. (2006) Thermal conductance of interfaces between highly dissimilar materials. *Physical Review B*, **73**, 144301.

[34] Costescu, R.M., Wall, M.A. and Cahill, D.G. (2003) Thermal conductance of epitaxial interfaces. *Physical Review B*, **67**, 054302.

[35] Hasselman, D.P.H. and Johnson, L.F. (1987) Effective thermal conductivity of composites with interfacial thermal barrier resistance. *Journal of Composite Materials*, **21**, 508.

[36] Nan, C.W., Birringer, R., Clarke, D.R. et al. (1997) Effective thermal conductivity of particulate composites with interfacial thermal resistance. *Journal of Applied Physics*, **81**, 6692.

[37] Nan, C.W., Liu, G., Lin, Y. et al. (2004) Interface effect on thermal conductivity of carbon nanotube composites. *Applied Physics Letters*, **85**, 3549.

[38] Choi, S.U.S., Zhang, Z.G., Yu, W. et al. (2001) Anomalous thermal conductivity enhancement in nanotube suspensions. *Applied Physics Letters*, **79**, 2252.

[39] Berber, S., Kwon, Y.K. and Tománek, D. (2000) Unusually high thermal conductivity of carbon nanotubes. *Physical Review Letters*, **84**, 4613.

[40] Kim, P., Shi, L., Majumdar, A. et al. (2001) Thermal transport measurements of individual multiwalled nanotubes. *Physical Review Letters*, **87**, 215502.

[41] Xue, Q.Z. (2006) Model for the effective thermal conductivity of carbon nanotube composites. *Nanotechnology*, **17**, 1655.

[42] Zhang, J., Tanaka, M. and Matsumoto, T. (2004) A simplified approach for heat conduction analysis of CNT-based nano-composites. *Computer Methods in Applied Mechanics and Engineering*, **193**, 5597.

[43] Davis, L.C. and Artz, B.E. (1995) Thermal conductivity of metal-matrix composites. *Journal of Applied Physics*, **77**, 4954.

[44] Nishimura, N. and Liu, Y.J. (2004) Thermal analysis of carbon-nanotube composites using a rigid-line inclusion model by the boundary integral equation method. *Computational Mechanics*, **35**, 1.

[45] Singh, I.V., Tanaka, M. and Endo, M. (2007) Effect of interface on the thermal conductivity of carbon nanotube composites. *International Journal of Thermal Sciences*, **46**, 842.

[46] Li, X., Fan, X., Zhu, Y. et al. (2012) Computational modeling and evaluation of the thermal behavior of randomly distributed single-walled carbon nanotube/polymer composites. *Computational Materials Science*, **63**, 207.

[47] Balandin, A.A. (2011) Thermal properties of graphene and nanostructured carbon materials. *Nature Materials*, **10**, 569.

[48] Shiomi, J. and Maruyama, S. (2010) Diffusive-ballistic heat conduction of carbon nanotubes and nanographene ribbons. *International Journal of Thermophysics*, **31**, 1945.

[49] Shiomi, J. and Maruyama, S. (2006) Non-Fourier heat conduction in a single-walled carbon nanotube: classical molecular dynamics simulations. *Physical Review B*, **73**, 205420.

[50] Zhang, Q., Rastogi, S., Chen, D. et al. (2006) Low percolation threshold in single-walled carbon nanotube/high density polyethylene composites prepared by melt processing technique. *Carbon*, **44**, 778.

[51] Sandler, J.K.W., Kirk, J.E. and Kinloch, I.A. (2003) Ultra-low electrical percolation threshold in carbon-nanotube-epoxy composites. *Polymer*, **44**, 5893.

[52] Thostenson, E.T., Ren, Z. and Chou, T.-W. (2001) Advances in the science and technology of carbon nanotubes and their composites: a review. *Composites Science and Technology*, **61**, 1899.

[53] Biercuk, M.J., Llaguno, M.C., Radosavljevic, M. et al. (2002) Carbon nanotube composites for thermal management. *Applied Physics Letters*, **80**, 2767.

[54] Clancy, T.C., Frankland, S.J.V., Hinkley, J.A. and Gates, T.S. (2010) Multiscale modeling of thermal conductivity of polymer/carbon nanocomposites. *International Journal of Thermal Sciences*, **49** (9), 1555–1560.

[55] Konatham, D. and Striolo, A. (2008) Molecular design of stable graphene nanosheets dispersions. *Nano Letters*, **8**, 4630.

[56] Konatham, D. and Striolo, A. (2009) Thermal boundary resistance at the graphene-oil interface. *Applied Physics Letters*, **95** (16), 163105.

[57] Shenogin, S., Bodapati, A., Xue, L. et al. (2004) Effect of chemical functionalization on thermal transport of carbon nanotube composites. *Applied Physics Letters*, **85**, 2229.

[58] Varshney, V., Patnaik, S.S., Roy, A.K. and Farmer, B.L. (2010) Modeling of thermal conductance at transverse CNT-CNT interfaces. *The Journal of Physical Chemistry C*, **114**, 16223.

[59] Roy, A.K., Farmer, B.L., Sihn, S. et al. (2010) Thermal interface tailoring in composite materials. *Diamond and Related Materials*, **19**, 268.

[60] Balandin, A.A., Ghosh, S., Bao, W. et al. (2008) Superior thermal conductivity of single-layer graphene. *Nano Letters*, **8**, 902.

[61] Bui, K., Duong, H.M., Striolo, A. and Papavassiliou, D.V. (2011) Effective heat transfer properties of graphene sheet nanocomposites and comparison to carbon nanotube nanocomposites. *The Journal of Physical Chemistry C*, **115**, 3872.

[62] Konatham, D., Bui, K.N.D., Papavassiliou, D.V. and Striolo, A. (2011) Simulation insights into thermally conductive graphene-based nanocomposites. *Molecular Physics*, **109**, 97.

[63] Cebeci, H., Guzman de Villoria, R., Hart, A.J. and Wardle, B.L. (2009) Multifunctional properties of high volume fraction aligned carbon nanotube polymercomposites with controlled morphology. *Composites Science and Technology*, **69**, 2649.

[64] Marconnett, A.M., Yamamoto, N., Panzer, M.A. et al. (2011) ACS. *Nano*, **5** (6), 4818–4825.

[65] Maruyama, S., Igarashi, Y., Taniguchi, Y. et al. (2006) Anisotropic heat transfer of single-walled carbon nanotubes. *Journal of Thermal Science and Technology*, **1** (2), 138.

[66] Zhong, H. and Lukes, J.R. (2006) Interfacial thermal resistance between carbon nanotubes: molecular dynamics simulations and analytical thermal modeling. *Physical Review B*, **74** (12), 125403.

[67] Kumar, S. and Murthy, J.Y. (2009) Interfacial thermal transport between nanotubes. *Journal of Applied Physics*, **106** (8), 084302.

[68] Yang, J., Waltermire, S., Chen, Y. et al. (2010) Contact thermal resistance between individual multiwall carbon nanotubes. *Applied Physics Letters*, **96** (2), 023109.

[69] Schroder, C., Vikhrenko, V. and Schwarzer, D. (2009) Molecular dynamics simulation of heat conduction through a molecular chain. *The Journal of Physical Chemistry A*, **113** (51), 14039–14051.

[70] Gates, T.S. (2010) Multiscale modeling of thermal conductivity of polymer/carbon nanocomposites. *International Journal of Thermal Sciences*, **49** (9), 1555–1560.

[71] Fang, M., Wang, K.G., Lu, H.B. et al. (2010) Single-layer graphene nanosheets with controlled grafting of polymer chains. *Journal of Materials Chemistry*, **20**, 1982–1992.

[72] Duong, H.M., Papavassiliou, D.V., Mullen, K.J. et al. (2008) Calculated thermal properties of single-walled carbon nanotube suspensions. *Journal of Physical Chemistry C*, **112**, 19860.

[73] Konatham, D., Papavassiliou, D.V. and Striolo, A. (2012) Thermal boundary resistance at the graphene–graphene interface estimated by molecular dynamics simulations. *Chemical Physics Letters*, **527**, 47.

[74] http://www.lammps.sandia.gov.

[75] Tersoff, J. (1988) Phys. Rev. *Physical Review B*, **37** (12), 6991–7000.

[76] Tersoff, J. (1989) Modeling solid-state chemistry: Interatomic potentials for multicomponent systems. *Physical Review B*, **39** (8), 5566–5568.

[77] Ong, Z.-Y. and Pop, E. (2010) Molecular dynamics simulation of thermal boundary conductance between carbon nanotubes and SiO_2. *Physical Review B*, **81** (15), 155408.

[78] Hertel, T., Walkup, R.E. and Avouris, P. (1998) Deformation of carbon nanotubes by surface van der Waals forces. *Physical Review B*, **58** (20), 13870–13873.

[79] Ong, Z.-Y. and Pop, E. (2010) Interfacial thermal conductance and thermal accommodation coefficient of evaporating thin liquid films: a molecular dynamics study. *Journal of Applied Physics*, **108** (10), 103502–103508.

[80] Hu, L., Desai, T. and Keblinski, P. (2011) Determination of interfacial thermal resistance at the nanoscale. *Physical Review B*, **83** (19), 195423.

[81] Marconnet, A.M., Panzer, M.A. and Koodson, K.E. (2013) Thermal conduction phenomena in carbon nanotubes and related nanostructured materials. *Reviews of Modern Physics*, **85** (3), 1295.

[82] Wang, Y., Vallabhaneni, A.K., Qiu, B. and Ruan, X.L. (2014) Two-dimensional thermal transport in graphene: a review of numerical modeling studies. *Nanoscale and Microscale Thermophysical Engineering*, **18** (2), 155.

[83] Gong, F., Bui, K., Papavassiliou, D.V. and Duiong, H.M. (2014) Thermal transport phenomena and limitations in heterogeneous polymer composites containing carbon nanotubes and inorganic nanoparticles. *Carbon*, **78**, 305.

Index

Advanced Computational Nanomechanics, First Edition. Edited by Nuno Silvestre.
© 2016 John Wiley & Sons, Ltd. Published 2016 by John Wiley & Sons, Ltd.